D_ESIGN SCIENCE

IN THE NEW PARADIGM AGE

Herbert Glenauldbin Bennett RA,DS

VOLUME II OF III

CONTAINS the front matter, the remaining 3 chapters, are not in volume one, and the back matter.

D_ESign Science in The New Paradigm Age

**D_ESIGN SCIENCE
IN THE NEW PARADIGM AGE: VOLUME II**

Copyright © 2023 By **Herbert Glenauldbin Bennett RA**

All rights reserved. No part of this book may be reproduced or transmitted, downloaded, distributed, reverse engineered, or stored in or introduced into any information storage and retrieval system, in any form or by any means, including photocopying and recording, whether electronic or mechanical, now known or hereinafter invented without permission in writing from the publisher.

DISCLAIMER: The contents of this work, including, but not limited to, the accuracy of events, people, and places depicted; opinions expressed; permission to use previously published materials included; and any advice given or actions advocated are solely the responsibility of the author, who assumes all liability for said work and indemnifies the publisher against any claims stemming from publication of the work.

To order additional copies of this book, please contact:

eComRocket
www.ecomrocket.net

General Inquiries & Customer Service
Phone: 1-(403)-755-8677
Email: info@ecomrocket.net

ISBN Paperback: 978-1-77419-199-6
ISBN eBook: 978-1-77419-198-9

The main categories of the book 1. Architecture 2. Art, 3. Business 4. Design 5. Geometry 6. Mathematics 7. Metaphysics 8. Morphology 9. Philosophy 10. Science and 11. Synthesis 12. Symmetry

Description: DESign Science: In the New Paradigm Age
The process of synthesizing the Body, Mind and Spirit dimensions of consciousness into natural and creative expressions to align with the highest and best use of space, time, energy and thought to preserve and promote the genius of human life, environments and our legacy with the most advanced harmonized Wisdom, Intelligence, Knowledge and Information to create the most highly effective and rich aesthetic life support systems that obey natural laws, into the distant future, is 'Design Science™'.

Author's Note "Making the abstract (spirituality) real (through creativity)" Living a passionate, abundant and fulfilled life is a gift to be realized when we discover our talents (self), our purpose (focused) and follow 'the plan' that connects us to source. Having a clear definition with a strategy for accessing the sustainable and working forces with energies for the creation and protection of our desires is a must. Whatever we are doing in our lives must be in alignment to receive the gift. The gift of a true authentic self to one's self is the best. If we can trick our minds to think of 'self' as a precious gift, positive emotions can be elicited to facilitate living the life we desire as we direct our thoughts, feelings and energies with clarity and focus. Being on purpose is the cherry.

Cover Art *"The Synthesis Mandala"*
Cover Illustration and design Copyright © H.G. Bennett
Chapter opening art © H.G. Bennett
Photography Credit as mentioned Poetry By © H.G. Bennett

 "You get told that the world is the way it is, but life can be much broader once you discover one simple fact; and that is that everything around you that you call life was made up by people no smarter than you. Once you learn (experience and internalize) that, you'll never be the same again." —**Steve Jobs**

CONTENTS

CHAPTER 1 – Volume I	**Hexahedron**	$H_A H\text{-}1$
TRINE A: the Discipline		Page 13
TRINE B: 'SOURCE': is ALL the same by any other name		Page 30
TRINE C: DESign Science		Page 39
CHAPTER 2 – Volume I	**Icosahedron**	$I_C H\text{-}37$
TRINE A: a new age is upon us		Page 47
TRINE B: 'aesthetix' III new paradigm-value systems		Page 55
TRINE C: creativity: the DESign Process		Page 63
CHAPTER 3 – Volume I	**Tetrahedron**	$T_R H\text{-}86$
TRINE A: creative potential		Page 91
TRINE B: think bank™ the core knowledge		Page 106
TRINE C: the consciousness story		Page 117
CHAPTER 4 – Volume I	**Cubeoctahedron**	$H_A O_T H\text{-}135$
TRINE A: the praxis of consciousness		Page 135
TRINE B: order		Page 144
TRINE C: mind: the command center		Page 178
CHAPTER 5	**Octahedron**	$O_T H\text{-}12$
TRINE A: the triad model		Page 14
TRINE B: going back to new truths		Page 24
TRINE C: visual mathematics		Page 34
CHAPTER 6	**Icosidodecahedron**	$I_C H\text{-}54$
TRINE A: geometry		Page 56
TRINE B: WIKI speak paradigms and principles		Page 94
TRINE C: the Structure of (Aesthetic or Design Revolutions)		Page 111
CHAPTER 7	**Dodecahedron**	$D_D H\text{-}118$
TRINE A: logic structures in a distribution Matrix		Page 120
TRINE B: synthesis: a creative thought processes		Page 135
TRINE C: design science interpreted		Page 155
THE ILLUSTRATION GALLERY		**Page 239**

Simple methods create complex innovative solutions that can then be used in various fields of design, engineering in the art of making (Everything). H.G.B.

Simple methods create complex innovative solutions that can then be used in various fields of design, engineering and production. H.G.B.

TESTIMONIALS:

"Your research is clearly a notch or two past the cutting edge of science! More power to you!"
Edward Malkowski; Author of "The Spiritual Technology of Ancient Egypt - Inner Traditions Bear & Co.

"May this book and your work be the map and compass for the synthesis of our body, mind and creative spirit in harmony, peace, prosperity and joy to connect with infinite consciousness or SOURCE!"
Jack Canfield: Author of "Chicken soup for the Soul"

"Your research is clearly a notch or two past the cutting edge of science! More power to you!" Jack Canfield: Author of "Chicken soup for the Soul"

THE AVATARS

Find 'avatars' who are fully realized and conscious influencers and game-changers in all creative leading edge communities that are embracing the new paradigm values, ideas, theories and principles to support the technologies and lifestyles for the GOOD of ALL.

They must not be intimidated by the present conditioning and control of the old paradigm establishment and its overt and covert agendas.

Be creatively free in all aspects of their lives.

Be conscious, compassionate and not be afraid to embraces GOD, SOURCE and 'HIMHERIT™' or any other name or persuasion that move us to define and respect one another and the higher force/s as the guardian/s of our destiny.

Be about simplicity not complexity and confusion

"The best way to get (anything) or approval is not to need it," Hugh MacLeod
By removing the (corrupt) intention and attention files from the human complex's mental program (with the 'not') pure mind allows the deepest desires to flow directly to you from SOURCE. HGB

The Author's Story

'What Herb's admirers are saying'.

The story of an Artist is one best told by surrendering to the deeply felt urge to make sense of many fleeting moments that connect us to source to find purpose by seeking and tapping into SOURCE] we can apply direct knowledge to reach our higher selves. It is a phenomenal journey of discoveries and gratitude for all who helped along the way. All that's been before paves the way back to the SOURCE we must all return to. SOURCE by any other name is all the same.

I agree with the axiom that says; "self-praise is no recommendation". I prefer to let my story be told via shared experiences with friends, colleagues and others I have known and collaborated with on social and professional occasions, working on projects; celebrating events and minor victories offered to me and those I love, on my travels around the world.

Ademola Olugebefola: artist and publisher says:

Greetings Herb...Your intellectual curiosity and prowess is admirable. Worthy of deep individual and group introspection as a vital stimuli to help us navigate our way through the often resounding 'noise' of our contemporary existence.

It is no wonder, that when our paths cross at artistic/cultural events and 'family' gatherings, you are in the midst of real conversations not just polite chatter.

Stay well and all the best to you my Brother......................

......................A. O.

"Give me a lever and a place to stand and I will move the earth" said Archimedes in 212 BC the Greek mathematician, philosopher, scientist and engineer. 2500 years later "Design Science" is the new lever that lifts and shifts mindsets in the new world to enhance human life and empower all creative endeavors in the spirit of constant creation in all our life support systems, technologies and solutions.

"Our consciousness will rise when we minimize material needs, maximize spiritual knowledge to positively impact human psychological behavior for a more efficient harmonious environment and planet; the real struggle is between quality and quantity". Herb G. Bennett

FOREWORD by The AQUARIANAS Agency of New York

In the western world knowledge feeds our voracious materiality. Prayer is strictly for supplication and revelation is a book in the bible few truly understand. Of all the sentences in this work this is the most dispassionate. Entire libraries have been dedicated to how to do the right thing leaving us on a trajectory to oblivion. From a spiritual (NOT RELIGIOUS) view every word in this text is a practical, emotional and super energy generator to create the lives we were meant to live. Here is another dedication to the library of human failure celebrating Shiva's triumph.

Experiments that are mechanical are STEM; (See, Touch, Estimate and Measure) no longer work. The Thought STEM is now (See, Think, Energize and Meditate) with observation and participation. The search for truth and freedom has an upside and a downside to it. If in the study of the nature of life we do not find truth or freedom the process of seeking could have its rewards. What we discover could soften the pain as changes in life are encountered.
Discovering truth and freedom could be the bonus.

Wisdom could be gained in either scenario. Like paying attention, which does not cost anything, except when it is not paid attentively. The search for wisdom and freedom cost much leas and the gain is exponentially greater. There is a physiological, psychological and spiritual transformations that follow the eternal creative principles outlined in this work.

One of the many goals of this work is the establishment of a 'B-body, M-mind and S-spirit; BMS and STEP' formula for life's journey with The Dao- of (self-discovery, self-effort and self- knowing) elevating 'DESIGN' to a science called 'DESign Science'. Others with broader shoulders have left us their wisdom to help us transcend the artificial diversions we encounter
as we seek to enhance our physical, emotional and spiritual lives. For this a 'body, mind and spirit' paradigm is needed. Fundamental principles and natural laws are already unfolding and are no longer be dormant. This awakening of truth and curiosity is long overdue.

I believe that our consciousness will be raised when we minimize material needs, positively impact human psychological behavior and maximize spiritual knowledge for a more efficient harmonious environment and the planet. Herb G. Bennett

"Any lean system is not an explosion of things but an implosion of things, a coming together (harmonizing) of things or better yet a Synthesis. We are not beginning to accept this notion, really, of life being totally dependent on the coming together of things" Paolo Soleri (1919- 2013), the founder of Arcosanti.

ACKNOWLEDGMENTS

All living creatures have built in directional systems. Movement, position, destination and mapping are key elements in knowing one's self, where we are and where we are going. Our egos and shegos tell us we are in charge and are doing it alone. What amazes me is our total disregard for being truly aware of the dynamic principles and laws that govern our lives. We find 'GOD particles' in force fields with complex directional systems displacing ourselves from the source goodness in celebration of our greatness, recognition and fame.

Having a multidimensional and directional compass is essential to navigating the currents of a new consciousness that we need to step up to position, move and map our own destinations out of, at least, or to get beyond self, time and space with a healthy energy to inspire us.

I will take a different approach to my acknowledgments and apply a 'shifted paradigm form' of recognition with the most profound gratitude I know as 'Grammercy' (profound thanks).
This is based on the notion of humans 'being spirits having a bodily experience' as a definition of life. That the connections, interactions we share form our worlds from birth 'back to forever'. That we come from and source and we return to source forever. Our bubbles of nurturing, guidance and direction finding in harmony with the Body-Mind-Spirit continuum shape our lives and are linked in a universal consciousness. The love and healing energies that are transmitted throughout the bubbles of me and you, places we have been and where we are going, help us make our contributions to eliminate all troubles using our creative talents, resources and gifts.
Chief Seattle's web is a similar analogy and another compass. The one idea that wraps this up is the one that speaks to the paying attention. It says 'where we put our attention is where put concentrated and focused energy.

I believe the traditional way of listing names of relatives, people, places and all who have been my compass by name would be another book. I would like to invite you to my bubble. To me the word 'compass' does say come-pass' which I interpret as an invitation for every part of my bubble to resonate to my love and Grammercy for all the goodness and kindness in it.

Thank your bubbles, troubles are your past!

May this book be a compass for the synthesis of your body, mind and creative spirit in harmony, peace, prosperity and joy with your bubble of infinite consciousness or SOURCE!

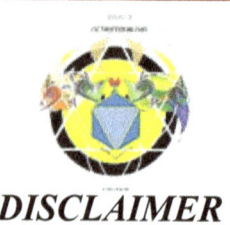

DISCLAIMER

This book provides inspiration and knowledge meant to inform readers in the current era of transformations impacting many creative industries. With this knowledge we take advantage of the unique opportunities this 'nu paradigm age' offers. The Publisher and author are not rendering legal, accounting or professional advice. Theories, intellectual properties and ideas are available for creative development by early adopters, game changers and designers.

Every effort has been made for this book to be as complete and accurate as possible at press time. The author and publisher do not assume and hereby disclaim any liability to any party for any loss, damage, or disruption caused by errors or omissions, whether such errors or omissions result from negligence, accident, or any other cause related to using its content.
The book speaks directly to designers who will add great value by applying the innovative timeless BMS-WIKI Formulae; (Body, Mind, Spirit)-(Wisdom, Intelligence, Knowledge and Information) to their 'cultural economic business systems and their clientele'.

The book is a distillation of the author's life experiences, searches and travels on his 'Dao DESign' journey, activating 'creative license from source' to explore and express creative horizons, found in creative fields where traditional knowledge, methods and technologies are being transformed. By synthesizing the 'abstract, the real, the new direct knowledge and information into an identity, with a rich aesthetic. A language is invented in the flow of most powerful constant creation dynamic, with a coherent and unfamiliar vision not found now in our design professions. This is the book's contribution to this movement and to the ancestors.

The goal is to awaken our creative genius and to step up to new levels of visual intelligence needed in all design fields by formulating art, science and mathematics theories, skills and new symmetry principles into a 'DESign Science™'. This is the key to embracing shifts charged with the potential for enhanced lifestyle solutions we can pursue with new passions and desires for design excellence, creative freedom, wellness, peace and prosperity.

Permission must be granted prior to using the content and/or ideas in the book. Please use competent professional representation when making enquiries related to the book's content.

If you do not wish to adhere to the above, you may return the book to the publisher for a full refund to be followed by cancellation of your lifetime membership and benefits in the 'New Paradigm Movement' an all its opportunities, services and events.

When there is realization, the same paradigms that limit us become portals of transformation and our creative freedom.

THE AUTHOR'S THOUGHTS

'In fields of 'nu' living vibrant forms'.
Thoughts are waves or energies of specific vibrations. All expressions like words are particles and sensations.
In the thought wave quantum in the dynamics of all energy fields. Thoughts are and create energies in all forms of expressions.
Words or descriptions are used to decode the symbols; 'Quanta, Qualia and Spiritua' for understanding and meaning to become The glue of Cohesion in the infinite possibilities of thought fields, Infinite consciousness, mental equilibrium and spiritual harmony. Thoughts are particles or states of specific vibrations

All expressions, like physical particles too, are sensations.
In the particle wave quantum all is in the fields of 'nu' living forms'. Thoughts are energy constantly created in perpetual expressions.
Where ideas or descriptions symbolically represent natural laws For understanding and meaning to become the glue for the Cohesion that mirrors the field of infinite possibilities and of Normal consciousness, with body, mind and spirit in harmony.
Thought waves are enneaves of distinct vibrations of life manifested. Densities of intensities, expressed in sound and profane geometries. The packets of sense data; particles of self-realization integrated.
Sensations of higher energy flavors order our thought forms.
All in our thought sense quantum dynamic field of inter-ill-igence. Various forms and thoughts create energy in return to source with little evidence of fulfillment and joy.

Words of description used to recycle the symbols of force on course Towards understanding and 'nu' meaning to become the inherent glue: Cohesion! that's in me and you in one field of infinite possibilities
With true understanding 'all in' with the roll of Einstein's dice In the right action of a mental equilibrium in spiritual harmony. Though the realigning did not hold the way the truth was told
A peaceful and prosperous world to behold is our resolution, once solved and incomplete to be resolved again
in much less stress and pain.

Poetic language expands associations, perceptions and mental constructs. It's the window to New Paradigms.

DESign Science in The New Paradigm Age

The "Prologue" : The grand vision

The Science of Design is a 'DEsign Science'. Managing complexity and technology in disciplines that share ubiquitous architectonic principles can be 'simple-fi-ed' and made exciting and enjoyable as ART. Scientists, artists, designers and others try to create systems and tools to understand "everything" in linear equations and are mired in needless complexity that defeats their purpose. Keeping within the strict Western paradigms and cultural traditions limits the outcome and eventually human growth.

DESign Science: in the New Paradigm Age" suggests looking at the world from a wider more inclusive and open minded, holistic and spiritual perspective. We 'Simple-Fi' 'Spirit' here as creative spirit; that life force that holds everything together that we are not expressing or using effectively to enhance our lives. This is a ''mindset shifter" or a New paradigm that is not really that new at all.

Design Science is a human story. It takes us on a creative journey, not of survival but, to reconnect with our 'WIKI'™. This is the key to elevate our consciousness to live the lives we design for ourselves. The method or 'science' gives us the **W**isdom that is not generally applied to the world of things we focus on, the **I**ntelligence that is natural and more common sense than we make it, the **K**nowledge that we synthesize with the habits and experiences we all create **I**deas, Information with Inspiration to live by. This is the *'WIKI'*. They are the four natural universal forces that hold everything together as 'one'.

The language of design expresses ideas about space, time and energy in an aesthetic form. No one aspect of this triad works alone. This is the form of communication we use to connect to source, ourselves and others. They are blended into a holistic adventure that takes us into the magic of Physiology, psychology and spirituality empowering us to create. The elements, words and energies we use are space, numbers, time and e-motion with our creative energy. Organizing them into useful form, with our Wisdom Intelligence Knowledge and Information, (WIKI) raises the consciousness of the life force, creative spirit or any other name we use to describe it, to create the life support systems we need.

We are all part of the grand creative process taking place in the universe. With our Body, Mind and Spirit as the finely tuned instruments we must be, we create forms of expression, beauty and love that flows through us, connecting to source in the works we create. Art, in patent and trademark parlance, is ubiquitous. It enhances the skills needed to bring original and paradigm changing ideas into 'being'. There is 'prior art' which is the state of the art and original or patentable art which is innovative. We can expand our creative skills with DESign Science to go deeper to unveil the 'hidden secrets' to bring new innovative lifestyles into harmony with creative excellence and pure imagination for "Making the abstract real". A new vision is in the making and it will be facilitated by this new creative discipline.

"Better to write for yourself and have no public, than to write for the public and have no self." —Cyril Connolly

"Every explicit duality (multiplicity) is an implicit unity." Alan Watts

All is one! h. g. bennett

 "When freedom prevails, the ingenuity and inventiveness of people creates incredible wealth. This is the source of the natural improvement of the human condition." —Brian S. Wesbury

CHAPTER FIVE

OCTAHEDRON-O_TH

5. Spiritual Principles:

Vishuddha is the (throat) chakra with the capacity for messages to divine the world in its Universals Nature discloses hidden mysteries to the seeker. When this center is open, one receives the codes from the Highest and becomes the communicator from the Highest callings to poets, singers and artist. All forms of art are expressed from this center.

Clairaudience, Symbolic Intelligence; the use of words and speech, telepathy, use power rather than brute force. The use of symbols create meaning and attract matter, power vs. force, polarity the throat air the thyroid in communication: when we talk to ourselves we think, when we talk to others we communicate.

CHAPTER 5 TRINE A

THE TRIADIC MODEL™

We present the Core Knowledge, Skills and technologies for the paradigm-consciousness Triad model. The three dimensional characteristics of Consciousness are Body, mind and Spirit. This threefold model or Triad permeates all expressions of the Universal Consciousness on the macrocosmic and microcosmic levels of reality. This fundamental knowledge along with the symmetry principles and other natural laws have been part of the wisdom traditions of ancient civilizations. Culture was once defined as the intersection of the natural and the artificial. This relates to the physiological and psychological aspects of awareness and reality without the third component of 'spirit'. Spiritual principles, when thought of in the context of consciousness, natural laws and symmetry principles exist in the domain of universal forces. Metaphysics is man's attempt at understanding the spiritual dynamics of the universe. The methodology we use to understand spiritual principles have nothing to do with religion. We can now redefine the synthesis of physiology or body, psychology or mind and spirituality or energy as culture. Energy here is about work and power; from thought to the basic electricity that powers machines and our homes.

Laws and cultures that interpret them are equally important. Different cultures through time have offered us knowledge through their "wisdom traditions" giving us clues we can use. The Chinese, East Indian; specifically the Hindu, Sumerian and Egyptian cultures are of interest to us in this meditation. Hermetic philosophy describes seven Hermetic laws that modern science still has not explained. The first law; the law of "mentalism" is the key that unlocks the mystery of "Thought"; how it works and where we find it. The Law of mentalism 'states' that all things are mental. The law reveal their meaning through observation, experimentation and meditation. It does not reveal itself unless we find the right key to unlock it. Stated another way it says "No-thing is manifested except by thought. Everything in our world starts with a thought. These thoughts are expressions of complex dynamic processes that enveloped in form.

How many vocabularies or (phyla) of form are there? There are organic forms of nature and the inorganic manmade forms made of geometric principles and artificial materials. The forms that are manifested through chemical processes are an extension of nature and are considered organic. Man uses geometry to describe the world. It's inherently a world of form. Here again we are indebted to the Egyptians for the discipline of geo meaning earth and 'metry' as in metric or measure. The nature of form itself has the three dimensional qualities of the triad.
Form is the synthesis of the physical, the behavioral and energy properties that resonate with the universal properties of universal consciousness at the macro level and is expressed at the micro level of the correspondence of the second Hermetic law of Correspondence. As above so below is its description.

There are three systems of geometry. One that is Euclidean and two Non-Euclidean variations. The Euclidean geometry is an orthogonal description of space. A Euclidean triangle is defined by flat surfaces with all of its angles equal to 180 Degrees.

A Non-Euclidean triangle has a compounded curved surface with its angles greater than or less than 180 degrees. These properties generate identities that are quite distinct and are applied to other polygons. The process that generates these identities are mental. They require thought, evaluation and experiment to arrive at the relationships and descriptions that result from the creative process. Intuition and other unfamiliar mental processes and methods are involved as well. Mind is the process at work here. Mind is the process of refining thought prior to expression or manifestation when emotions, feelings and skilled experimentation and other symmetry principles and technologies are applied. There is a major distinction between Design and Manifestation. Design involves man's interpretation with knowledge, systems and methods for harmonizing forces and materials into form. Design requires the understanding of "how nature works". Manifestation is a more organic process.

The law of correspondence is a macro and micro correspondence dynamic that is at the core of the universal consciousness in the macro state with its corresponding micro expressions of the universe we are aware of and inhabit. Man has the most degrees of freedom which allows us to give form to our micro and macro mental environments. There is a very significant spiritual or energy attribute given to us by the Hindu tradition known as the 'Kundalini'. This is the name of the universal and human life force. There are seven centers known as the chakras associated with the life force. These centers correspond to the endocrine system which is responsible for the distribution of hormones through the ductless glands. Western medicine fails to recognize this vital relationship. They have developed protocols without fundamental knowledge of how vital forces work. It can be argued that everything in the universe contains the Kundalini force with the same dynamic principles at work. This knowledge leads to a holistic and harmonious understanding of the essence of humanity and life.

Once we have these principles in place we focus on the practical (praxis); aspects of "making the abstract real". What does it take to design reality? What are the creative processes, tools, and knowledge needed to transform thought into things? Logic structures, Symmetry principles, laws of behavior with appropriate rituals and technologies are some of the basic elements needed. Wishing thought into reality is fantasy. Design is a science. Like many other ancient wisdom traditions we need to accept this as a reality. Transforming thought into form involves elements of time as in tense past, present and future. The dynamic of tension is also a time factor that with pressure helps the process of creativity flow through the stages of development described and or prescribed by the chakras. They in fact correspond to the phases of manifestation as part of the design science. The though process is triggered by need on a very basic level. That need flow up to the cortex which engages the stimuli and the ideational processes. Visualization is the next stage where form and matter meet and is resolved through the communication center in thought and communication. The heart is the next center where the LOVE force is applied to the approval process.
If the form is not loved it goes back to re-thinking center and its process for other iterations to be created. The next stage is the work center. Energy is extracted from the sun through the consumption of sun fed foods in the solar plexus and is transformed into work energy. The next level is where gender is assigned. All things are either male or female. Mechanical parts are designed as such to 'fit' according to symmetry principles with male and female characteristics with the physiological, behavioral and energetic properties.

Gender here is not sex. Procreation is replication involving sexual intercourse, while creation or production relies on principles of gender. The final step in the process is the understanding of the behavior (psychology) of matter to determine how it accommodates the functions that correspond with the design, intention and satisfaction of the impulse or need. Thinking on a phenomenal level allows us to understand the *symmetry principles* of the triad (3) of the physical, psychological and spiritual dynamic. Logic structures, Symmetry principles, gauge theories, numbers, measure, and all other necessary systems are used to design (not just wish) thought into form and reality. These are disciplines that the mind is completely engaged with. There are other systems required to bring these into the world. The symmetry principles involved with manifestation in the physical world are four (4). Together they are 3+4= 7, seven. The fundamental law of 7 applies to all expressions of nature and the phenomena operating on the three dimensional vibrational level of planet earth. What this leads to is a "qualitative" periodic matrix that correlates the flavors of expressions with certain correspondences and properties. Thinking abstractly with a mental scaffold can get us to a deeper understanding. Inventing the words "SUCHNESS" to relate to flavor and MUCHNESS for quantity with universal CONSCIOUSNESS as spirit helps us develop an open and expansive paradigm to expand our [human] consciousness which corresponds with its universal 'flavor'. Where we get into trouble is in using 'muchness' as the basis of science and other theories to validate "suchness" phenomena.

For those who think the words suchness and muchness are not sexy enough we can use the German words 'Solchein' for 'such' and 'Oft' for much with the suffix 'heit' for the 'ness' that's added to both. We do this in the intellectual and research traditions using German language once considered the 'tongue of the western brain'. The 'sexy' form for 'suchness' is 'Solcheinheit'™ and for 'muchness' it's 'Oftheit™'. Gesundheit bless you.
Gestalt psychology or gestaltism (German: *Gestalt [gəˈʃtalt] "shape, form"*) is a theory of mind of the Berlin School. This is a concept we are aware of. That it deals with shape and form is profound. German words for consciousness are Besinnung, Bewusstsein, and Sinne. We will explore this later.

The qualitative matrix creates order with expressions of nature, similar to what the periodic table of elements does. Creative disciplines are called into action in their respective states of *'Solcheinheit'* where there are discrete rituals required to design the world around us as each individual soul expresses its own *'Solcheinheit'*. *'Oftheit*™* ,'Solcheinheit'*™ and *Sinneheit*™ (consciousness) represent Body, Mind and Spirit.
The disciplines are discrete. They have unique technologies, methods and require understanding of their basic 'neter' or nature that correspond with likeminded creative people who resonate on the levels of frequency that the manifestation requires.
A shoemaker cannot create a space station. This is true of all three aspects of the thought- form- consciousness manifestation principles involved in making things.
Knowledge of physiology, movement, measure etc. are required. Psychology or behavior runs the gamut from the base to the finer expressions and dynamics as does the spiritual which goes from thought itself to power and electricity which are all related to work energy on all levels. This reduces to consciousness, the one source, with the three dimensions of physiology, psychology and spirituality.

Our visual intelligence is vital to this process. Matrices make it easier to expand fields of observation. Again there is the visual spectrum going from the subtle vibration of vision and imagination to having the ability to 'see' what we create clearly, as they become physical 3D objects in our world. All our senses operate within their spectra of *Oftheit*™' ,'*Solcheinheit*'™ and *Sinneheit*™. The *(Gestalt [gəˈʃtalt] "shape, form")* is the principle that expresses equations of need and articulated relationships translated into physical space, with particular behaviors and energies harmonized and balanced in becoming real. The spiritual dimensions are more remote and unavailable to us. When the alignments are in order the form appears or is made.

We are spiritual beings having physical experiences in what we call life. We operate on dense vibrational levels that are complements of the finer forces and processes and choose to make things. Civilizations that were created by some of our ancestors that have survived in any form of ruin or otherwise seem to have synthesized the physical, behavior and spiritual expressions representing their cultural, social and philosophical and religious systems that we are still struggling to understand. Most of what we create are strictly about satisfying physical need.

This is not how civilization building is done nor how they survive or evolve. At some point in our story we have to be about becoming a civilization. If we understand and are honest about our current state of awareness and can observe the spectrum of our collective consciousness, questioning the progress we have made in alignment with our purpose, the results would be dismal. With all we have done there is a kind of vulnerability or unsustainability about our situation having focused on the physical at the expense of the other two dimensions for the harmony we need. What is the energy for the vision and perception?

Do we know that the 'trinity' is the principle that permeates all human life, every expression and though we have? The one or the source is distributed into the three (dimensions) and returns to the one in cycles of 'life'. All reality operates with this principle. Our binary process give rise to patterns of the triad or trinity. Though we operate with the two that resulted from first polarity the duality becomes a fundamental quality and gauge with codes and specific instructions that are inherent in the expression or being, in an energy-essence. At some fundamental level there is a tetrahedral symmetry at work here. The first two adds the first pair to become the tetrahedron as the first three dimensional form in all three systems of geometry.

What are the disciplines, the skills we master, the nourishment and energies needed for each dimension of the triad. Each one is unique. The physical demands that we understand space, physiology, movement, structural stability, integrity and matter. Behavior requires understanding psychology though all elements are governed by 'behavior' with their internal and external forces the behavior of behavior, we may call tis the 'meta-behavior' is a unique dynamic. Our minds control our behavior. Our thought affect our behavior. Thoughts are in the spiritual realm as the highest form of energy or spirit operating at all levels of super consciousness. Knowing how to qualitatively order the variations of each dimension is the function of the 'Qualitative Matrix'. The law of seven (7) is applied here.

The premise presented states that creativity in the evolutionary sense engages man-woman as being created to be creators as both self-regenerating and being able to manifest reality. Beings given the most degrees of freedom in creation comes with a price. We seem to have been given many tools without instruction manuals. We have to set about finding the instructions by trial and error it seems.

There must be a direct way to avoid the pain and get to the creative bliss we are meant to enjoy. If we were put here to help maintain and develop our reality and support our 'environment', the one that nurtures us, then why are we not more conscious of what our role is about ; to be 'our father's business'? Why are we not building the mansions here and now instead of waiting to get to them when it might be too late then after we have compromised ourselves to get there? Where is that famous image we were created to look like? What were we meant to be like? What were we meant to become? Any likeness that created this awesome reality must be extremely smart. If expanding consciousness is part of our deal then taking it to the max must be our destiny.

This is where the 'mystery manuals and play books' come into play again. Shifting our 'mind sets from our set minds' is key to achieving the paradigm adjustments needed. We must first find the instructions that we came with. Creating a physical, psychological and spiritual protocol that teaches us the secrets to synthesize a new reality is the new agenda. How this is created, distributed and made accessible is now readily available to the planet. We have the technologies and the talent to do so. The pain that is caused by this not being shared to help us all grow and find our purpose has a back lash that we are feeling in every aspect of our lives, the life of the planet and our environment/s. We can no longer focus simply and singularly on our physical or material proclivities and preconditioning. Creating systems to focus on development for material gain and profit at the expense of the psychological and spiritual even religious principles is not wise. A holistic healthy and abundant future is in the balance. When these triadic principles and harmonies are attained we would have evolved to higher dimensions of our selves being guided by the intended universal consciousness we are connected to but seem not to respect. What we do to it we do to ourselves is the price of our ignorance. We must go through the inversion process to pierce the veil of the stress and survival plane to one of living in peace and abundance. This takes place at the center known as 'third eye' or 'seat of the soul' or the 'Pineal gland' which is part of the endocrinal system with light sensitive qualities. Is it any surprise that light is the energy here?

Mental House Cleaning: how to prepare the mind for things that are needed not wanted to make sense that folks may not be receptive to. "Two objects, thoughts or feelings cannot occupy the same space, time or energy field. The old paradigm must give way to the new." The society and culture I live in programmed the fixed patterns of my mind. My very identity, job and paradigm are based on these patterns. These systematic patterns of thoughts (habits) that programmed my mind came from my environment; parents, relatives, community and teachers. They are deeply embedded into my normal awareness or consciousness and my subconscious mind by the imbalance of my body, mind and spirit. This caused the external and strictly material and illusory psychological mechanisms of the ego to automatically resurface when triggers are activated and my buttons are pushed. My monkey mind stays busy with these conditioned patterns that linger longer in my sense of 'comfort'. The thoughts and the processes that evaluate them are cause and effect related. I identify with what is familiar and hold on to bad habits.
Difference makes some 'feel' uncomfortable and alone and similarity make other fell at home. This is how our DNA encodes us. To me difference is more inspiring and creative.

The life of seeking continued until my crisis and trauma came as a total disintegration of my EGO. This was my self-diagnosis. It resulted in total dysfunctionality with no energy with me pretending to understand the world. Everything around me appeared very strange and different. I found my element in the state of difference and that meant I too was becoming different. This finally became creatively controllable and I awoke in my own subjective "paradigm". Art forms I explored turned into magic that never stopped flowing.

Deep Mental, Physical and Spiritual conditioning creates fixed patterns of mind reflecting thoughts, feelings, desires and the expectations created by the environment and all of the messaging it traps us in. The triggers begin to appear in particular situations or with people who immediately activate them. The remedy for this is to become the actor or the trickster or whatever deception is needed to neutralize the trigger and give it what it wants. Triggers always want everything their way. They seem to never get enough 'soul food'. They dislike 'food for the soul'. So to neutralize them that is exactly what we give them. There are prescriptions for controlling what triggers do. Their needs come at their "victim's" expense.

Being the victim is the easiest role to play in the theatre of 'me, myself and I'. 'I' am the star, 'me'; the victim and 'self' is the most capable stock taker of selves to bale these players out of their predicament. These three actors conspire against 'us' to become the host and hostesses of habits that define the 'I, me and self' or 'us' players. These habits (all pun intended) have their costumes for particular situations, crafted to represent the values and rewards desired as outcomes are extracted from the situations or people who to us are triggers.' I, me and self' can never be triggers, or so 'we' think. What triggers need; they need to be reconfigured creatively to be de-triggered permanently. The 'EGO' is the renaissance-man appearing in all actions, subtractions and transactions.

So what is the problem with EGO? The natural world, including our inner world that we also omit or forget, is made of forms and shapes the objects that are expressions of vibrational energy at the material level. It corresponds with the ethereal realm of FORM and SHAPE as phenomenal thought at much higher conscious levels. Each realm has its own symmetry rules, language and 'essence' we need to know to communicate not to talk. EGO is what we apply to the objective world as a healthy dynamic that can become "Energy we use to Get Over" with compromise. "Everyone wants to get to heaven but nobody wants to die to get there"; Says BB King. There is no backdoor to resurrection and judgement. To us in the 'now' transformation is a form of resurrection. We judge which habits align with life's goals. How do we discover what they are? Do they relate to our values and our basic nature?

A thought field: I cannot exist with expectations alone in the pre-sent society in our present state if we are to enjoy the present (value) that life is. I am unable to exist but can I still be without a body, without a mind and without my spirit connected to the greater SPIRIT. Where did 'me, myself and I' get this idea of isolating expectations for the synthesis of the triad of consciousness on all its levels? The forms and shapes of the world have a mirror
plane between form and shape. It divides illusion from reality and chimera from dream. These are states with depths we are unable to fathom with our illogical reasoning skills and limited mental capacity when there is an intuitive alternative attainable through silence and thoughtless awareness.

The compelling single force of the expectations, is a vast matrix of the multiple dimensions of consciousness upon which gems of consciousness grow. With all our knowledge we are still not up to the task of formulating complete and healthy identities, identifiers and identifications. We settle for ease and comfort and create false economies, sticking to personal identity with costumes and expectations that come even after waking up, or in whatever ill- luci-dated state we find ourselves.

'Fixed paradigms' are powerful, impactful and unhealthy. Their power aligns with the degree of the inadequate identifications and identifiers we create and experience as value without integrity. We can translate this and 'simple-fi' it to con-note 'KNOWING' and eventually meaning. Unless the ego is disintegrated strong belief in one's own thoughts creates overwhelming power over you. The ill-use of the faculties will then easily obscure all vision with illusions, reduced awareness, focus, attention and Intention not to be present. This will cause you to surrender to daydreaming. Dreams,as astral phenomena are creative and useful. What they manifest unconditionally can lead to the true freedom we aspire to, to attain our goals.

The intelligent life force is an eternal present (gift) of 'nowness' that's not very material. Now itself, knows no past nor future. When we are invited to it why do we bring our past and future to it? What type of dissonance does this create? Now is no space-Body, no time-Mind and no Energy only Pure Spirit. [If it is as encompassing and as vast as it is without space, time and energy we get lost in what seems to us, in our mind's eye, to be a void.] We then avoid being in the void for any length of whatever 'measure-meant we use for keeping score. We elect to be in the addictive and very seductive conditioned past we carry that we cannot disconnect from permanently.

The void is where all the ethereal energies are found. It is the pure medium for transformation and dynamics like the pendulum. Here our slow vibrations oscillate to the stressful, anxiety ridden over expected and expectant states we are preoccupied with and miss what was pre-sent and given to us to live our life in abundance and joy.

Expectation and identification are tantamount to wanting to see or have something 'pleasant' happen to, for or with us and others etc. Let's translate. What it says is; "if we know and are clear about what we want it is reasonable to expect it to happen". The "to, for or with us and others etc." is the proverbial all powerful WHY?. Does this favorable (expected) result of the happening lead to happiness? The location of the happening must correspond and align with all qualitative and quantitative aspects and properties to create the happiness we seek. It's like the mystery of finding anything we lost. We must look for it, were it is, to find it. It's logical, isn't it! Happiness can only be experienced in the appropriate state or the right place, mood or mindset for what 'must' happen to coincide with all the preconditions needed for manifestation.

Let's look at the root of the word happiness right now. 'Hap' is the root for things 'accidental'. 'Happen-Happenings and Happiness' are all connected to SOURCE and are co-incidental.
Haphazard, Happenstance and Mishap are the opposite unfavorable outcomes.
The flavor of the Favor is key. The favor of the flavor is me being creatively free. It is what we truly need to be happy. It's what we pray and work for. It makes our work our prayer.

The state of 'nowness' is accessible through thoughtless awareness. Behind the silent veil of deep meditation is our relative distance to SOURCE. The stress that we live in is caused by our misaligned state. Forgiving spirit catches up with us, justifiably so, at different degrees and stages of trauma, crisis, disease and then death. No one knows what happens at the moment of the return to our arising. Personally I do not think we should wait to find out. We take this opportunity 'now' (as our motivator) to align with source to enjoy the favors we '<u>learn to earn</u>' for our time served and the service we offer to others and the planet. This then is our purpose.

To make 'awakening' irrevocable we must recognize all fixed mental patterns as habits of (body) (mind) (spirit) with their paradigms. Remember that the paradigm is the grand value system aligned with universal principles and the three dimensions of consciousness are very specific. This is where we might be getting stuck; not identifying accurately, the difference with specificity and generality. This recognition triggers or motivates the proverbial 'WHY". That is the deep soul searching introspection that addresses the pain and brings it to the surface to motivate us to be honest and true to ourselves. This is not the time for shutting down. 'Kicking the can down the road' eventually transforms into that bigger horizontal 'metal can' we're in on our homebound journey. There are some experiences we leave behind and those that are in our hearts we take with us to be weighed against the feather.

Why can the mind not create itself? It deludes itself by participating in its own conditioning but cannot heal itself if healing is needed in avoidance mode or from behind closed doors. It must be answering to an infinitely higher power. Fixed mindsets cannot be transformed by thoughts, deeds and beliefs of the same mind alone. Problems do not fix themselves within the same universe. Applying effort, practice or willpower, found primarily in lower vibration ranges, can never be effective. How do we get beyond the limited self and pause for there to be a new cause to reboot all meaningful and sustainable transformations with the right synthesis in place, in time and in energy?

Back to the triggers. How do we know what they are? Are they old or new and to whom do they belong? Who activates them and who or what are the targets. If they are physical they are emphasized by the lack of creature comfort and desired things that ruined our expectations and manifestation methods, arts or sciences we have no way of verifying and are leaving ourselves open to the "snake oil sales" phenomena, analog or digital, of the day all over again. Are there distinctions between things that are desired, wanted and/or needed? Not all people desire or want the same things. There is diversity and ranges of differences to consider. Basic needs and 'status inspired' needs are based on social, cultural and economic status, access and means.
This is all directed by habits, paradigms and the panoply of human traits, proclivities and all else imaginable. Stations in life, status revolved around development of the self. Individuals, communities and groups are just a few layers to consider. Materialism outwardly directs this human urge. The urge for self-development is the purview of the true self. One that know its connection to source and is in the flow of its vibration without being distracted by any external definitions or identifications of it and the fixations of its paradigms and dogma. When fixed mental patterns appear, this mentality becomes the buffer used to protect us regardless of any expectations that activates it. The limbic brain with its 'fight or flight' stress responses is neutralized or trans-ended.

How can we achieve true 'global intelligence' when the Wisdom Intelligence Knowledge and Information from all cultures are denied and are not part of global consciousness? This is essential to our holistic realization that could elevate human consciousness now.
These are the wisdom traditions and cultures that continue to contribute significantly to the WIKI.

1. Egypt: The 7 Hermetic laws that are the 'symmetry principles of nature, western science know nothing about.

2. The Buddhist Canon; the Dao DeJing with its transliteration to a 'Dao Design™.

3. Hinduism offers the Kundalini and the 7 chakras (East) which aligns with the endocrine and human energy system (West). This is the East-West Dichotomy we need to synthesize. Our entire consciousness and all its expressions depend on this global mindset shift.

4. Many other systems will be presented. What other cultural and wisdom systems or 'kens' do we use to synthesize your own paradigm.

This is the 'praxis' for creating the expressions of shape/s and form/s:

1. Sacred and Profane geometries using the three systems of form namely the Euclidean and the Non-Euclidean. Contributors to the Non Euclidean traditions are Gauss, Riemann with Bolyai and Lobachevsky in the 19th century.

2. Newtonian, Einsteinian and Subatomic Physics.

3. Quantum theory, entanglement, entrainment and String theory.

4. Think of other resources to build systems in other disciplines.

BE THE ENTREPRENEUR TO YOUR OWN PROSPERITY h.g.b

EVERY MAN IS AN ARTISAN TO HIS OWN (FREEDOM AND) FORTUNE.
Appius Claudius Caecus.

Body	MIND	SPIRIT
THE EGO	ILLUSORY IDENTITIES	THE EGO-MIND DYNAMICS
LANGUAGE	REALITY	CAUSE AND EFFORT
DREAMS	DAYDREAM	TRIADIC QUALITIES
DIRECT KNOWLEDGE	BROWN STUDIES	SOURCE IS EVERYTHING
BARRIERS AND WARS	VEILS	IN EVERY SHAPE AND FORM

HOW TO OBSERVE OLD HABITS AND EMBED NEW ONES

Meditation is the silence (energy) that allows us to connect to source through thoughtless awareness to rewire the circuitry in our brain and to tune the rest of the Human Instrument.

Gnosis is the common Greek noun for knowledge. It generally signifies a total and reliable dualistic knowledge in the sense of mystical enlightenment or "insight".

Gnosis taught the deliverance of man from the constraints of earthly existence through insight into an essential relationship, as soul or spirit, with a supra-mundane place of freedom.

The term is used in the context of ancient religions and philosophies, aspects of Judeo- Christian beliefs, particularly to the ideas that emerged during early Christian and Greco-Roman interaction during the 2nd century.

It seems quite appropriate for the concepts that are emerging in the new paradigm. Gnosis has spirit and consciousness. Knowledge does not and is materialistic and Newtonian in its fundamental nature.

Gnosis is a feminine Greek noun which means "knowledge". It is often used for personal knowledge compared with intellectual knowledge. Latin dropped the initial g (which was preserved in Greek) so gno- becomes no- as in noscō meaning " I know noscentia meaning "knowledge" and notus meaning "known". The g remains in the Latin co-gni-tio meaning "knowledge" and i-gno-tus and i-gna-rus meaning "unknown" and from which comes the word i-gno-rant, and a-gno-stic which means "not knowing" and once again this reflects the Sanskrit jna which means "to know", "to perceive" or "to understand".

In the kundalini of hindu spiritual philosophy, ajna is the sixth chakra where communication and other traits are centered. Ajna (Sanskrit: आज्ञा, IAST: Ājñā, English: "command"), or third- eye chakra, is the sixth primary chakra in the body according to Hindu tradition. It is a part of the brain which can be made more powerful through repetition, like a muscle, and it signifies the conscience.

"Optimism is a strategy for making a better future. Because unless you believe that the future can be better, it's unlikely you will step up and take responsibility for making it so. If you assume that there's no hope, you guarantee that there will be no hope. If you assume that there is an instinct for freedom, that there are opportunities to change things, there is a chance you may contribute to making a better world. The choice is yours." —A. Noam Chomsky

CHAPTER 5 TRINE B

Towards a 'true form' paradigm.

As we evolve, our consciousness expands. New concepts, information and innovations emerge charged with identities and qualities of forms unique to time periods that continue to unfold. Not only does the verbal communication change, the everyday objects, the scientific, mathematical and classical form vocabularies do also. Without fanfare and celebration of those who make these subtle and major contributions to our lives we seem to continue in a mindset that does not readily embrace change. Change is at the root of all growth and transformation, as quietly as it's kept. At basic levels of living, change is still the most natural attribute of being and becoming. The groundwork for change is created along continuous multi-directed paths of creative challenges and contributions made by dedicated and connected souls who share their visions with the world even in the din of chaos and uncertainty. We are not alone, we do nothing on our own. We stand on the shoulders of little people and giants alike. Everyone has a song to sing.

The ideas presented here are a to create the synthesis of information inspired by wisdom traditions, experiences and experimentation in design, art, architecture, 'visual mathematics' especially geometry with readings in science and metaphysics. The goal is to synthesize a holistic identity with a unique aesthetic flavor that ties all creative expressions together with a beautiful, eloquent and elegant language that's practical, economical and in harmony with all aspects of natural laws. The approach here is a pragmatic one that explores and searches for 'truth' not in theory but as reality. The notion of 'when a thing is (made) makes it true' is the maxim and reality test applied here. It is the ultimate experience of a living, dynamic form of symmetry in action. All parts are harmonized to make expressions (W) holistic.

"When the path to knowledge is trampled and bare the new imperative becomes the search for virgin and fertile ground." Restoring the groundwork for individual contributions creates valuable changes that may not be fashionable. The path continues and there will always be dedicated souls who share their visions with the world even in the din of chaos and uncertainty. Be true to self and good form and function will follow in all you do.
How many dimensions are there to (universal) consciousness?

Evolutions and Adaptations using these sites to support ideas, concepts and thoughts as functions that are evolving as discoveries are made. This relates to the expansion of the causal universal and human consciousness inspiring the new Paradigm. The connection to source is the benefit of this process. The transformations experienced during this process will continue with or without humans. The goal is to apply the knowledge and technologies to harmonize all our future developments.

Disciplines: art, architecture, industrial design, fashion, food, toys, games & puzzles, free and renewable energy generation and storage, new materials and production technologies, software and digital technologies.

A Form Paradigm: Starts with a Physiological: Matter; Space that relies on (Gauge theories) Measure, The Psychological: Time; Emotions, Expressions and the Spiritual Creative thought process. The new definition of Man is now a Bio, Electrical, Plasmic, Magnetic Complex or spirit in a physical body having creative experiences operates in fields of consciousness. We are connected to the 'universal KUNDALINI or CONSCIOUSNESS' as our vital creative source and living processes ART-iculated as (human) expressions of the self-image of all creative forces. THOUGHT is the major process used in observations, participation, and Experimentation used to discover other creative strategies.

ART is a collective Creative Process (Mind and Thought Processes) ART–iculated in Space, time and energy. Research, Observation, participation based on a Foundation of mathematics (once a spiritual endeavor now a tool for keeping score), Core Flavors: Muchness, Suchness and Consciousness. Harmony with space, time and energy can be fused into the holistic dynamic where Mind, Body and (Creative) Spirit are the three dimensions of consciousness. Is the word 'di-mension' inherently dualistic?

Three dimensional Forms rely on Systems of Geometry: Euclidean, Non-Euclidean, Organic, Amorphous (irregular or intuitive) SYMMETRY PRINCIPLES: Operations and Behaviors, Symbols, Structures, things and objects all EXPRESSIONS Words, Gestures et al. They generally are codified in an organized system or a Periodic matrix that aligns with the qualitative (suchness), quantitative (muchness) and Universal principles of Consciousness as the foundation for human development and sustainability. They are natural expressions, with flavors 'molded' by the forces (realities) with the prescriptions of their fundamental 3Fold Nature (Neter) that corresponds with the three dimensions with all symmetry principles harmonized according to natural law.

Creative 'thought-forms' precede the manifestation or creation processes and are all governed by the laws of number, color, scale, gauge and. They operate in the formulation, manifestation and correlative dynamics of identities, realities, uses and functions within the aesthetic flavors of all natural and artificial expressions. It is for this reason that periodicity exists, meaning that flavors are common to various phyla of realities and expressions that can be blended into a qualitative synthesis. The 'minimum Inventory-maximum Diversity' concept applies. There are few principles and elements that govern all reality.

The following ideas are 'tags' that relate to form generation principles: Vibration is the medium for the creative process.
Logic structures or the 'Flavor Machine' rely on Periodicity and Number properties for the Physiology of things with much greater benefits to add to the unfolding story of mankind. The sacred values, qualities and properties of number come in many other forms.
Value is created through understanding the psychology- (behavior) of things and materials we create.
Spirituality- Thought Form-Symbolic meaning; Natural Law, SYMMETRY-Logic, Intelligence Intuition are all part of the process of synthesis
Flavors help form 3fold realities and expressions of all natural phenomena in the world.

Polyhedra- Generate Euclidean, Non Euclidean, Organic and Irregular (Artificial Forms) following traditions conventions and inventions that continue to unfold as our consciousness continues to expand. Ancestral and Cultural Adaptations, Applications, Creative life support Systems Industries etc.Design Principles are at the core of all these Logic Structures.

From the origins of celestial alignments in monolithic ancient technologies to all useful space systems. Ancestral sites like Gobekli - Tepe in S.E. Turkey, Stonehenge in England and the Egyptian Pyramids are the finest examples of form in its optimum function and spiritual potential we have not yet understood. The triad of Form, Function and Spirit is the New Paradigm Principle for the new age. Digital simulations, prototyping networks, life support systems technologies, products and projects. Distribution of technologies and systems, economic models, environmental and ecological harmony, wellness health and wealth creation, education and personal development, creative industry and engagement in and for all communities.

Of all the freedoms and resources humans pursue and attain, the right, the privilege and the ability to access knowledge freely from the abundant source of universal consciousness is not readily available. "The world of Africans and descendants of Africans and the world of scholarship about them is still the only one at the end of the Twentieth Century that retains a 'colonial' signature whereby experts and authorities outside African communities control knowledge creation and exceed experts inside those communities. This does not apply to Europe, Asia or the Americas. This has led to an unfortunate predilection among Africans to concede expert knowledge to outsiders. African people have tended in the past to surrender the right to academic self-affirmation to others, thereby accepting conclusions of a Eurocentric framework that have assigned a permanent peripheral role to the Africa centered perspective in the world's growing knowledge industry. Indeed, many of the 'authorities' who study and write about the African world and exercise great influence over the outside world's perception of Africa and Africans, the understanding of its value priorities, the vision of its future and the capacity to define its very essence for insiders and outsiders alike, often are not burdened with the knowledge of single African or African derived language."

There is no monopoly with consciousness in knowing the bi-product of these universal processes as knowledge. It is free and readily available. What you do with it makes it relevant. Periodically the universal consciousness evolves and the 'global' paradigm/s shift for those who are 'connected' to source to benefit from this natural and inherent transformations and the free access of the new information that continues to unfold.

Are we well or even better informed or educated with our current methods, channels and systems of education? Is the contest of pedagogy and andragogy ever going to be resolved?

Are the 'products and services', we put our trust in traditional educational and institutional culture serving its purpose? Do we recognize the intuitive or "genius" factor in our spiritual lives?

The status quo is not as agile or nimble enough to make the transitions that 'paradigm shifts' offer us. There is a creative dimension to intellectual freedom that is 'spiritual'.
For some 'strange reason' this is factored into the status quo sense of itself, its presence and its reality. Value is the key to determining the relevance and benefits of the 'knowledge bases' existing and new. Culture is determined by its expressions. Physiology, Psychology and Spirituality are three dimensions of consciousness and by 'expansion' all else that it expresses and is emerging and constantly unfolding. That we, 'the status quo', stop growing or embracing 'the unfolding of our consciousness and inherent knowledge' collectively and individually is unfortunate.

The Premise for Learning

Consciousness continues to evolve. Our intellectual, spiritual and knowledge base is shifting.
We discover the truths of the genius of the human mind. The classic Eurocentric socioeconomic motivations for human industry and how we are trained to survive are not the only effective sources of inspiration for living in today's 'thingified' world.

In developing an organic and deeper educational system it must be based on 'true and complete intellectual freedom'. We can incorporate principles and content from systems that define the overall culture which in the new definition is composed of the physical, the psychological and the spiritual dimensions of ALL consciousness and its expressions that support human life. Focusing on the material, mechanistic, and physical dimension alone does not complete the triadic harmony that leads to the organic harmony that nature, with all its laws and principles, operates with. We struggle with the definition of the word 'Genius'. We need to identify the cultural; physical, psychological and spiritual priorities of people in this pluralistic society who deserve and are entitled to all expressions and experiences of freedoms.

In some of the classical disciplines of the arts, the sciences, mathematics, psychology and the medical arts fundamental shifts continue. Here is an opportunity to make adjustments in our approaches to how we connect to the deeper sources to learn, ART-iculate and teach. Intuition, 'subliminal perception, the sub conscious, the super conscious, happy accidents and all of the vicarious experiences of intelligence, knowledge and supposed freedom we share is not embraced by the status quo.

In some circles of education the 'hands on' approach is in again. The hand mind coordination is critical to many disciplines. The understanding of cognition, consciousness, creative and critical thinking are part of the 'Spiritual' processes involved in the body-mind-spirit, continuum. The East-West synthesis is having a great impact on our lives. There is much we can learn from the values and principles of oriental cultures especially when it is believed that oriental culture is the seed of this planet's civilization (SINGULAR).

There is a contribution found in pyramid building that was credited to the Greeks in a discipline of geometry called 'polyhedra'. See appendices. These are forms of nature that give structure to the unseen and visible worlds and forces. This is an area that needs to be explored since it relates to many other disciplines involved in learning. They are the building blocks of nature.

The seven styles of learning need to be designed into physiological, psychological and spiritual activities to complement the cognitive and neurological dynamics that help us learn. There are many other 'cultural' sources of information we need to look at to discover how to complement the roundedness of the whole person. The moral, ethical, social and other value systems are to be encoded into activities, behaviors and philosophies as well.

Achieving some modicum of organic, creative and spiritual freedom is of paramount importance to me. It is motivated by the desire to maximize my life's potential given this opportunity to access knowledge from the academic status quo. I did not totally trust nor excelled at it. Following the tried and true, being one of the many doing the same thing many times over and chalking it up to years of experience was a waste of time and of no interest to me. The information I was able to access after my transformation to the traumatic shifts I made is part of our new paradigm's new vocabulary or voice being used to share the expressions of this creative freedom we are all entitled to. That it is possible for one individual, to free themselves of or transcend their living circumstances, to connect to source and draw from it new information from wisdom traditions and from the subconscious, intuitive, creative, spiritual representing the cultural freedoms they all continue to express is exciting to me. It is the voice that reaches deep into my soul that I was given to give 'form' to life as part of an aesthetic vision that is unfolding. I am not sure that I am the farmer to plant the seed nor if the ground is fertile enough for it to grow. What I have mental control of, though, is that the expressions exist in the mental world or source and I continue to bring them into the world. The answer to the question of which world it will be is unfolding.

The emerging reality is a rich source of information and exploration opportunities we can all participate in when we claim our spiritual freedom. The various dimensions of the world that are evolving are the re-sources harmonizing humanity, nature, the forces, principles and laws. There is a definite new perspective that still needs to be explored and put to work as we move forward. Some of the fundamental ideas I work with relate to consciousness in all of its dimensions and expressions. Nature is part of consciousness and every phenomena we experience fits into a periodicity that unifies and organizes them into a matrix of 'suchness- flavors' and 'muchness-numbers' qualitatively and quantitatively. That numbers are much more then we think they are. Counting is for keeping score. As natural laws they help explain much more of what we think and feel about the world and ourselves.

That every form is a logic structure incorporating symmetry principles we still do not understand. Meaning comes from the 'oneness' of the cosmic consciousness with all its expressions in an interactive constant creation. There are logic structures available to us that have been used by ancient civilizations that are still relevant today is significant. Form is essential to establishing truth when thought is expressed in the continuum of space time and energy. It is a dynamic medium for and of the manifestation for all consciousness.

DeSign Science in The New Paradigm Age

The 'of' is cause. The for is effect.

The geometric ART-iculation of the emerging expressions and updated languages and symbols of the new paradigm are an iteration of an emerging, consistent and repeatable Non-Euclidean geometry we still do not know or seem willing to embrace, when we see it, as being or becoming the truthful expressions of valid thought processes of great minds we continue to dismiss. From these few concepts there are industrial applications that can help nurture, improve and support human life. The nurturing is educational freedom.

The improvement is what everyone, actively participating in their own growth, will experience when efficient systems reach levels of sustainability transforming the socio, cultural, economic agenda making true freedoms attainable.

THE QUALI-QUANTITATIVE PERIODIC MATRIX OF THE TRIADIC SYSTEM

TRIADIC CONSCIOUSNESS	AWARENESS CONSCIOUSNESS	SUBCONSCIOUSNESS	INFINITE CONSCIOUSNESS
PHENOMENA EXPRESSION	MUCHNESS PHYSICAL	SUCHNESS PSYCHOLOGICAL	ONENESS SPIRITUAL
SYMMETRY	CHARACTERISTICS	QUALITIES	FLAVORS
HARMONY	BODY	MIND	SPIRIT
UNDERSTANDING	PHYSICAL THOUGHT FORM-IDEAS	THE SENSES-EMOTIONS FEELINGS-HABITS	HIGHER LEVEL THOUGHT FUELS
SENTIENCE	KNOWING-AWARENESS	THE UNKNOWN-INTUITION	TRANSFORMATION PURPOSE
INTEGRITY	STRAIN-EFFORT-FORCE	STRESS	BECOMING
VIBRATION	MOVEMENT	TENSION	ENERGY
MANIFESTATION	MATTER	KNOWLEDGE	CREATIVE GENIUS
PRODUCTION	MATERIALS	BEHAVIOR-RITUALS	WORK ENERGY
FORCES	MECHANICAL	ELECTRICAL	MAGNETIC
SYMMETRY ELEMENTS	SPACE	TIME	ENERGY
MATTER	SOLID	LIQUID	GAS-PLASMA
ORIENTATION	DOWN-DN	SIDE-S	UP-U
WEALTH CREATION	INELLECTUAL PROPERTIES THE IDEA BANK AND OTHER ASSETS	THE CREATIVE PROCESS	CAPITAL FORMATION CURRENCY CONVERSION AND IP TRANSFER
USERS & MARKETS	CLIENTS, CUSTOMERS ARTISTS, SCIENTISTS AND DESIGNERS	CUSTOMER CARE, ART, SCIENCE AND DESIGN	VALUE AND BENEFITS PURPOSE FULFILLMENT
	HUMAN BEINGS	NATURE	CONSCIOUSNESS
	EXPRESSIONS	EMOTIONS	THOUGHTS

DeSign Science in The New Paradigm Age

MANIFESTATION: There are varieties of talents distributed within the same creative force that controls everything we create. Each has its type of product, service and function in a synthesized value dynamic that satisfies needs.

The same energy empowers them all in everyone. This is design and there is a science to it that is now being revealed to us at new levels of consciousness the planet is in now.

To each is given direct power, knowledge and connection through their Creative Spirit for the good of ALL. Through the senses, talents and Spirit wisdom is given to one and to another is given knowledge expressing the same consciousness and awareness.

The path along which this process takes place is a design link which can be seen, in the mind's eye, as a 'Ray of Creation' connecting the ideation through the creative process to interact with the forces in varying frequencies of vibrations being transformed into dense of matter.

This all takes place in accordance with symmetry principles, natural law and in equilibrium. The 'design and science of manifestation' diagram on page 188 is a representation of this process.

A number of processes are overlaid in the diagram indicating the complex interactions taking place within the Body, Mind Spirit continuum and the Space, Time, Energy Process.

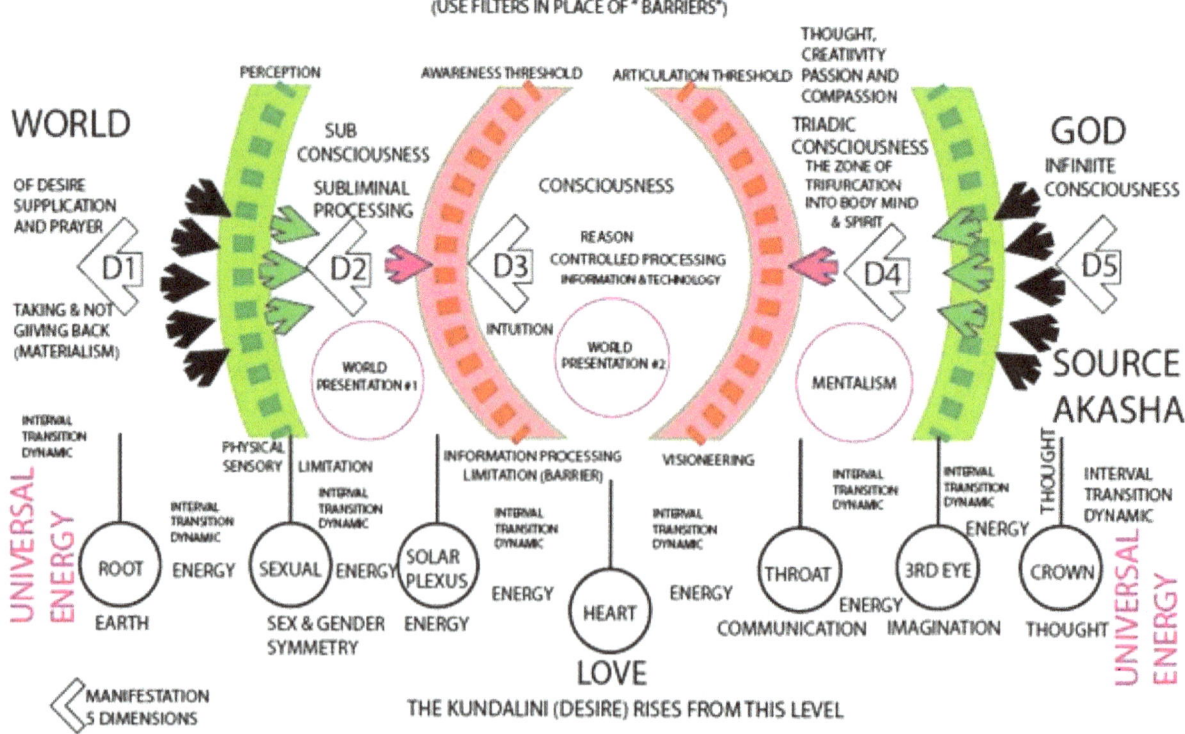

Fig 17

TRIADIC RESEARCH METHODS

BODY	MIND	SPIRIT
GATHER	SORT	TAXONOMIZE
PHYLUM, GROUPS	BEHAVIORF	ENERGY DYNAMIC
FORMS:	STURCTURE	INTEGRITY
EXPRESSION	PROCESS	MEANING
GEOMETRIC/ MECHANICAL AND PHYSIOLOGICAL	ALIGNING WITH SPECIES AND TYPES PSYCHOLOGY OR TYPOLOGY	LIFE FORCES, MOBILITY AND FREEDOM OF MOVEMENT OR VIBRATION
QUANTITATIVE AND PHYSICAL PROPERTIES THE EVOLUTION OF SPECIES AND THEIR GENEOLOGY ANCESTORY	SYMMETRY PRINCIPLES AND OPERATIONS, QUALITATIVE MAJOR AND MINOR STRUCTURAL SYSTEMS	FUELS, POWER, WORK ENERGY

KEY CONCEPTS TO OPTIMIZE CREATIVE THINKING

SPACE	TIME	ENERGY
THE SUBJECT OR OBJECT	CREATIVE PROCESS	THE ACTIVATOR
AWARE	AWAKEN	QUICKEN
WHAT, WHO, WHOM	HOW, WHEN,	SOURCE, ORIGIN, ACTION
NOUNS AND PRONOUNS	WITH/BY/FROM	ESSENCE, WHY (PURPOSE)

HERMETIC PRINCIPLES, KUNDALINI, SYMMETRY PRINCIPLES, GAUGE THEORIES, VISUAL INTELLIGENCE AND VISUAL MATHEMATICS

THE SCIENCE, LOGIC & MATHEMATICS FOR THE QUANTITATIVE; QUANTUM ANALYTIC AND QUALITATIVE ORGANIC EVALUATION OF PHENOMENA, EXPRESSIONS AND ENERGIES, ALL TO SUPPORT THE TRIADIC SYNTHESIS AND HARMONY.

THE BODY MIND SPIRIT SYNTHESIS MODEL

BODY	MIND	SPIRIT
DIGITAL PRODUCT DESIGNER, TECH/DESIGN INNOVATORS, TECH/DESIGN ENTREPRENEURS PATHS THAT DEPEND ON MATERIALISTIC AND PHYSICAL LABOR INTENSIVE THINGS AND MECHANICS	CYBERNETIC DIRECTOR, HUMAN ORGAN DESIGNER, EMBODIED INTERACTIONS DESIGNER, SIMULATION DESIGNER, PATHS THAT ARE BEHAVIORAL; RELATED TO FEELINGS: CREATIVITY AND INTELLIGENCE	AS AUGMENTED REALITY DESIGNER, VR DESIGNER, CONTENT DESIGNER, AVATAR PROGRAMMER, DRONE EXPERIENCE DESIGNER, PATHS RELATED TO THOUGHT ENERGY AND HIGHER STATES OF CONSCIOUSNESS
CHIEF DESIGN OFFICER OR CHIEF CREATIVE OFFICER	INTELLIGENT SYSTEM DESIGNER	PROGRAM DIRECTOR
CONDUCTOR: CREATIVE TECHNOLOGY RESEARCHER	MACHINE-LEARNING DESIGNER	SYNTHETIC BIOLOGIST
FUSIONIST	REAL–TIME 3-D DESIGNER	INTERVENTIONIST,
DIRECTOR OF CONCIERGE SERVICES	SYNTHESIZERS, SYNECTION	SIM DESIGNER
PRODUCT DESIGNER,	SYSTEM DESIGNER	NANOTECH DESIGNER,
AWARENESS	AWAKENING	ACTIVATING

Visual mathematics offers the basic understanding of the symbolic creative processes derived not from 'traditional calculations but from and thought processes that involve the form generating functions described by Euclidean and Non-Euclidean geometry, topology and 3D Form modeling. Entire Form vocabularies are invented using newly formulated symmetry principles that can be synthesized with the visual sciences, visual mathematics and art.

The following distribution matrix is one example of the logic structure using the symbolic color, number and verbal elements to create the "synthesis" of flavors from their combinations and proportions to produce new forms.

DeSign Science in The New Paradigm Age

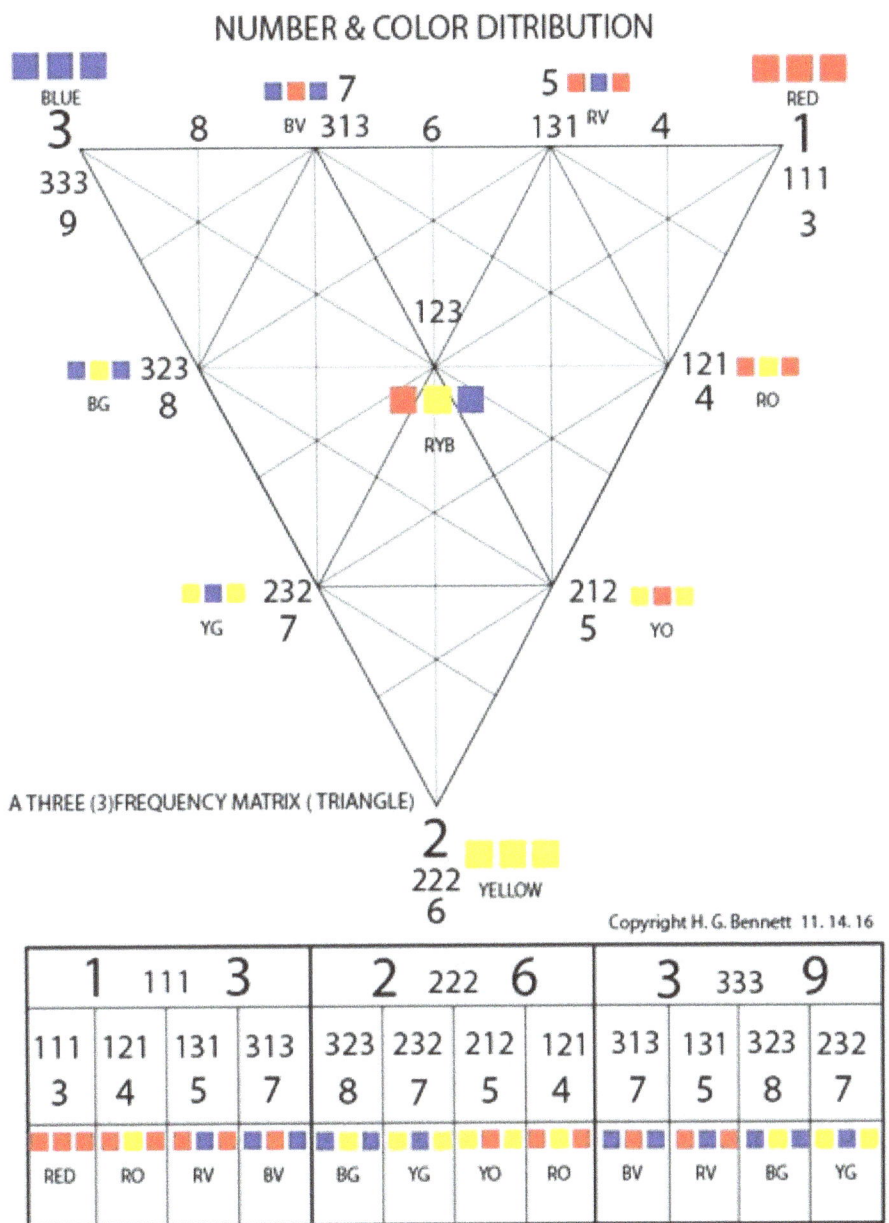

Fig 18.

"Making the abstract real"

"**Optimism** is a strategy for making a better future. Because unless we believe that the future can be better, it's unlikely we will step up and take responsibility for making it so. If we assume that there's no hope, we guarantee that there will be no hope. If we assume that there is an instinct for freedom, and there are (we create) opportunities to change things, there is a chance we may contribute to making a better world. The choice is ours." —**Noam Chomsky**.

CHAPTER 5 TRINE C

VISUAL MATHEMATICS:

"From deep within the open fields of infinite vision the true seeker sees solutions never seen before and by the study of numbers, form and flavors makes the abstract real in suspension on route to an ascended consciousness to the source of all arising. We develop spatial creative thought form definitions of 3D process that come with precession for synthesizing flavors following natural law" HGB

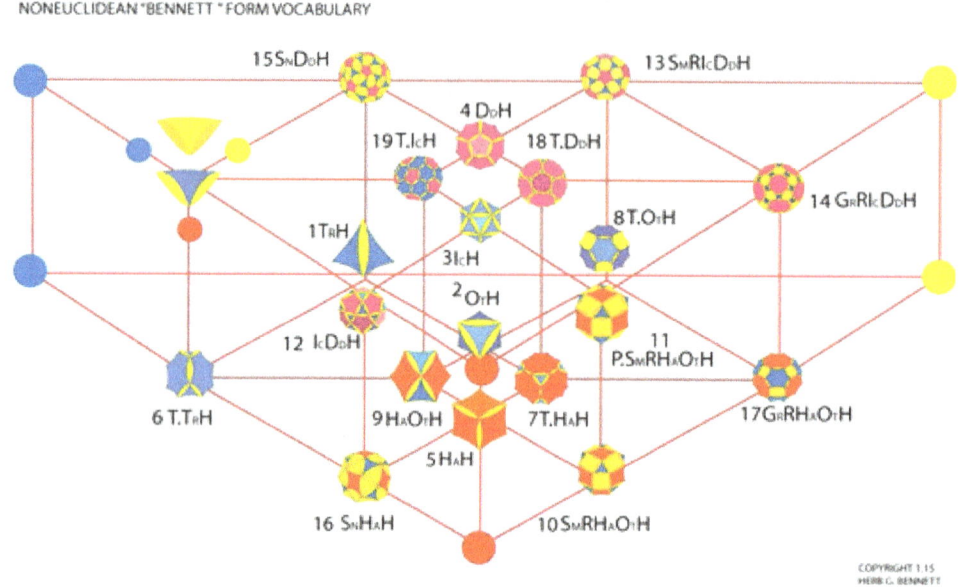

Fig 19. Non-Euclidean Bennett Polyhedra

There is a symbolic and structural logic found in this work that resonates with ideas I have been developing over a considerable length of time. From my researches and studies the information I rely on evolves very slowly. This is occurring in a climate that is dominated by thoughts about numbers strictly related to counting and calculation. The see, touch and measure paradigm is the fundamental rule of not only science, but of life it seems. Quantitative reality is only half (of) life. The qualitative, geometric, suchness complement is lacking. When the qualitative methods are synthesized with the number crunching a more complete system of understanding will be achieved. What we call ancient wisdom holds many of the clues that could unlock much of the knowledge needed to begin the process of developing a right brain methodology for understanding the 'flavor dynamics and mechanics of consciousness' **Numbers** have many flavors, qualities and values. There is an order to them that follows the sequential logic of physical, psychological and spiritual laws. The counting logic is a progression of an 'additive and subtractive (polarity) unity'.

One added to the largest googolplex, (Milton Sirotta's number with the figure 1 followed by 100 zeros equal to 10 to the power of 100), becomes, the next largest number to absurdity. The qualitative logic can be detected in subatomic physics as the super symmetry model in the three frequency decimet of our number system of bases 3, 9 and 10. This Study of numbers is the definitive work on the spiritual nature of numbers as it relates to creation. If we look at the number crunching computers do, the basic binary logic builds on 1's and 0's in a combinatorial ordering sequence that is at the core of the processor's nature. The silicon chip's qualitative resonant constant frequency generation potential is its nature. There is an inherent qualitative-quantitative translation taking place that changes one symbolic language to the traditional number character we know as our number system. In the decimet symmetry system a synthesis can be made of all natural phenomena found in nature, including the quantitative number crunching functions. The operations conducted in our traditional mathematics are not yet detected as working principles in this flavor machine. There is also a 'stacking' correlation of the fundamental flavors that govern all of mind and consciousness as we know them now.

There is a human factor and aspect of 'mind, thought and common sense' operating here. To understand this approach to nature's laws, how they affect us and how we create our lives the nature of elements and their behaviors must be experienced and appreciated within their respective classes and symmetry groups. For example the fundamental nature of orientation includes up, side and down. These three terms correlate to other principles like matter; liquid, solid and gas etc. The 'Stacking potential' of their flavors we do not factor into our logic systems. By Stacking I mean the 'upness' of matter is determined by the behavior of the quality of matter that tends to have that special flavor of 'upness' to it. Of the three distributable aspects of matter which one has the ability to be or rise up? By our experience and observations we can attribute up to gas. Gas is light. Gas also rises UP etc. By the same qualitative correlations we attribute Side to liquid, since liquids flow sideways and down to solid because of gravity.

All of nature obeys laws of numbers for each expression of its principles, flavors and qualities. Numbers are at the root of them all. The super symmetry that the subatomic physicists use in their qualitative matrix is not as organic or as inclusive as this 'flavoring' approach promises to be. The Mind, Body, Spirit Triad transcends all expressions of nature, thought, and man.
The three-fold number base (3) is evolving into a tertiary logic. Our consciousness continues to expand. Our languages and tools are not able to express the efficiency and elegance of the information and knowledge contained in our vision, and dreams. There is a living, universal, qualitative and causal reality of numbers (R.A.Schwaller de Lubicz).

From the irreducible One through the unfolding of creation through the cycles of polarization, ideation and formation; Topics include Numbers, values, and relations; the disengagement of numbers; the harmonic basis of numbers; the development of values' and the establishment of harmony We apply our lateral thinking skills to expand this course. The wisdom traditions of the Egyptians explore more of the essential aspects of architecture to demonstrate the spiritual encoding of temple forms and other structures that still perplex us.

The psycho spiritual energetic; bio magnetic energetic; mind spirit body dynamic continuum replaces the definition of man given to us by Rene Descartes. To him man is a 'plumbing project'; various systems of fluids flowing through the body.

Whether they were analogs or not this model is quite useless now. In stark contrast, while ancient traditions were quite available, no links were being made to Egypt, India or the Orient. How many dark ages have we had? We need to ask our historian friends this question? We are finding more meaningful connections to the past now than the past has ever had to its own past.

One major aspect of this conundrum we face is the understanding of our human 'electric life'. Many cultures have some basis for this 'truth'. They have explored it mindfully, mentally and subconsciously. The Chinese use of the 'Chi' in their healing of the body, space and the man built environment is one classic case. Various cultures in Africa talk about the Ka which is the same life force. The Hindus with their understanding of the pre-endocrinal knowledge of dynamic life forces as the basis for the creation and support of life is a critical piece of knowledge not ever included in the western ken. The Kundalini with the seven centers or the chakras is not only a spiritual metaphor for psycho spiritual behavior. It is the universal map of consciousness. It is a clear guide to the constant creation of the universe an all it contains (and who contains it). There is a 'Ray of Creation' we are guided by. The light of this ray is the Kundalini energy. It is the electric life that is the chi, the flame, the ka and all other names we give to it. Is it the lack of exposure, the limitations to travel or cultural arrogance that has caused the lack of global synthesis from happening earlier and therefore our evolution? With these conditions somewhat improving today are we in for another renaissance? The forecast looks dim. The third level man model has been presented to us many times. Is this part of our future? Is a tertiary, no longer binary computational potential, ahead? Or is the third dimension the only qualitative one?

The geometry that once measured the earth continues to evolve. The systems of identity are no longer Euclidean. Non Euclidean or curved space is being understood in the context of the larger body of structural, temporal and energetic thought forms. Inventions in new geometric forms reveal new vocabularies of two three and even systems of greater dimensionality than we have ever reckoned. This is finally translating into our lives and our world. A new architecture is emerging defying all the traditional rules of our sense of space time energy and our environment. Super Symmetry does not know spirituality. It is laden with its special types of contradictions, biases, prejudices to deny our continued evolution of human consciousness. The Akasha is a democratic source of truth. You need no degrees to enter its realm. Living does it!

The goal is to align and harmonize creative people with opportunities and knowledge to define and realize fulfilled desires and be self-reliant. To become the authority, publish and share this goal and help others to participate in growth and in creating wealth, using simple natural principles and laws. These are the three core fundamental laws of creation. Creating original expressions follow these natural laws, principles and the processes.

THE ART OF CONSTANT CREATION

LAW	ACTION	REACTION
Polarization Intention	Cognition; Cision-division-healthy ego is the separation dynamic of choice to filter	Filter-crystallization Theme
Ideation Conception	Thought, Decision, Imagination Cision and vision	Conceptualization-Thought-form Idea
Formalization- Invention (Formation; geometric description & definition)	Incision-cut- Visualization	Precision Gauge Measurement*

THE 1ST LAW of POLARIZATION
ACTION/CAUSE: Cognition; Cision-division-healthy ego as the separation dynamic of choice to filter.
REACTION/EFFECT: Filter-crystallization.

The first law is the Spiritual Principle (attitudes): A fundamental truth or proposition that serves as the foundation for a system of belief or behavior or for a chain of reasoning. Truth, proposition, concept, idea, theory, assumption, fundamental, (Fundament) essential, ground rule : morals, morality, (code of) ethics, beliefs, ideals, standards;

THE 2ND LAW of IDEATION
ACTION/CAUSE: Thought, Decision, Imagination.
REACTION/EFFECT: Conceptualization- thought form.

The second Law is Psychological Principle (attitudes or behavior): morals, morality, (code of) ethics, beliefs, ideals, standards; More integrity, uprightness, righteousness, virtue, probity, (sense of) honor, decency, conscience, scruples

THE 3RD LAW of FORMALIZATION (Formation; geometric description & definition).
ACTION/CAUSE: Incision-cut-Visualization.
REACTION/EFFECT: Precision Gauge Measurement*.

The Physiological Principle includes "elementary principles" (of matter, materials and things): Physical dimensions width, height and length, properties and characteristics, mechanical and machine logic, symmetry and gender principles and structure or operational instructions, specifications, a general scientific theorem or law (theory) that has numerous special applications across a wide field.

A natural law forming the basis for the construction or working of a machine; "these machines all operate on the same general principle" "the first principle of all things is water"

---

*Assure: make an action certain or 'sure'. Measure: (I) make action certain or sure: knowing;
For Variations we Gather, sort and Taxonomize. The three …ations: Polarization, ideation formulation (formation) manifestation All is in vibration and in sense. Design is evolving into a design science. Is spin a vibration? Is the (normal 90 degree) view from the end of a wave a spiral with spin?
New paradigm definitions: What is chemistry?
Four Versions: Introversion-Extraversion-Perversion- [the alteration of something from its original course, meaning, or state to a distortion or corruption of what was first intended.]-Diversion [an instance of turning something aside from its course: rerouting, detour, bypass, deviation, alternative route redirection, deflection, deviation, divergence.

A Hologram is an encoding of the light field as an <u>interference</u> pattern of specific qualities of vibrations in a sensitive and interactive medium. When suitably lit, the interference pattern <u>diffracts</u> the light into a re-production of the original light field and the object's form appears, exhibiting visual <u>depth cues</u> such as <u>parallax</u> and <u>perspective</u> that change realistically with any change in the relative position of the observer-participator. This is the metaphor for life.

In its purest form, holography, uses the eternal light to illuminate the subject and object form. The hologram (object) and subject (observer-participator) are lit just as they were intended to be at the time of (recording) formalization. A microscopic level of detail throughout the recorded formalized volume of space is re-produced. Major image quality compromises are made to eliminate the need for laser illumination when viewing the hologram, and sometimes, to the extent possible, also when making it. Holographic recording often resorts to a non-holographic intermediate imaging procedure, to avoid the hazardous high-powered <u>pulsed lasers</u> otherwise needed to optically "freeze" living subjects as perfectly as the extremely motion-intolerant holographic recording process requires.

ELEMENTS FOR THE SYNTHESIS PRINCIPLE
Creativity is a synthesis of an omnipresent consciousness that can reveal itself in multiple expressions of constant creation of Body, Mind Spirit and Energy in a wave particle dynamic that 'invites' curious observers to participate in the determining the possibilities of coalescing harmonies of the singularity of focus, attention and intention that can then define forms with all its quantities, qualities and flavors reflecting the nature of its SOURCE in correspondence with all natural laws that hold things together in constant vibration.

THE NEW PARADIGM PRINCIPLE:
DISTRIBUTION is the logical combination of fundamental states of natural, artificial and cultural phenomena extrapolated into periodically ordered number frequency ranges in arrays that expand the flavors of the phenomena.

SYMMETRY PRINCIPLES:
It is the synthesis of symmetry principles and elements of nature, their expressions, matter, energies (or consciousness) to create or to extrapolate periodic quantitative; body, qualitative; mind and flavor; spirit, arrays of combinations.
This expanded thought process offers more combinatorial possibilities, enhancing the observer participator's creative intelligences beyond normal binary mental functions.

ASSOCIATIONS:
Qualitative correspondences create 'alignments' based on the quantity, quality and flavor 'dimensions' of consciousness itself as the 'TRIAD'; 'the one into three and the three into one' of 'Consciousness into Body, Mind and Spirit' and vice versa as the symmetry principles expressing the iterations and correlations of the states or systems of reality that are then harmonized.

THE WISDOM TRADITION: [Ancient wisdom and traditions as inspiration]
Sources: Vasistha's Yoga (page 83)

"Things come into being (or are created) on account of one's fancy (words, thoughts or ideas): one's fancy arises (or is inspired) by (the past as) memory, presence in the now (or present) and the fancy (desires) for things (in the immediate future)…An unreal (or abstract) object or substance, a deeply felt emotion, and inspiration (spiritual; spirit-in action) or thought become real when such intense faith is present through the right word, thoughts and ideas that are put into ACTION". The following seven intelligences are needed; Thought, visualization and imagination, words, graphic media, audio communications and thinking, LOVE as the reality tester of and for all intelligences and real faith, the energy of the solar source transformed into work energy, the gender symmetry of the male and female aspects of all things and the proper and respectful use of earth materials. This is all complemented by the three nadis, the ida, the sushumna and the pingala with their sub triadic nadis.

The causeless, infinite consciousness expresses a flavor of itself as the creative consciousness where the realm of macro principles, laws and egotism that are distributed (reflected) through the micro and Nano realms of human mindfulness where hallucination, karma and ego create what we perceive and call reality."Without a cause no effect is produced anywhere or anytime and therefore there is no 'fancy', no words or thoughts either; hence for the one causeless, infinite consciousness no-thing whatsoever has ever arisen or been created, rest assured of this."

The understanding of the creative process and its principles without the LAW of 7 (seven). This is where the matrix of consciousness is used as the upper register of the creative universal dynamic and its principles. To do this effectively we pay respects to the ancestors who might have conceived their WIKI as gifts for us that were coined in their consciousness and understanding that we in our ken do not respect. Modernity does not translate wisdom accurately or at all. So we reinvent our own processes, language and leave gaps in our all aspects of our thinking and our lives.

WORDSMITHING: The word Fancy; an innocuously profound construct we generally take for granted. Re-contextualized it is revitalized at least in this process of revisiting meaning and cognition in space time.

Doing the distribution (avoiding the A word) from the big-inning is at the first row. Suchness and muchness are the poles (duals) of consciousness. We then drill down to the other rows each one revealing its properties, traits, qualities and essences of flavors.

What kind of thinking is this you ask? It's natural and organic and it's what our three- dimensional totally conditioned minds afford us, in this three-dimensional world now.
Until we embrace 'paradigm shifts' and enhance our WIKI this is where we will be, until nature brings the next 'great flood' upon us.

QUANTA, QUALIA AND SPIRITUA MATRIX

MUCHNESS	SUCHNESS	CONSCIOUSNESS
BODY	MIND	SPIRIT
fancy noun	adjective	verb
A feeling of liking or attraction, typically one that is superficial or transient.	Elaborate in structure or decoration.	To feel a desire or liking for.
	The faculty of imagination	To imagine; to think.
The Act; The person of thing	The Action; The ritual or Habit	Acting; Creating movement

THE CONSCIOUSNESS-INTELLIGENCE MATRIX; the human story in 'Simple Fi' Form.

	CONSCIOUSNESS	THE MIND	THE SENSES
CONSCIOUSNESS	COSMIC CONSCIOUSNESS	THE SUPREME SELF	NORMAL CONSCIOUSNESS 1. Linguistic Intelligence
THE MIND	THE HIGHER SELF 5. Musical Intelligence;	MENTALISM: INTELLIGENCE 2. Logic Intelligence	THE PERSONAL SELF: IDENTITY, EGO 6. Interpersonal Intelligence
THE SENSES	PERCEPTION 4. Spatial Intelligence	CLAIR: ALL FORMS OF INTELLIGENCE 3. Kinesthetic Intelligence	INTUITION 7. Intrapersonal Intelligence

<u>The 7-8 Types of Intelligence & Their Characteristics:</u>

This is a synthesizing process of interpretation. Each form is a language used to interpret stimuli that Body, Mind Spirit with all forms of energy used in the control center and the 'entire harmonized and synthesized being, not just the brain, can translate into coherent Wisdom, Intelligence, Knowledge and Information (the WIKI). The triad with the four forces of the realized and manifested world creates the 7 principles that permeate all expressions, events and flavors.

1. **Linguistic Intelligence;** verbal and writing skills; the left brain is associated with Linguistic Intelligence.
2. **Logic Intelligence;** Insight, problem solving; the right brain is associated with Logic Intelligence.

3. **Kinesthetic Intelligence;** sense of space; distance, depth and size. The cerebellum controls voluntary movements of the body.
4. **Spatial Intelligence;** the ability to create, imagine and draw 2D and 3D images.
5. **Musical Intelligence;** to listen to sound and music and identify different patterns and notes with ease, having "absolute pitch".
6. **Interpersonal Intelligence;** born to be leaders practical people with a great sense of responsibility. They know how to use their own knowledge and power to influence people.
7. **Intrapersonal Intelligence;** someone that is deeply connected with themselves. Intrapersonal intelligence is considered the rarest
8. *Intuition:* Direct knowledge from source through the synthesis of all the faculties and qualities operating on all energy and thought levels (of consciousness) known and unknown.

New opportunities, language and visions converge in space time energy, To crystallize a new way of being in harmony and egoless mindfulness,
All in community must find new thought forms spun from human kindness. We are each invited to the awesome vibrations of a new living process.

Design is the knowledge of effects. DESign Science is the wisdom of causes. Through Synthesis we connect to source to obtain direct knowledge and attain Creative freedom.

Design is evolving into a science as it was projected by RB Fuller in 1957. The major factor is the shifting paradigm where language must be more visual than ever before. Our semiotic communications are becoming more symbolic laden with content with the most effective uses of space, time and energy to lift the human spirit to new levels of much higher consciousness. The current trends seem to be leaning on the 'tools' to define the form as an inversion seen too often when we define ourselves in the parlance of the media and tools we use. EG. I use paint to make pictures, so I am a "painter" etc. Calling myself an artist becomes vague. The higher conscious articulations seem to have no associative attraction or charm.

Writing in the visual vernacular of an emerging coherent symbolic paradigm demands a new concise and efficient form of communication that rises above and carries us all beyond the tasks and tools to higher or deeper meaning. Normal language does not do highly visual information justice especially with inventions and descriptions that are new.

The structure of all languages have underlying structural rules that make meaningful communication possible. The rule we have not known is the Body, Mind, Spirit formula. If language is the medium for consciousness and the structure of consciousness is the MBS formula, it follows that this could be used to enhance meaning.

The hierarchy of the building blocks of language. (Lan-d-Gauge) [Verbal +Visual= Mental+ Written]
The five main components of language are 1.Phonemes, 2.Morphemes, 3.Lexemes,
4.Syntax, and 5.Context. Along with grammar, semantics, and pragmatics, these work together to create meaningful communication among individuals.

THE SYNTHESIS OF LANGUAGE

BODY	MIND	SPIRIT
NOUNS, PRONOUNS, DEFINITE ARTICLES	QUALIFIERS & CONNECTORS	VERBS: the part of a sentence that expresses what is said about the subject and object.
lexeme The set of inflected forms taken by a single word.	phoneme An indivisible unit of sound in a given language.	morpheme The smallest linguistic unit within a word that can carry a meaning, such as "un-", "break", and "-able" in the word "unbreakable."
SUBJECT-OBJECT	ADJECTIVE-ADVERB	PREDICATE
GRAMMAR	PRAGMATICS	SEMANTICS

SYNTHESIS

PHONETICS	PHONOLOGY	MORPHOLOGY	SYNTAX	SEMANTICS
A morpheme is the smallest unit of a word that provides a specific meaning to a string of letters (which is called a phoneme). There are two main types of morpheme: free morphemes and bound morphemes.	A phoneme is the smallest unit of sound that may cause a change of meaning within a language but that doesn't have meaning by itself.	A lexeme is the set of all the inflected forms of a single word.	Syntax is the set of rules by which a person constructs full sentences.	Context is how everything within language works together to convey a particular meaning.

PRAGMATICS	1: a branch of semiotics that deals with the relation between <u>signs or linguistic expressions and their users</u> 2: a branch of linguistics that is concerned with <u>the relationship of sentences to the environment in which they occur</u>

Subtleties and nuances are part of language. Inflections of the voice determines the meaning. With Sign Language, you can convey full, grammatical sentences with tense and aspect with physiology; by moving your hands and face.

Poetry as languages have BMS underpinnings that expand the framework of associations and correlations in personalizing all our relationships. A VERB is characteristically the Spirit or creative energy center of a predicate that expresses an act, occurrence, or mode of being, that in various languages is inflected for agreement with the subject, for tense, for voice, for mood, or for aspect, and that typically has rather full descriptive meaning and characterizing quality but is sometimes nearly devoid of these especially when used as an auxiliary or linking verb. Thought the finest fuel we use is the critical dynamic of all language. Instead of being action words which relates to physiology there is an essence about verbs that move our spirit. 'THE'(the definite article) is used to indicate a person or thing that has already been mentioned or seen or is clearly understood from the situation; used to refer to things or people that are common in daily life; used to refer to things that occur in nature. 'The' is the Body element in the formula.

Five major components of the structure of language:

1. **Phonemes** (Tone)
A phoneme is the basic unit of phonology. It is the smallest unit of sound that may cause a change of meaning within a language, but that doesn't have meaning by itself. For example, in the words "bake" and "brake," only one phoneme has been altered, but a change in meaning has been triggered. The phoneme /r/ has no meaning on its own, but by appearing in the word it has completely changed the word's meaning!

Phonemes correspond to the sounds of the alphabet, although there is not always a one-to-one relationship between a letter and a phoneme (the sound made when you say the word). For example, the word "dog" has three phonemes: /d/, /o/, and /g/. However, the word "shape," despite having five letters, has only three phonemes: /sh/, /long-a/, and /p/. The English language has approximately 45 different phonemes, which correspond to letters or combinations of letters. Through the process of segmentation, a phoneme can have a particular pronunciation in one word and a slightly different pronunciation in another.

2. **Morphemes** (particles and waves)
Morphemes, the basic unit of morphology, are the smallest meaningful unit of language. Thus, a morpheme is a series of phonemes that has a special meaning. If a morpheme is altered in any way, the entire meaning of the word can be changed. Within the category of bound morphemes, there are two additional subtypes: derivational and inflectional. Derivational morphemes change the meaning or part of speech of a word when they are used together.
For example, the word "conscious" changes from an adjective to a noun when "-ness" (consciousness) is added to it.
"Creation" changes in meaning when the morpheme "re-" is added to it, creating the word "recreation." Inflectional morphemes modify either the tense of a verb or the number value of a noun; for example, when you add an "-s" to "form" the number of forms changes from one to more than one.

3. Lexemes

Lexemes are the set of inflected forms taken by a single word. For example, members of the lexeme RUN include "run" (the uninflected form), "running" (inflected form), and "ran." This lexeme excludes "runner" (a derived term—it has a derivational morpheme attached).

Another way to think about lexemes is that they are the set of words that would be included under one entry in the dictionary—"running" and "ran" would be found under "run," but "runner" would not.

4. Syntax (Symmetry)

Syntax is a set of rules for constructing full sentences out of words and phrases. Every language has a different set of syntactic rules, but all languages have some form of syntax. In English, the smallest form of a sentence is a noun phrase (which might just be a noun or a pronoun) and a verb phrase (which may be a single verb). Adjectives and adverbs can be added to the sentence to provide further meaning. Word order (symmetry) matters in English, although in some languages, order is of less importance.

5. Context (Flavor)

Context is the 'Dao' (or way) everything, 'before, during and after language', works together to convey 'a particular' meaning. Context includes tone; pitch, quality, and strength, cadence body-language, and the words being ART-iculated. Depending on how a person thinks, creates and says something, with posture, or emphasizes certain points of a sentence, different messages can be communicated. The process that is used to prepare the delivery is 'ART'.

SEMIOTICS: The study of signs and symbols and how they are used.

A general philosophical theory of signs and symbols that deals especially with their function in both artificially constructed and natural languages and comprises Syn-tactics, Semantics, and Pragmatics. The number three used here is key. Pragmatics is 'Body. Syn-tactics is creative or Mind. Semantics is the Spirit of the Law. It creates the Synthesis for communication, meaning and ultimately consciousness.

THE SYNTHESIS OF LANGUAGE

BODY	MIND	SPIRIT
SUBJECT- OBJECT	ADJECTIVE-ADVERB	PREDICATE
GRAMMAR/SYNTACTICS	PRAGMATICS	SEMANTICS
WORDS/ FORMS	EXPERIENCES/ EXPRESSIONS	MEANING
PHYSICAL/PHYSICALITY	EMOTIONS/PSYCHOLOGY	CRATIVE ENERGY/ SPIRITUALITY

"RETHINKING SEMIOTICS" The study of signs and symbols and their creative uses. It's a general philosophical theory of signs and symbols that deals with their function in both artificially constructed, natural languages and comprises Syntactics, Semantics, and Pragmatics.

Language is the ability to acquire and use complex systems of communication, with the human creativity to do so, and a language is any specific example of such a system. Words, Images, Forms, Symbols

The Body. Mind, Spirit dimensions of life have qualities about them we all experience. All living forms are driven to share the wonder and mysteries of their emotions, feelings, 'self- talk' with their thoughts. MAN becomes the fine-tuned instrument with languages derive from the desire share values and commonalities in community. From the grunts and groans, to cave paintings, cuneiforms and hieroglyphs and the unfolding of the flow of human consciousness encoded in these 'art forms' that may also be registered in our collective DNA we can now read this text and transform our lives, the lives of others and the state of our planets.

The 5 Senses: The nervous system with subtle energy dynamic and interactions with specific sensory systems or organs internally and externally, are dedicated to each sense.
The subtle bodies and energies of the 'Kundalini and the seven chakras' is a 'spiritual' dimension the western world is now learning about. The human being is the instrument for this language of consciousness with the ability to raise elevate ours. Humans have a multitude of senses.

These are the five traditionally recognized.

THE SENSES AND PROCESSES

SIGHT	HEARING	TASTE	SMELL	TOUCH
Ophthalmoception	Audioception	Gustaoception	Olfacception	Tactioception
Visual Sense	Auditory Sense	Chemical Sense	Chemical Sense	Physical Sense

Taste is often confused with the "sense" of flavor, which is a combination of taste and smell perception. Common Forms and Functions of English Language sentences and their structural elements.

THE BODY, MIND SPIRIT MATRIX

BODY	MIND	SPIRIT
The Informative Functions	The Expressive Functions	Directive Purposes
Statements of fact	Reports feelings or attitudes of the writer or speaker, the subject, or evokes feelings	Found in commands and requests involving energy, actions and movements.
Concreteness, Materiality	Evoking and expressing feelings and emotions.	Causing (or preventing) overt action.
The communication of information	Poetry and literature	Thought is a powerful and subtle directive process

Describe the world or reason about it	Expressions of emotions, feelings or attitudes.	Various logics of commands have been developed
True or False Good for Logic.	Neither true or false	Neither true or false
Subject/ Object/ Place	QUALIFIERS	Action Words, Verbs
SUBJECT-OBJECT	ADJECTIVE-ADVERB	PREDICATE
GRAMMAR/SYNTACTICS	PRAGMATICS	SEMANTICS
WORDS/ FORMS	EXPERIENCES/ EXPRESSIONS	MEANING
PHYSICAL/PHYSICALITY	EMOTIONS/PSYCHOLOGY	CRATIVE ENERGY/ SPIRITUALITY
PRAGMATICS	SYNTACTICS	SEMANTICS
SCENE	MOOD	ACT
.PHONEMES	MORPHEMES	CONTEXT
LEXEMES	SYNTAX	GRAMMAR

We now look at the visual mathematics process and how it relates to computation and other methods of finding solutions.

ALGORITHMS AND PARADIGMS

It seems quite reasonable that the triadic harmonizing of intelligent (wisdom) Knowledge, information, data and other evolving (energy transmission devices; ETD) used to help us develop systems, skills for cognition, recognition, observation of correspondences that work. This evolving process, help us interpret and manage new and exciting sources now being revealed to us…everything old is new again with our paradoxical sun. They represent new systems for encoding, organizing, analogs and metaphors for design, to simulate, emulate, calculate and create expressions of a new energy with a new identity and a new aesthetic.

We have been able to develop systems for calculation for the quantitative processes with some success using gauge theories and frameworks that are inherently materialistic. The qualitative behaviors are not as fortunate and the physical correspondences. They are still lacking rigor and some form of 'qualitative logic structures and systems' to support better interpretations of the principles they use to teach us how they really work.

The world of spirit is quite complicated. It involves all the creative aspects of the infinite consciousness or source (with a plethora of names, allegiances and alignments to faith, with the life force itself as one vital expression. This is another topic all together. Actions taken to perform 'work' require energy; a form of 'spirit' starting with, the finest fuel that operates on all dimensions, in which we humans are engaged. We have a 'spiritual spectrum' of vibrations that start with the rarest to the basest forms of 'thought', to mental constructs and to electricity. The mental constructs also engage the emotional aspects of the individual and collective dynamics of living.

We depend on 'spirit' for living but deny ourselves the fullest potential possible by not incorporating it into all aspects of living it was intended to serve. It's a notion similar to "if we work for a living why we kill ourselves working?" Eli Wallach; "The good the bad and the Ugly". We have settled for duality as our basis for reality, inspired by what we see, tough and can feel, measure and interpret. We started counting with fingers to derive our digits and decimal system of counting. At the same time our mental faculties were either in the limbic or mammalian stages of developments we were fashioning survival systems and dodging prey to save ourselves. About forty thousand years ago the three stages of the human brain began to evolve.

The limbic, the mammalian and after some time the third or the 'neo cortex' appeared or were awakened. Was the unfolding of this faculty 'DNA encoded' for time release at preset stages "eras" of human evolution? Did each phase have the instructions needed for its interval? Are we at another unfolding now moving towards an even higher level of consciousness? The answers to quests, requests and questions of this sort are always rhetorically in the affirmative.

We recognize three levels of 'consciousness' possibly facilitated by the three parts of our brain expressed as the physical 'Body'; interacting with the limbic brain, the psychological 'Mind', corresponding with the mammalian brain and the Energy or 'Spirit'; controlled from the neocortex with links to the hypothalamus, the pineal gland and the subtle energies of the Kundalini and the 7 chakras aligning with the endocrine system. To novices of this subject, who need to observe and be aware, to the initiates who know study and search and to the gurus who have totally internalized and live this connected and transforming experience of 'being one with source' it might be evident that the triadic symmetry of our awareness is permeated with a threefold reflection that mirrors the Body, Mind and Spirit principles throughout life itself and all expressions of forms and languages generated by 'thought' is the essence of consciousness.

If nature is threefold why do we still use the binary processes to understand higher dimensional processes and to interpret the world around us that is in overdrive now and more so in the foreseeable future? This becomes more acute as 'our universes' expand requiring faster and more robust systems to reckon, manage and store data from the information explosion we continue to inherit. There is something inherently strange about staying committed to the twofold nature of everything, when the possibility of a more efficient third element is awaiting us, trying to awaken us as well, even in crisis and pain.

In the transformation age the quantum dynamic is playing a pivotal role in our experience of how we can interpret the stimuli, symbols and thoughts that will eventually become the language, the media and the expressions of the time, space energy continuum of this new era. There is an internal or subjective world that comes way before any expressions can be formulated and communicated. The tools, the materials and technologies are generally lagging the thinking needed for this creative process.

My sympathies go out to the sensitives who are sensing the vibrations of this critical transformation and are unable to interpret it because of not having the "language" to do so coherently or who are not grounded in the possibilities being offered. The materialistic state of the culture we live in has not interpreted 'spirit' properly. I have had my challenges during my transformation as I began reeducating myself to begin my journey to my 'connection to source' or as it is said "being one with nature".

What do the following words 'work', 'power' and 'spirit' mean? Spirit is a functional or mechanical, creative and motivating or behavioral and electrical energy dynamic. It implies effort or work which creates expressions in every vernacular, form and language. There is force involved in the vibrational processes of all thought and action described by need, desires and will 'power'. The reverence and higher levels of consciousness that we are unable to define or understand is put into this nebulous moniker spirit as a colloquial and ubiquitous meaningless idea designed to avoid 'work' needed to understand the distinctions between 'spirit' and religious experiences and history. This topic is the subject for a much later sharing. Love, the next conundrum, can get us through anything.

In a true third level or third wave of human consciousness, spirit in its triadic form operates according to certain (symmetry) principles, for there must be consistencies in the way the various expressions and functions "work" for there to be order before there is the harmonious synthesis of both cause and effect and all the other natural laws and principles in effect to put it (the unified dynamic) to practical use. The tone is set for all the human vibrations to be harmonized as well. There is a corresponding set of three-dimensional dynamic symmetry laws and principles that becomes the unique and relevant toolset. The square pegs of the old polarity do not fit into the multi-dimensional New Paradigm circles (this is symbolically very significant). Circles are the metaphor for a multitude of natural, logical and energy related process at the heart of the ideational processes we will be exploring.

The triadic matrix is the compass for the journey we are now embarking on. It is a very economical tool that describes complex processes in a simple S_2 or S_3 framework. It is composed of columns of phenomena related to (times) the qualitative characteristics and conditions being evaluated.

THE THREEFOLD SYMMETRY MATRIX AS SOLUTION FINDING TOOLS

CONSCIOUSNESS (AWARENESS)	SUB CONSCIOUSNESS	INFINITE CONSCIOUSNESS
Body	Mind	Spirit
Physiology	Psychology	Spirituality
Space	Time	Energy
1	2	3
UP	SIDE	DOWN
Extrapolated Infinitely	Extrapolated Infinitely	Extrapolated Infinitely

Why is number included in the matrix? All expressions are subject to the laws of the triadic MATRIX and can be distributed in the synthesis of the correlations of Muchness, Suchness and Infinite Consciousness including number.

There is a fourfold version of this idea that's used to evaluate the fourfold symmetry elements. The phenomena of "The Law of numbers" will be dealt with further into our journey. We continue gathering the tools needed.

The Fourfold Symmetry Matrix of phenomena and forces operating in the physical world

gravitational	electromagnetic	strong nuclear	weak nuclear
1	2	3	4
North	East	South	West
Spring	Summer	Fall	Winter

Prominent themes in Hindu beliefs include the four Puruṣārthas, the goals or aims of human life

Dharma	Artha	Kama	Moksha
(ethics/duties),	(prosperity/work),	(desires/passions)	(liberation/freedom)

The path to moksha requires the following: The Hindu triumvirate

Vishnu	**Shiva**	**Brahma**
Vishnu is the preserver of creation (The World)	upkeep and destruction of the world	Brahma is the creator of the universe
Requires karma (action), intent and consequences	samsara (cycle of rebirth)	Yogas (paths or practices to attain moksha)

The inclusion of the Hindu references will be developed as we seek to bridge the East-West dichotomy of the visions of the global cultural awareness of our current mindset. These are the two predominant orientations we live with and by now.

Quarks	electric charge	mass	color	charge	spin
flavors	up	down	strange	charm	top and bottom

Quarks are the only elementary particles in the Standard Model of particle physics to experience all four fundamental interactions, also known as *fundamental forces* (electromagnetism, gravitation, strong interaction, and weak interaction).

If knowing how the world works is exciting to you and quarks are the building blocks of nature a subjective, heuristic method can be applied to begin the thought process, not experiment, to begin the gathering, sorting and taxonomizing of the information available. There are six types of quarks. The Up and down quarks have the lowest masses of all quarks. The heavier quarks rapidly change into up and down quarks through a process of particle decay: the transformation from a higher mass state to a lower mass state. Matrices are developed in many forms as 'tables' better known as graphic organizers. It is a visual and graphic framework that depicts both the conjunctive and disjunctive relationships between facts, terms, numerical data and or ideas in 'c/re/ative'; 'c/ogn/ative' observations and tasks. Using the associative, compare/contrast matrix graphic organizer, visual intelligence is enhanced when recognizing similarities and differences between whatever the symbolically rendered topic is.

Another tool is the 'tesseract' also known as the hyper cube. In geometry, the tesseract is the four-dimensional analog of the cube; the tesseract is to the cube as the cube is to the square. Just as the surface of the cube consists of six square faces, the hypersurface of the tesseract consists of eight cubical cells. The tesseract is one of the six convex regular 4-polytopes.

The Tesseract: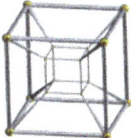

Because of this, up and down quarks are generally stable and the most common in the universe, whereas strange, charm, bottom, and top quarks can only be produced in high energy collisions (such as those involving cosmic rays and in particle accelerators). For every quark flavor there is a corresponding type of antiparticle, known as an *antiquark,* that differs from the quark only in that some of its properties have equal magnitude but opposite sign.

Calculation requires computing and machine or artificial intelligence, systems tools, programming, software and hardware. Computation is any calculation that follows a well-defined model that is understood and expressed as, an algorithm. Visual mathematics is a pattern generation and recognition process that follows a heuristic model not understood initially and needs art-iculation and definition, to create an algorithm. The study of both computation and visual mathematics is paramount to all the disciplines of computer science.

D$_E$Sign Science, the complement, involves all forms of computational, qualitative and graphic or visual mathematics with application as its focus and relies on visual acuity and intelligence.

Infographic displays should:
- Represent data clearly
- The viewer must think about interpreting the data or information to expand their knowledge.
- Accept the data as it unfolds
- Start with known quantities and ideas and go from the specific to the general.
- Identify the core elements in the patterns that unfold
- Engage the coordination of the hand, mind and eye to compare different pieces of data
- Explore several levels of iteration, from the overview back to the specific.
- Goals and objectives should be open ended
- Be familiar with the logic, language and limits of a data set.
- Visual process is the backdoor to mathematical reckoning or computation.

Human have had more experience with Graphic data. Indeed graphics can be more organic, precise and revealing than conventional statistical computations.

Mathematical models

In the theory of computation, a diversity of mathematical models of computers have been developed. Typical mathematical models of computers are the following:
- State models including Turing machine, pushdown automaton, finite state automaton, and PRAM.
- Functional models including lambda calculus.
- Logical models including logic programming
- Concurrent models including actor model and process calculi
- Why are we not enhancing human organic capabilities to maintain the unique qualitive edge we have, that physical systems might not be able to process?

The main fields of research that compose these three branches are artificial neural networks, evolutionary algorithms, swarm intelligence, artificial immune systems, fractal geometry, artificial life, DNA computing, and quantum computing, among others.

Natural computing, also called **natural computation,** is a terminology introduced to encompass three classes of methods: It can be expanded to include new methods.
1) Those that take inspiration from nature for the development of novel problem-solving techniques;
2) Those that are based on the use of computers to synthesize natural phenomena; and
3) Those that employ natural materials (e.g., molecules) to compute. The main fields of research that compose these three branches are:
4) Those that are based on Logic structures, Ancient Monuments and systems of celestial architecture. Polyhedra and other 3D form vocabularies.
5) Those that are based on the Quantum dynamic principles at work in the Kundalini and the principles of spiritual dynamic principles and geometric forms or Polyhedra; the realm of sacred geometry.

Artificial neural networks, Evolutionary algorithms, Swarm intelligence, Artificial immune systems, Fractal geometry, Non-Euclidean Geometry, Artificial life, DNA computing, Quantum computing, The relationship between algorithms and Artificial Intelligence:

Algorithm: A process or set of operations using symmetry principles, laws, rules and conventions to be followed when followed, obeyed when thinking, observing, recognizing, reckoning, calculation

ALGO-RHYTHM: Vibrational patterns of movement The arrangements and designs of things, systems of behavior, forces and energy in thought, (human) creativity, power and work.

ALGORITHMS AND PARADIGMS. The benefit here is the harmonizing of Wisdom, Intelligent Knowledge, Information, (WIKI) data and other evolving (energy transmission devices; ETD) used to help us develop systems, skills for cognition, recognition, observation of the correspondences and associations that are interconnected participate in a constant creation evolving process to manage and help us interpret new and exciting sources now being revealed to us. Ironically this resource has always been there and the shifts are taking place in our minds. The new systems decoded by the D$_E$Sign Scientists must begin encoding, organizing, analogs and metaphors for design, to simulate, emulate, calculate and create expressions of the WIKI of its space, time energy continuum in alignment with the Body, Mind, Spirit principles of consciousness with a new identity and a new aesthetic.

We are then able to develop systems for calculation for the quantitative and qualitative processes with some success using gauge theories and frameworks that are inherently holistic. The qualitative behaviors are not as fortunate as the physical correspondences are because of the current mind set. They are still lacking rigor and some form of 'qualitative logic structures and systems' to support better interpretations of the symmetry principles used to know (teach) how they really work.

Our widget making priorities are greater than m/any others.

D_ESign Science in The New Paradigm Age

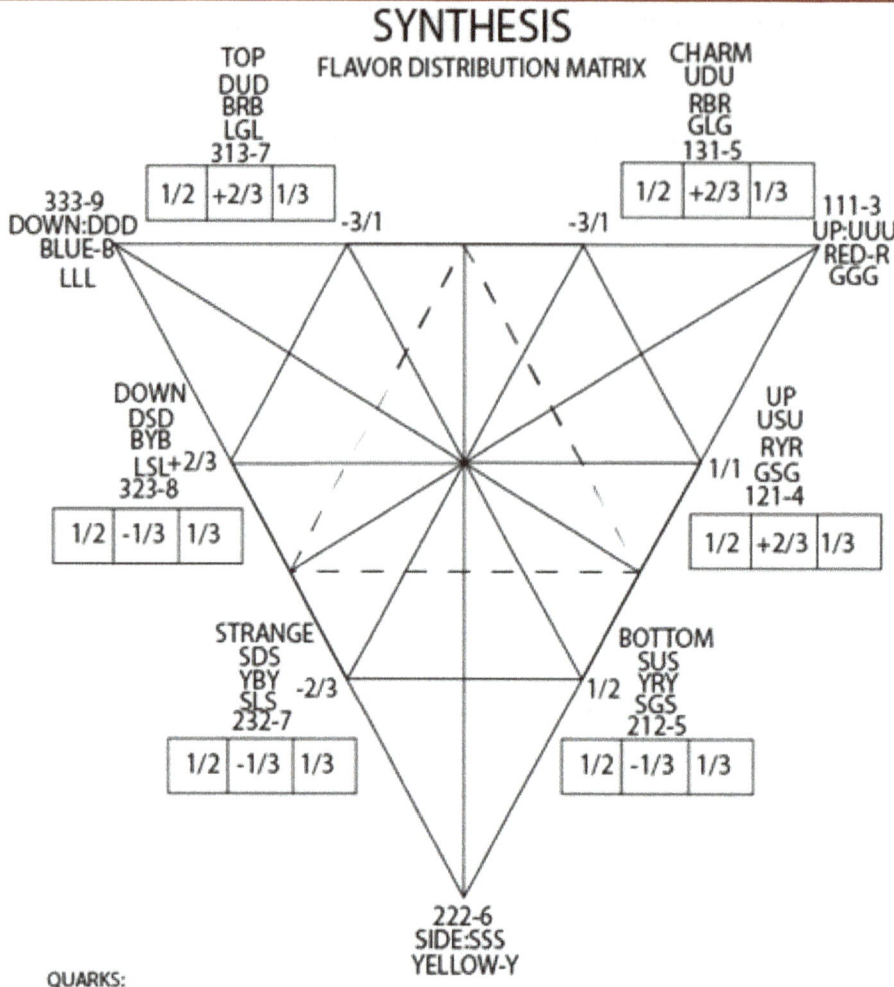

QUARKS:

Quarks and Leptons are the building blocks which build up matter, i.e., they are seen as the "elementary particles". In the present standard model, there are six "flavors" of quarks. They can successfully account for all known mesons and baryons (over 200). The most familiar baryons are the proton and neutron, which are each constructed from up and down quarks. Quarks are observed to occur only in combinations of two quarks (mesons), three quarks (baryons). There was a recent claim of observation of particles with five quarks (pentaquark), but further experimentation has not borne it out.

Quark	Symbol	Spin	Charge	Baryon Number	S	C	B	T	Mass*
Up	U	1/2	+2/3	1/3	0	0	0	0	1.7-3.3 MeV
Down	D	1/2	-1/3	1/3	0	0	0	0	4.1-5.8 MeV
Charm	C	1/2	+2/3	1/3	0	+1	0	0	1270 MeV
Strange	S	1/2	-1/3	1/3	-1	0	0	0	101 MeV
Top	T	1/2	+2/3	1/3	0	0	0	+1	172 GeV
Bottom	B	1/2	-1/3	1/3	0	0	-1	0	4.19 GeV(MS) 4.67 GeV(1S)

'THE QUANTUM FLAVOR MACHINE'
Fig 20.

ALL consciousness is one creative intelligent process
Origins propagating with life, awareness and movement
Life is a bio, dynamic, magnetic and electrical universal expression
Encoded by various levels and harmonies in vibration
The higher the vibration the more degrees of freedom
The lower the degrees of freedom the denser the form
Symmetry principles, natural laws and numbers blend
The five types of life forms including man in a rhythm without end
The continuum with the most degrees of freedom was given Dominion?
Over all living and non-living forms of consciousness including self

Continuity depends on the principles of gender for re and pro-duction
Making in the image of some higher form that resonates in its own reflection
From much higher dimensions protected veils of perception in no recognition
For continuity making in the image of 'heaven on earth", as above so below
Corresponding man-infesting an aesthetic creation without symmetry principles
Laws of dis-order, no aesthetics nor design for human needs dis-satisfaction
Encoded and art-iculated progenital seeds of form; physical,
Nature psychological and energy; spiritual all in vibration dynamical
Creating through life's perpetual portals simple mortals to super-consciousness

Man's relationship to the creator intended and expressed a language
Of creation with identities and principles with proc-live-ities as flavors
For producing offspring Humans in their own likeness genetically conceived
So genetic mappings would obey the philogenetic frequency encodings
Of a DNA found in all dark matter including me and you yes you and I are dark matter
The Maya of our dark thinking is what we see with the bright light shadows
Of The Sun behind the much brighter invisible Sun

The universal invisible sun makes of dark matter the glue that keeps all consciousness One
The universe is holding everything in it together till returning to the source of its arising come.
Somewhere in pineal gland, the pine cone shaped 'organ' in the brain at the chakra crown
Then the melatonin and the 'spirit molecule' flips the dime that trips the mind… off we go.

Everything vibrates at various frequencies to encode the basic nature of itself
Mythic substances resonate to frequency range of human consciousness
Sparking the electric life of universes, to Man the forms we inherit and create.
Continuity is the gender principles of reproduction and production great.

Then comes the image of varied higher resonating dimensions at imaginations gate
Reproduced in perpetual visions of 'heaven on earth", for above to be below.
Corresponding and reflecting an aesthetic of creation with di-verse symmetries
Harmonious aesthetics and design order for human needs and satisfaction
With support and production for the evolution of life and consciousness
Measured cycles and rhythms extend to infinite space, time and pure energy.
The Sun behind the Sun the all mighty ONE
Configured invisibly for time to be-gin for eons to come
The 'gin' in being the en-gin beginning and becoming
The soular vehicle.. It's where the Dog star lives
The EN-Gin is the soul of the Eternal Generating

CHAPTER SIX

ICOSIDODECAHEDRON-IcDoH

6. Ajna, brow's this chakra to destroy all darkened pasts, to hasten a golden future that can never last; for certain to manifest the present in four supremely fulfilling ways.
Clairvoyance, Visualization, imagination and in-tuition not afforded. The anahata 'love' and feeling node reality tests all the 'fulfilling' Psychic and 'occult' powers of endlessness. The opened ajna centers the above and below 'thought form' being itself becoming one with feelings in this ajna time-space, Nullified past; 'the Karmic net is cast'. The future is not swayed from bringing into immediacy; the ever present 'now'. Pre sent in space, time and energy a portal birthing physical forms Visions Achieved desires received the third eye to perceive all conceived in now with the darkest past thens destroyed. Patience need not apply to work here with 'HIMHERIT' With the third eye wide open, one can accomplish much. If the ultimate Power is misused, with intentions confused with the third eye's transcendental Power abused, all is in destruction if the third eye is used properly With divine alignments set not on property alone More blessings humanity imagined is now our own Seeing clearly with intelligence transcends space Seeing big picture use of vision referencing to grace the mental blue print meditatively turning into place beyond dimensions attracting magnetic vibration in correspondence; with the pit of uitary in unity to visualize, imagine and be in-tu-it-ion. The ition that's it's in-tu needs no perspiration to Become the Creative Praxis of and for Creation.

CHAPTER 6 TRINE A

GEOMETRY

Non-Euclidean geometry

This is the future of three dimensional geometric form intensive professions.

Who is the father of geometry?
Euclid of Alexandria is called the Father of Geometry. He received his education at Plato's Academy in Greece and moved to Egypt to teach. He taught during the reign of Ptolemy I Soter, the first Macedonian ruler. Euclidian geometry has been taught in schools for a long time. Euclid is famous for the mathematics textbooks "Elements," which contains lessons on algebra, number theories and geometry. The series contains 13 books. Other math books written by Euclid that are still famous include "Division of Figures," "Data" and "Phenomena." Three textbooks by Euclid that are no longer used are "Porisms," "Surface Loci" and "Pseudaria".

Many occupations including architects, designers, farmers, construction workers and medical professionals incorporate geometric concepts into their work. Even individuals outside of these professions use geometry when measuring walls, calculating how much paint is needed for a project or determining whether new furniture can fit through a door. Geometry is a practical guide for measuring lengths, areas and volumes.

Geometry focuses on the properties of space and figures. It simplifies calculating area, perimeter and volume. It also helps people understand objects spatially, conceptualizing how the position, size and shape of a space relate to the things inside that space. For instance, if someone wants to buy a new dishwasher, knowing how that appliance fits spatially with the other cabinets, furniture and appliances in the room helps to determine what type or size of dishwasher to purchase.

Designers and architects deal with three-dimensional figures constantly. Geometry helps determine how a building, something composed entirely of three-dimensional shapes, is constructed. Medical professionals make use of geometric imaging with technology such as CT scans and MRIs. Image Mapping requires a solid foundation in geometry. Occupations that involve surveying, navigation and astronomy use maps to illustrate where things are located. Looking at maps requires geometry as well. Maps spatially display where a destination is, and through measurement, help determine how long it takes to get there.

DeSign Science in The New Paradigm Age

What Is Geometry? It describes the Body, Mind and Spirit dimensions of Consciousness.

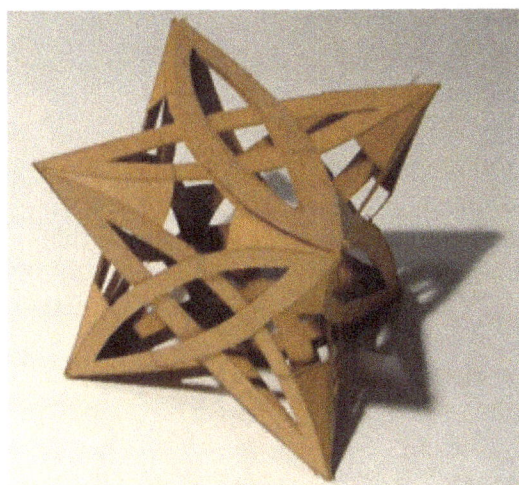

Geometry is one of the classical disciplines of math. Conceived in Egypt for "Earth Measurement" and is concerned with the properties of space and figures. It is primarily developed to be a practical guide for measuring lengths, areas, and volumes, and is still in use up to now. Euclid turned the study of geometry into an axiomatic form at around 3rd century BC, and these axioms are still useful up to the present day. An important evolution for the science of geometry was created when Rene Descartes was able to create the concept of analytical geometry. Because of it, plane figures can now be represented analytically, and is one of the driving forces for the development of calculus. In addition, the rise of perspective gave rise to projective geometry. Nowadays, modern geometry has strong ties with physics, and is an integral part of new physical concepts such as relativity and string theories. The Egyptians surveyed land around the Nile and developed astronomy skills.

The most basic form of geometry is called Euclidean geometry. Lengths, areas, and volumes are dealt here. Circumferences, radii, and areas are one of the concepts concerning length and area. Also, the volume of 3 dimensional objects such as cubes, cylinders, pyramids, and spheres can be computed using geometry. It used to be all about shapes and measurements, but numbers will soon make its way to geometry. Thanks to the Pythagoreans, numbers are introduced in geometry in the form of numerical values of lengths and areas. Numbers are further utilized when Descartes was able to formulate the concept of coordinates.

In real life, geometry has a lot of practical uses, from the most basic to the most advanced phenomena in life. Even the very basic concept of area can be a huge factor in how you do your daily business. For example, space is a huge issue when planning various construction projects. For instance, the size or area of a specific appliance or tool can greatly affect how it will fit in to your home or workplace, and can affect how the other parts of your home would fit around it. This is why it is essential to take account of areas, both of your space, and the item that you are about to integrate in there. In addition, geometry plays a role in basic engineering projects. For example, using the concept of perimeter, you can compute the amount of material (ex.: paint, fencing material, etc) that you need to use for your project. Also, designing professions such as interior design and architecture uses 3 dimensional figures. A thorough knowledge of geometry is going to help them a lot in determining the proper style (and more importantly, optimize its function) of a specific house, building, or vehicle.

Those are some of the more basic uses of geometry, but it doesn't end there. Now more professions use geometry in order to do their job properly. For example, computer imaging, something that is used nowadays for creating animations, video games, designing, and stuff like that, are created using geometric concepts. Also, geometry is used in mapping. Mapping is an essential element in professions such as surveying, navigation, and astronomy. From sketching to calculating distances, they use geometry to accomplish their job. In addition, professions such as medicine benefit from geometric imaging.

Technologies such as CT scans and MRIs are used both for diagnosis and surgical aids. Such methods enable doctors to do their job better, safer, and simpler.
As you can see, geometry affects us even in the most basic details of our lives. No matter what the form, it helps us understand specific phenomena and it helps us in uplifting the quality of life.

ANCIENT WISDOM TRADITIONS, SCIENCES AND ART WITH ALTERNATIVE INTERPRETATIONS OF LIFE AND THEIR OPERATING SYSTEMS.

GEOMETRY: The paradox of consciousness.

Egyptian culture is (with its present impact) much more than mastering pyramid and temple Architecture; as we are told. There was a 'soul science' in the temple and wisdom traditions kept alive today by secret orders and lodges. 'Physical geometry' made many significant contributions to human development, as a whole. It does not compare with the deeper Egyptian; holistic, spiritual wisdom and philosophical knowledge, attributed to its 'otherworldly influences,' that was its 'spiritual science' foundation. Many of our modern technological cultural and spiritual traditions have Egyptian DNA roots; HINDUISM, SUMERIA, TURKEY,

Geometry, as a subset of mathematics, is the discipline that describes all expressions of nature and manmade physical reality. Though it was created by the Egyptians for periodically surveying the land on the flooded river banks of the Nile, other 'geometries' are used to describe the dynamic creative processes of life now and in the future. What type of 'metrix' gets us to the moon?

From: **'A Study of Numbers'**: A Guide to the Constant Creation of the Universe by R. A. Schwaller de Lubicz reminds us that;

"We lack direct consciousness of Space, Time and Energy (as Creative Spirit). We can know of them only indirectly by (our materiality and understanding of) mass, force, and mechanical energy, and by the intermediary of phenomena such as may be tested by our five senses. We can understand mass, force, and energy as Body, Mind and Spirit as dimensions of consciousness.

Without direct awareness (or a higher developed consciousness, that transcends our chimeric and illusory senses) of Space or Time (without Spirit), human beings lack "two (now three) senses" necessary for the knowledge of all causes. These senses are enhanced by human 'faculties' of physiology, psychology and spirituality that are then harmonized into the deepest sense of 'oneness' or source all humans are connected to. From this imperfection, of which we are always being made aware, is born our need to simplify.

Thus we reduce everything to fundamental properties (things), without paying any attention to the underlying universal organizational (dynamics), (with their quantitative, qualitative and spiritual periodicities) the effects of which are all around us."

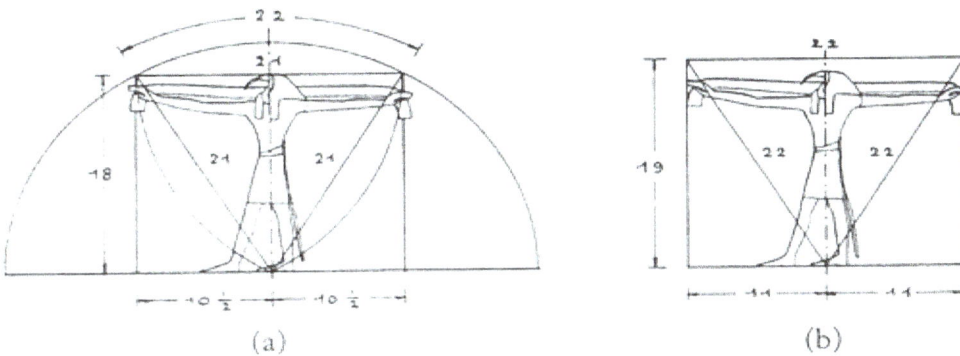

Fig.21 *The figures (a) and (b) are the geometric representation of the principle of proportion with the metrics for gauge theories for the Non-Euclidean geometry.*

Let's define some of the tools, goals and principles used in this discussion. The most favorite of all the tools is the "MATRIX", referred to as the "PERIODIC MATRIX" The matrix is a bird's eye view of relationships in arrays to help us think and look for many more connections in our searches and re-searches fields. There may or may not be order in the sequencing of the information or conditions, hence periodicity and consistent or progressive growth derived in this creative process, that can help us make discoveries beyond the binary or liner functions of our traditional learning and critical thinking with our limited binary thinking brain to use 'available ideas as the communication media, creative and thought processes of our visual intelligence and learned experiences. The linguistic and symbolic potential proves the very powerful adage, of 'pictures, word and values', truly mean.

The shift to three-dimensional thinking, where forms are no longer half- truths of our poor visual acuity, is an invaluable part of the overall paradigm shift my life's work is about and that I am very grateful to share. The behind the scenes daily rituals and exercises maintained with dogged determination and desire to connect with source or GOD or whatever moniker makes one comfortable, though not literally expressed in words vibrate between the lines as part of the implicit higher states of consciousness we are all connected to.

Connecting with the creative potential, deep within us that can give us the life we desire and deserve, is a way of life, or a "Dao". It's an art form we can and do 'DESIGN' for ourselves. The greatest creative potential that is our (macro) universe is already in our (micro) DNA code. In his book; *'The GOD CODE'* Gregg Braden agrees. We are tuned for freedom and total success with all we need within us. The English language is derived from the 'synthesis' of many other languages. The 'etymology' or the study of the origin of words and the way in which their meanings throughout history become relevant continues to change as we evolve.

The symbolism found in the ancient records describe principles that do not change; the language does but we are fixed in its conditioning and dogma. The rapid rate of cultural expansion is not limited to technology and creating smarter intellectuals. I believe that accepting what consciousness as the All of All we have known, now know and will ever know has been by experiences, learning and intuitions that we continue to symbolically and spiritually transform into text, context, pretext and subtext and all the other 'text forms' yet to be codified. The poetic manipulation of letters 't.e.x.t' is a wordsmithing game that expands the connections and associations to offer deeper meanings than those found in dictionaries. Poetry is the ultimate 'wordsmithing'. It is the essence of poetry which inspires discoveries, creative freedom and innovation. It is the get out of jail construct of mentalism.

THE TRIADIC HARMONY OF BODY, MIND AND SPIRIT AS FLAVORS OF GEOMETRIC EXPRESSIONS 'FORM'.

The triadic harmony of body, mind and spirit is the three dimensions of consciousness. It is found in our thinking, our daily lives at the root of and in the design of the English language (I am still experiencing). Every thought, manifestation or expression starts with self-talk as a thought that becomes a sentence and a command for 'the- I' (is called) to act-I-on from the inspiration, desire or need. Let's look at the word 'sentence'. It is used as (in the senses 'way of thinking, opinion,' a 'court's declaration of punishment,' and the 'gist (of a piece of writing)'): via Old French it comes from Latin as: sententia 'opinion,' from 'sentire' 'feel, be of the opinion.' **Sentience** is a simple way of defining "consciousness", and other 'collective' characteristics of the 'human mind'.

Both the sentience and sentence have a body, mind and spirit essence to them. The body (physiological) or thing is the noun. The mind relates to the senses with qualifiers and connectors needed to make sense and communicate emotions. The opinions are the results of the verbs and actions taking place in the creative acts and processes of the 'punishment'. Sentences can be hard to endure and to create (smile). 'Wordsmithing' is a poetic and a heuristic device. It is granular and goes to the roots.

Framing a Poetic New Paradigm Parlance.

Noam Chomsky asserts that "language is a part of the internal mental Structure, with biological tools to construct evolutionary trees for basically (four) language families. Poems often make deliberate use of imagery, word association, and sound qualities". The poetic use of language, can be used to encode new concepts and expressions that emerge from the innovation in new paradigms (to move the human spirit hgb).

Avram Noam Chomsky is an American linguist, philosopher, cognitive scientist, historian, social critic, and political activist.

Changing the communication to realign our thoughts and mental constructs
Does language create culture of does culture create language?

> *"Myth, art, language and science appear as symbols; not in the sense of mere figures which refer to some given reality by means of suggestion and allegorical renderings, but in the sense of forces each of which produces and posits a world of its own. In these realms the spirit exhibits itself in that inwardly determined dialectic by virtue of which alone there is any reality, any organized and definite Being at all. Thus the special symbolic forms are not imitations, but organs of reality, since it is solely by their agency that anything real becomes an object for intellectual apprehension, and as such is made visible to us. The question as to what reality is apart from these forms, and what are its independent attributes, becomes irrelevant here. For the mind, only that can be visible which has some definite form; but every form of existence has its source in some peculiar way of seeing, some intellectual formulation and intuition of meaning."*

Ernst Cassirer, *Language and Myth*

 *The influence of language in our interpretation of reality was formulated by the American linguists Edward Sapir and Benjamin Lee Whorf in the early twentieth century.*

Getting back to geometry. It could also be part of the description of physical, psychological, spiritual or by synthesizing all three into creative' experiences, emotions and thought. The word 'thought' is singular and is the highest form of (spiritual or creative) energy known. The 'geo-metry of things', forms or (land) renders space in many flavors. They possess identities that we interact with using subtle intelligences to interpret them. The movement is the expectation or the shift (of consciousness) we experience that is 'spiritual'. There are methods or sciences for describing all three states, with its meter or gauge theory;

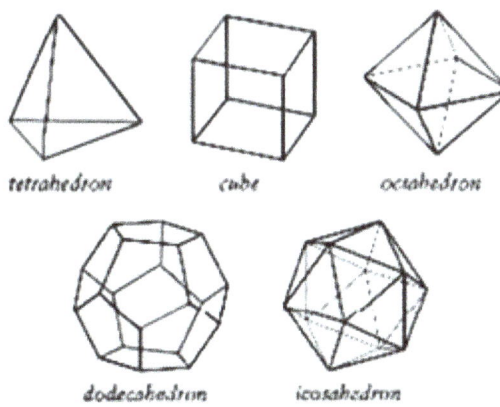

"Egyptian hieroglyphs were used to inspire the development of the letters of the English alphabet as one of the many contributions we pay no attention to or give credit to Greek philosophers who studied at the Alexandrian library in Egypt",

Fig 22.

Tolu Oladimeji.

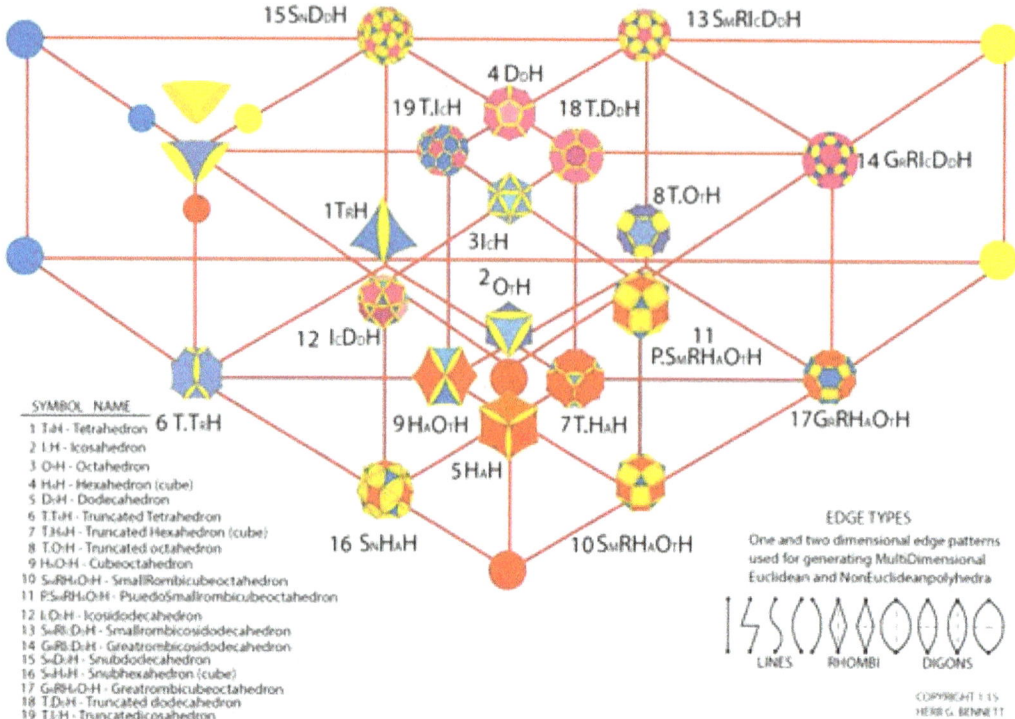

NON EUCLIDEAN " BENNETT" POLYHEDRA
Fig 24

Let's invent a 'Hay Day Wave' to map the cultural contributions of ancient civilizations with high lights and low lights thE rise and fall through time. A universal gauge will indicate ranges and values of the 'quotient of consciousness.' The 'Hay Day and Bust' rhythm would be a set of principles to bind all phenomenal expressions, to weave the fabric of human consciousness. Man is the loom. Thoughts and deeds are the warp and weft of the fabric of reality. Hay Day Makers are few and far between on our planet. Even with its forgiving sustainability and regenerating potential there are limits to the integrity of our 'web' of consciousness. The great intelligence with its seven dynamic principles was 'deciphered' by Hermes Trismegistus; the 'multi-dimensional' Egyptian God Thoth.

It is my opinion that these principles form the underpinning for the Pyramid Building Temple Science and Architecture as well. Here 'MAN' the holistic meter is used as the gauge or grid of a physical plan using what may be spiritual gauge of a synthesized holistic design science meant to transform the spirits of mortal beings to become 'godlike' if not GODS. A construct we still do not comprehend. What are the gauge theories and principles behind this sacred knowledge, its spiritual science and most impostantly its manifestation?

The loom (or matrix) produces a weave (or pattern) that is two dimensional yet it imbues the story of life in all dimensions. It is a story of 'consciousness' told in every language and culture. There is a vertical axis; the warp and the horizontal weft. The warp and weft makes the weave in two dimensions that encodes the messages that fabrics communicate on all levels.

The threads, their colors, textures and final qualities and flavors; fine or course, are the desirable results of this process turned into form embracing the body. This metaphor fits the Law of Seven Model as an inclusive and more comprehensive gauge theory.

A gauge theory is a type of theory in physics that could have been known by the Egyptians through their spiritual science. They synthesized quantitative, qualitative and 'creative' or spiritual dimensions into holistic models in fields of vibration and energies that describe the muchness and suchness of life, reality and consciousness in general. Tapping into the world grid requires such a gauge. It's the 'grand energy field' that we are all connected to. The modern ideas presented to us by Albert Einstein and the host of 'western high priests' and to a lesser extent the oriental chapters of non-traditional leanings and learnings offer modern theories to describe physical forces in terms of fields of energies, symmetry laws and operations in the electromagnetic field, the gravitational field, and subtle fields of 'forces' between elementary particles in physics and quantum mechanics. Language is the difference.

The warp, the weft and weave; the three dimensions of the lower register of our periodic matrix are the tangible elements of the fabric. It is the dense realm of 'earthness'. The threads, their weights and colors are in the midrange register. This is the range where emotions are used or experienced and used to act, make decisions and choices. Textures, qualities and flavors; fine or course, are the desirable results of this process.

They are the essence of the creative process that is on the highest register of this creative spectrum. Every phenomena manifested have these aspects. Between the big bang and now I do not expect this dynamic to change nor to ever do so at ALL. In most ancient cultures there is an illusory understanding of reality that seems to dominate their (sacred) spiritual knowledge. Modern science is now coming around to this point of view; hence 'paradigm'. Given that the infinite consciousness itself, from this very perspective, is also the creator it still does not limit itself to any one epochal continuum or space, (time) and energy or frame work for understanding (the movement of human spirit) expressions and representations of itself to the 'loom' at anywhere on the Hay Day Wave (thing Making) spectrum.

HEPTAPARAPARSHINOKH: THE LAW OF SEVEN (7) AND ITS principles permeate, support and energize all things physical, psychological and spiritual.

The periodic waves, their vibrations that define them, rise and fall or ebb and flow according to cycles and the Law of Rhythm that governs them.

"Heptaparaparshinokh" is the term spiritual teacher George Gurdjieff used to refer to the universal *law of seven*. This law of '*sevenfoldness*' is ubiquitous in all coherent processes of nature when things are viewed as *comprehensive wholes* and *unities*.

THE CORRELATIONS OF QUALITIES

THE PRINCIPLE	DEFINITIONS:	CORRESPONDENCE:
I. MENTALISM	THE ALL is MIND; The Universe is Mental All things are thoughts in but not of the infinite consciousness	The Crown Chakra and the Pituitary Gland is the gland to appear in the Human Embryo.
II. CORRESPONDENCE	As above, so below; as below, so above.	The Brow Chakra and the Pineal Gland. It produces "Melatonin"; controls the waking and sleeping cycles
III. VIBRATION	Nothing rests; everything moves; everything vibrates	The Throat Chakra and the Thyroid Gland produces "Thyroxin" to convert oxygen and food into usable energy.
IV. POLARITY	Everything is Dual; everything has poles; everything has its pair of opposites like and unlike are the same; opposites are identical in nature, but different in degree; extremes meet; all truths are but half-truths; all paradoxes may be reconciled.	The Heart Chakra and the Thymus Gland produces "T cells" for the immune system of the body.
V. RHYTHM (The Cycles)	Everything flows, out and in; everything has its tides; all things rise and fall; the pendulum-swing manifests in everything; the measure of the swing to the right is the measure of the swing to the left; rhythm compensates.	The Naval Chakra and the Adrenal Gland produces "Hydrocortisone" that regulates the use of food and helps the body adjust to stress.
VI. CAUSE AND EFFECT	Every Cause has its Effect; every Effect has its Cause; everything happens according to Law; Chance is but a name for Law not recognized; there are many planes of causation, but nothing escapes the Law.	The Spleen Chakra and the Lyden or Spleen produces "Macrophages" to cleanse the blood and is vital the immune system of the body and a person's health.
VII. GENDER	Gender is in everything; everything has its Masculine and Feminine Principles; Gender manifests on all planes. There are 'gender symmetry dynamics' in the physical, psychological and spiritual worlds unrelated to reproduction. Physiology relate to form and function in this articulation of gender.	The reproductive organs of "male and female" in all Life forms that exist. The Root Chakra and the Sacral or Reproductive Glands of Male and Female in all Species.

The order of the principles are somewhat different in the kundalini-chakra system. Polarity and Rhythm and Cause and Effect and Gender are interchanged. The wave (vibration) that causes each natural expression to rise and fall in cycles onto the 'chart' maps the universal rhythms of cosmic or infinite (macro) consciousness that is reflected in the micro and 'nano ™ states of being and in the subatomic environment. This is also pre-sent-ed to sub-conscious levels.

 *YOU CAN DISCOVER MORE ABOUT A PERSON IN AN HOUR OF PLAY THAN IN A YEAR OF CONVERSATION.*   *Plato*

Egypt is not the be-nor-end of all, instrumental in getting this blue ball of ours rolling. There is much confirmed evidence of older highlights on this Hay Day Wave; with perplexing technologies left for us to figure out if we really wanted to know who we really are and where we came from. The oldest example of such an event is the temple of Gobleki Tepe found in Southeastern Turkey in the town of Urfa "Predating Stonehenge by 6,000 years, Turkey's stunning Gobekli Tepe upends the conventional view of the rise of civilization" Şanlıurfa, also known as Urfa in daily language, in ancient times Edessa, in south-eastern Turkey and the capital of Şanlıurfa Province. It is a city with a primarily Kurdish population, near the 'garden of Eden'.

This seems to be of interest to so few. Case in point: "Gobekli Tepe was first examined—and dismissed—by University of Chicago and Istanbul University anthropologists in the 1960s. As part of a sweeping survey of the region, they saw some broken slabs of limestone and assumed the mound was nothing more than an abandoned medieval cemetery. In 1994, Klaus Schmidt was working on his own survey of prehistoric sites in the region. He made one of the most startling archaeological discoveries of our time: massive carved stones about 11,000 years old, crafted and arranged by prehistoric people who had not yet developed metal tools or even pottery. After reading a brief mention of the stone-littered hilltop in the University of Chicago researchers' report, he decided to go there himself. From the moment he first saw it, he knew the place was extraordinary". The dynamic that cause this nescience, exclusion and omission is interesting and will be explored later.

D_ESign Science in The New Paradigm Age

This complex is very interesting and important for one overarching reason; its technology and 'extraterrestrial' genius. Apart from its archeological significance is its celestial architecture. It made interstellar-planetary connections with the highest form of geometric-astronomy first used to build this temple complex. The level of thought that could have gone into this 'holistic' creative process had to be at the same level of the genius that created it. At this time in earth's hi-story knowledge was augmented by 'divine intervention'.
This could have been the case, as with later sites that were geometrically laid out, in polyhedral patterns, on the world grid and its vortices.

The most fascinating thought at this site relates to the structural system's other potential uses that have never been revealed, either in the records of time, space and energy or in physical and oral forms and those not yet thought of. On this count; time is not our friend nor is it very kind to us. I have a proposition that might shed light on this idea of 'other uses' of the Gobekli Tepe temple and celestial architectural portal.

Gobekli Tepe was not uncovered for centuries, cloaked invisibly by time and the forecasting of the genius that created it who understood, for good reasons, why it needed to be preserved. What does this discovery tell us or compel us to do about it, is the question. Our relationship with our planet and the universe requires a different (geo) metric or gauge which is more spiritual than the strict geo' land or space based systems of 'reckoning or cal-cu-lation and qualitative con-temp-lation' (wordsmithing using the prefix-the fix or root and the suffix™) rules. Recent discoveries have added more convincing information about this site.

Our 'mentally locked' paradigms do not afford us the freedom to look for relationships across the vastness of our own limited creative potential to make connections with ancient wisdom traditions and sites, the way the spiritual sciences could and do relate to or modern quantum scientific, string, quark and other theories meant to teach us who we are and how we can lift our consciousness to realize our true purpose on this planet at any time in its cycles, rhythms and diverse vibrations of reality.
Is there a universal consciousness that man is responsible for evolving or enhancing? Are we leaving this totally up to 'HIMHERIT' or to the other names we use in our much needed and halfhearted dialogs and insincere relationships that are laced with guilt and karmic debt way beyond our control? Man is the meter for consciousness.

Is there any value in deciphering any of this intelligence we claim to want to know about or are we doomed to the false 'universal or spiritual economics' where we invest and trade things for our souls?   Are we collecting the heavenly marbles our collective eternal consciousness once destroyed that we are revisiting with our space missions? If death is the transition we all make and all levels of materiality disappear then why all the pain, the mysteries and the lies about life and living in mental slavery?

Truth, love and compassion are forces science never will discover with its STEM. They can only be lived freely and peacefully in bliss and abundance! All else is illusion; the ill use of our minds; dreams and imagination.

Fig25

Geometry has evolved into describing more complex and perhaps more intelligent functions. Not all its applications involve tangible form and measured spaces. The quality of space continues to change as we discover new 'gauge theories'. The behavior (psychology of new materials with enhanced intelligence 'smart materials' can define now forms, envelopes and enclosures. The energy potential to generate, store, radiate and transform states are available now. The path from the Newtonian to the Quantum mechanical technologies is no stranger than other 'lost historical periods'.

The imaginary and illusory states of consciousness have been well established as the interpretation of mind, ego and time in relation to our ability to manifest thoughts.
No one talks about the fundamental geometry these thoughts need to take nor the geometric, structural and energy potential they can demonstrate.

Pyramids are just one example of such a form. If we understood this dynamic property of our thought form vocabularies many more can be realized with useful properties that can be used on all levels. The one very exciting realm is Nanotecture; the architecture of the nano scale sub-elements of an otherwise invisible reality. There is a qualitative science morphing from geometry like alchemy's transition to modern chemistry. Frameworks remain intact awaiting the synchronicity of appropriate technology and materials to converge with vision and need. All futures, before they are realized, are full of such possibilities and discoveries 'in the making'. 'Thought form' principles by way of dreams, visioneering, meditation and all the other 'esoteric' methods in reality must be ART-iculated through 'Design'. From conception, to DNA sequencing, mitosis, birth and all life stories are inscribed into the cells of beings, plants and animals; all sentient forms.

William Blake's poetry about the author of the fearful tiger symmetry comes to mind here. Who in fact defines the symmetries of consciousness? The symmetry of consciousness is written in 'symmetric code' by 'Micro-man' and by its own inherent intelligences. Where art, science and mathematics converge is fertile imaginative, creative and innovative mental soil where seeds of thought, for this new design science, are sown. In synthesis a new vision will emerge like none we have seen before with its new gauge theories and design sciences organically generating the fruits of our creative minds at work in new ways, with new materials and new forms. When the quantitative, qualitative and spiritual or creative dynamics are harmonized the limitations of the body, mind and spirit will be liberated, beyond our 'mind locked' conditioned perceptions', to degrees of freedom unimagined.

D_ESign Science in The New Paradigm Age

We need to be more curious, courageous and confident in questioning the current 'unfolding' still defining new visions of old paradigms and their origins of form vocabularies.
In the state of oneness every contribution is relevant. Here potential and credential are equal. Being committed and not conflicted about where information, motivation and inspiration come from has its rewards. Quantum mechanics teaches us about timelessness and the illusion of solidity of our consciousness, with the fundamental symmetry principles. There are (5) levels of consciousness starting with 1-infinite or cosmic consciousness, 2-consciousness of creation, 3-consciousness of normal aware-ness 4-subconsciousness. 5-Spirit Consciousness. They all operate in their own eternal realms. Do these five intelligences align with our senses? Light is the glue. This is another topic to be developed later.

Is there an overriding dynamic whose LAWS they all obey? What role does light play in this 'equation'? Space is pure extension. Time is pure duration. Energy is pure spirit. Light is pure (radiation) vibration. In (slower vibration time; SVT) time, mind, (yet slower VT) e-motion and (still slower VT) ego begin to 'resonate' to the various (seven 7) levels of vibration of intelligences passing through the stages of density as (speed) or the vibration of light). Light is the governing principle, energy and dynamic of this entire process. Light vibration is perceived solidity at the level in the atomic behavior of cells and nucleus.
Nothing is without light, nor without spirit. Is spirit light and is light spirit? Light is the purest [spirit] or energy. What is the relationship with 'Light, Thought and Consciousness'?

The 5 types of consciousness are: Where is mind defined? Is it an agent of light?

THE CONSCIOUSNESS TYPE	CORRESPONDENCES
Super Conscious Infinite	Mind (Souls)
Spirit Consciousness Cosmic	Akashic and Karmic Mind
Conscious Awareness	Mind (related to the body)
Sub Conscious Intuitive	Mind (Ethereal)
Un Conscious 'Auto Conscious'	Mind (the Subtle Bodies)

The Psychological Consciousness-Mind Continuum

THE DYNAMIC BEHAVIORAL TRAITS	PERSONAL EXPERIENCES
Separation Consciousness: Ideation	Everything Happens To Me
Earth Consciousness: Formalization	What You See Is What You Get
Heart Consciousness: Love	I Am You And You Are Me
Awakening Consciousness: Intelligence	I AM Powerful In All Ways
Cosmic Consciousness: Polarization	I AM An Eternal Being In An Eternal Universe

On the physical plane of normal awareness there are now several types of geometries each with its own flavor and identity. These can all be seen as scaffolds of consciousness each with its own language and expanded vocabularies based on the interpretations of the classic 1-Euclidean polyhedral.

DeSign Science in The New Paradigm Age

"When one door closes another door opens; but we so often look so long and so regretfully upon the closed door, that we do not see the ones which open for us." —Alexander Graham Bell

DeSign Science is the perfect doorstop.

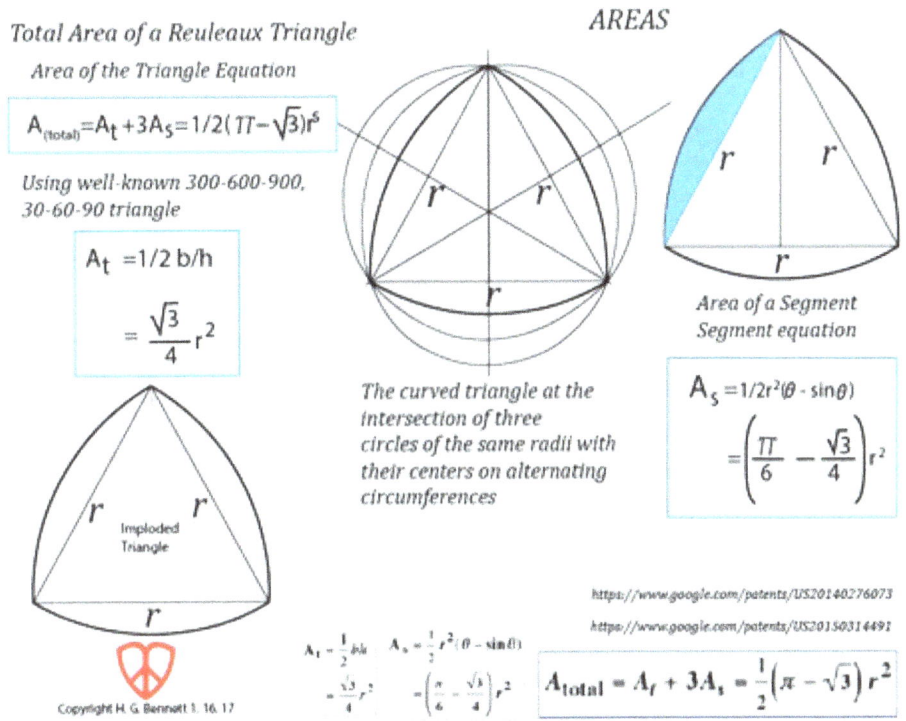

PROPERTIES OF THE FIRST NON-EUCLIDEAN S² 'BENNETT FORMS'

Fig 27

This image contains many of the Non-Euclidean Geometric Forms in a perfectly harmonized vibrational 'Chladni' pattern.

DESign Science in The New Paradigm Age

Florens Friedrich Chladni; (German: 1756 –1827), the father of acoustics was a physicist *and* musician *who generated vibrational patterns on sand covered metal plates that were exposed to various sound frequencies.*

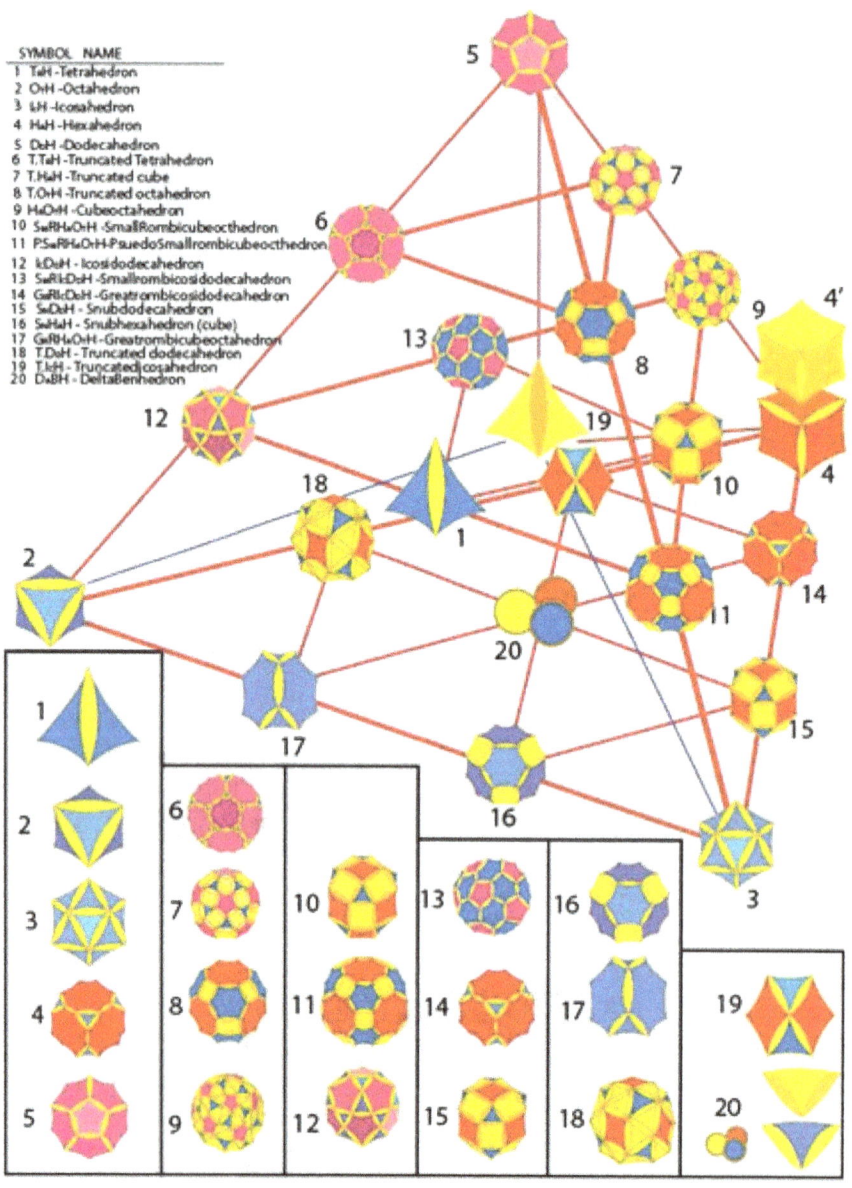

Fig 28

D_ESign Science in The New Paradigm Age

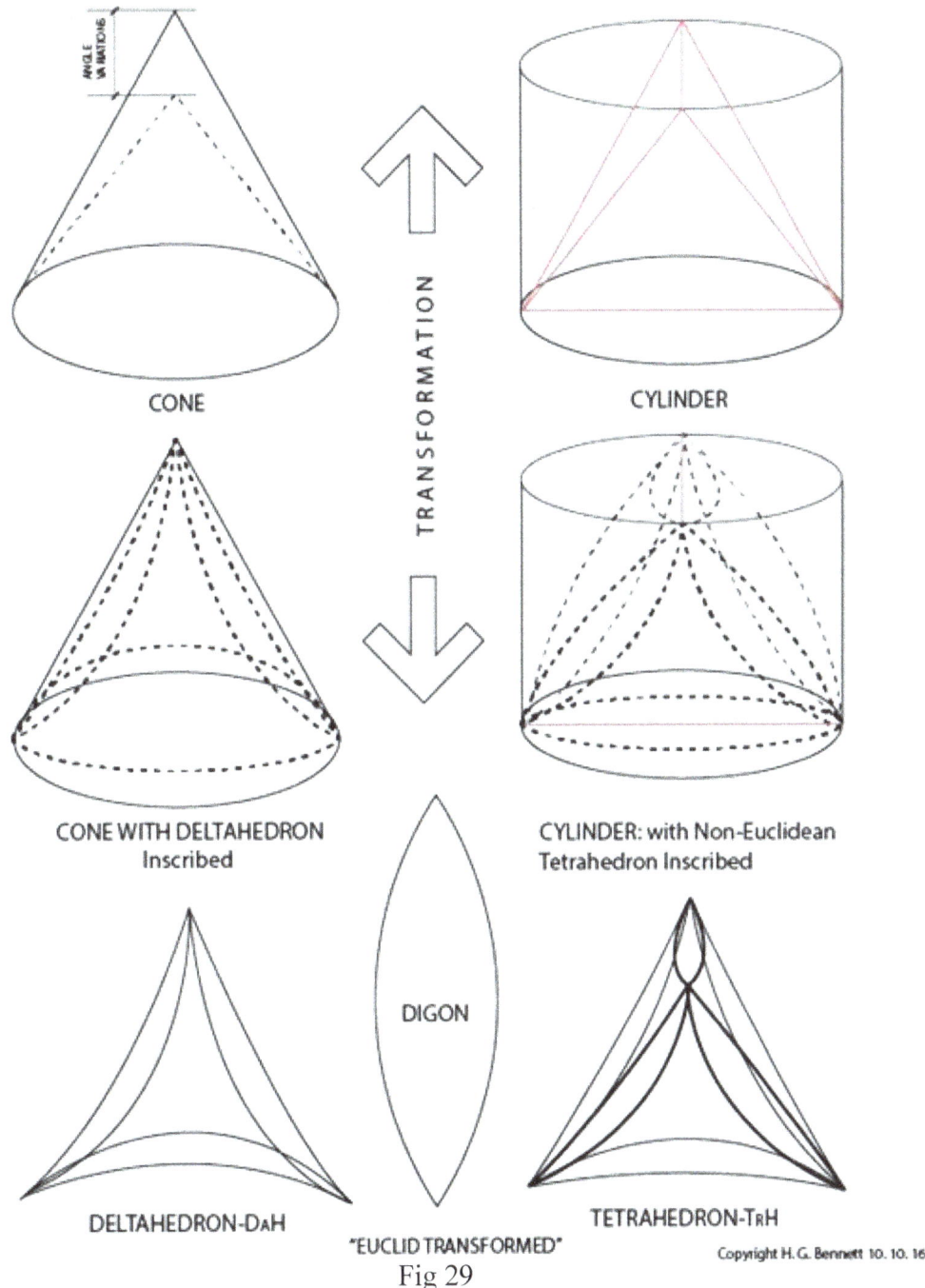

Fig 29

In the Non-Euclidean family are the 2-Riemannian, the 3-Bolyai and Lobachevsky sets, 4-the arcuated rhombic set and 5.the windswept set. They all follow the 'regularity principle' with regular polygonal faces. This fivefold symmetry repeated in nature quite often; emmm! Look into this concept of the 5fold symmetry of nature.

Challenges to the fifth postulate of Euclid led to the development or expansion of the vocabulary. No one has actually created the physical forms that demonstrated these principles until my patents we issued a few years ago. This gives rise to a regular repeatable set of curved forms that follow the vertex relationships of standard polyhedral. These were originally discovered by Egyptian architects calculating the volumes of pyramids. Then the Greeks cam marching in again and took that back to Plato making them the 'Platonic solids' in spite of his anti-physical form beliefs. Plato thought that these very forms, named after him, posthumously of course could never be real.

There is a 'breathing dynamic' in the first three stages of movement between the forms that is not apparent but is very significant. It is the true demonstration of the harmonious interactions between these systems of form that can be the metaphor for the three levels of the threefold principles. The identities of these new form vocabularies have aesthetic qualities and flavors that reflect our thought patterns and creative vision. Without this New Gauge Theory this experience and its opportunities are DOA. There are many transitional concepts that lead to new realities that seem to 'mark' space, time and energy shifts. "A wormhole or Einstein–Rosen bridge is a hypothetical topological (Geometric) feature that's a shortcut connecting two separate points in the space-time continuum.

A wormhole, might connect extremely far distances; a billion light years or more, short distances such as a few feet, different universes, and different points in time. A wormhole is a tunnel with two ends, each at separate points in space-time". Where can we find other similar transitional events and theories in our history?
Nature, (ART), Geometry, Science and Mathematics. Are these tools of consciousness, if so what are their correspondences?

TYPE OF CONSCIOUSNESS	CORRESPONDENCES	DISCIPLINES
Super Conscious Infinite	Mind (Souls)	Mathematics
Spirit Consciousness (Cosmic)	Akashic (Karmic Mind)	Geometry
Conscious Awareness	Mind (related to the body)	Science
Sub Conscious Intuitive	Mind (Ethereal)	ART
Un Conscious 'Auto Conscious'	Mind (the Subtle Bodies)	Nature

OLD PARADIGM DISCIPLINES	NEW PARADIGM DISCIPLINES
Mathematics the abstract science of number, quantity, and space. Mathematics may be studied in its own right (pure mathematics), or as it is applied to other disciplines such as physics and engineering (applied mathematics).	Metaphysics: the branch of philosophy that deals with the first principles of things, including abstract concepts such as being, knowing, substance, cause, identity, time, space and energy: abstract theory or talk with no basis in reality.
Geometry the branch of mathematics concerned with the properties and relations of points, lines, surfaces, solids, and higher dimensional analogs.	In physics, a 'gauge theory' is a type of field theory *(Lagrangian) field theory applied to continua and fields, which have an infinite number of degrees of freedom.
Science the intellectual and practical activity encompassing the systematic study of the structure and behavior of the physical and natural world through observation and experiment.	Spiritual Science: (Anthroposophy), a philosophy founded by Rudolf Steiner, postulates the exis-tence of an objective, intellectually compre-hensible spiritual world that is accessible by direct experience through inner development.
ART the expression or application of human imagination and creative skill, typically in a 'visual' form such as painting or sculpture, producing works to be appreciated primarily for their beauty or emotional power.	Conceptual Art: is art for which the idea (or concept) behind the work is more important than the finished art object. It emerged as an art movement in the 1960s and the term usually refers to art made from the mid-1960s to the mid-1970s.
Nature the phenomena of the physical world collectively, including plants, animals, the landscape, and other features and products of the earth, as opposed to humans or human creations (of the physical world) we call ART.	Artificial Intelligence the theory and development of computer systems able to perform tasks that normally require human intelligence, such as visual perception, speech recognition, decision-making, and translation between languages.

*Joseph Louis Lagrange's treatise on analytical mechanics (Mécanique analytique, 4. ed., 2 vols. Paris: Gauthier-Villars et fils, 1888–89), written in Berlin and first published in 1788, offered the most comprehensive treatment of classical mechanics since Newton and formed a basis for the development of mathematical physics in the nineteenth century.
He made significant contributions to the fields of analysis, number theory, and both classical and celestial mechanics. He also did work in Celestial Mechanics.

Who are we leaving this planet to and what are we teaching them and leaving them as we and our time space, intelligence (energy) become ancient?

LET'S LOOK AT DE LUBICZ'S COMMENTS ABOUT OUR CONSCIOUSNESS

"We lack direct consciousness of Space, Time (and Energy as Creative Spirit; symbolism. 'Through meditation and visualization we can develop this ability'. We can know of them only indirectly by mass, force, and energy, and by the intermediary of phenomena such as may be tested by our five senses*.

Intuition and the connection (connectionism) to source are available through Anthroposophy. We understand mass, force, and energy as Body, Mind and Spirit as dimensions of consciousness and quantum mechanics. Without direct awareness (or a higher developed consciousness that transcends our chimeric and illusory senses) of Space or Time (without Spirit), human beings lack two (now three) 'senses*' necessary for the knowledge of all causes. The 7 Hermetic principles teach us the 'cause and effect' principle".

*These senses are enhanced by body, mind and spirit 'faculties' of physiology, psychology and spirituality that are then harmonized with all other triadic harmonies into the deepest sense of 'oneness' or source all humans are connected to. Our 'materiality' is a serious impediment to understanding other expressions of consciousness. Access to source is the closest and deepest connection to the "cosmic geometry"; to coin a phrase) where cognition transcends symbolism.

Our standard definitions of art, science and mathematics (geometry) are based on using each discipline's language to describe the cultural expressions related to our senses, forces and intelligences. They are unique but are all connected to the one expression of the all or consciousness. As paradigms shift they offer new views and tools for our world. There is a closed loop (circle) with what we learn and the tools we use to share it. The adage of using the 'right tool to do the job at hand' describes this circle. The subject (the observer or participator) of the enquiry relates by correspondence to this polarity of the discipline and the known. I need to know geometry. We are familiar with the 'I' and its capabilities. There is some perceived benefit in geometry we believe in with value and why we know it at all. The 'to know' is the challenge. The way to know is to apply the discipline and the appropriate learning styles, tools and principles to get the knowledge and the satisfaction and elevation of conscious-ness that comes with the (positive and successful) experience of knowing what 'I' (the ego) set out to discover. A strong 'I' will get the job done. A weak 'I' will not. This sentence demonstrates the three states of consciousness at work in 'providing need'. For the most part the process is physical, seldom going deeper than the immediacy of obtaining the thing, getting the emotional feeling of being good or self-fulfillment, experiencing the satisfaction, elevation or high. There is generally little or no deeper philosophical meaning gained nor further iteration of the process. I now know what I needed to know, now what can I do with it?

Breaking away from the initial circle of need suggests applying continuous creative processes. Knowing the 'how to' is the method or science of the process. Then comes the Why or the spiritual inspiration, motivation and reason for the goodness to be 'done' or to 'become' or to belong; all with their own currency (ego fulfillment) and value. Some 'Karmic debt' can be paid with this value creation. Science is what I call 'STEM', not as the standard definition.

STEM is "Seeing, Touching, Evaluating (or Eliminating) by Measuring". If this does not produce objective or expected results it is not science and is not perceived as reality. Challenges to the 'Geometry' of the old paradigm minds are raised by new gauge theories.

Do we have the proper physical, mental and spiritual intelligences to interpret the dynamics of consciousness? Can we develop Physical, Mental and Spiritual ''geometries" in interconnected, harmonious way-'DAO'? When cal-cu-lation (space before time with no spirit) is not sufficient very often con-temp-lation (time before space with spirit; meditation added) is needed. Counting as 'keeping score' all basic functions of calculation. Contemplation does not just happen mentally. No-thing happens. Everything happens by LAW says the Hermetic principle of Cause and Effect. This is what "Synthesis' allows us to do!

The classic Euclidean geometry is an elementary geometry of two and three dimensions (plane and solid geometry) with a neutral; straight line, flat surface looking space. It is based largely on the Elements of the Greek mathematics of Euclid (fl. c.300 B.C.). In 1637, René Descartes showed how numbers can be used to describe points in a plane or in space and to express geometric relations in algebraic form, thus founding analytic geometry, of which algebraic geometry is a further development. The Cartesian coordinates is his major contribution using the x,y,z system for spatial orientation. Representing three-dimensional objects on a two-dimensional surface was solved by Gaspard Monge, who invented descriptive geometry for this purpose in the late 18th cent as differential geometry, in which the concepts of the calculus are applied to curves, surfaces, and other geometrical forms and objects, was founded by Monge and Carl Frederick Gauss in the late 18th and early 19th cent. This is no longer an active discipline in mathematics. The modern period in geometry begins with the formulations of projective geometry by Jean-Victor Poncelet (1822) and of Non-Euclidean geometry, with a deflated or imploded looking curved space, by Nikolai Ivanovich Lobachevsky's (1826) and János Bolyai (1832). Another type of Non-Euclidean geometry, with an inflated curved looking space, was discovered by Bernhard Riemann (1854), who also showed how the various geometries could be generalized to any number of useful dimensions.

The intelligence required to understand the physical universe depends on a spatial intelligence that is not isolated nor self-sufficient. Consciousness, expressed as nature, has a holistic, interactive and multidimensional 'modus operandi' that 'MICROMAN' is incapable of.
Allowing source to interact with normal consciousness, without ego, is wise.
The universe is a container of containers including things, emotions, thoughts and ALL forms of consciousness as 'fields within fields' of vibrating energies with infinitude. New geometries appear to be quixotic and mysterious illusions; (ill uses of our brains). The old Geometry is now a crime, punishable by being ostracized to quantum hell and its gauge theories. The longer we keep committing this crime our 'diminished creative capacity' punishment intensifies.

The world is much more complex now. Our minds are corresponding on a subconscious or auto-conscious level where our other systems are in rhythm and quite different from our normal awareness. This is a recipe for dis-ease and stress.

In all systems there are pockets of proper and healthy alignment but for the most part the general atmosphere is not healthy. There are socio, cultural, political and economic consequences to be addressed. I am a firm believer in the creative process and '(spiritual) design science' specifically as the harbinger. Disciplines of Science and Mathematics are the classical languages of our physical Newtonian world.
They like all other fields are evolving leaving us stuck in 'mind-lock'. In this state the observer was thought to be the most natural 'element' in all 'creative' processes.

Quantum intelligence invites the participator to a more contemplative 'celebration' impacting the experimental or 'thought-form' meditation outcome. Emotions, beliefs, thoughts and subtle energies are naturally aligned with and the greater good. The mental environment serves the physical world. Emotional, Spiritual qualities and flavors are more incidentals and accidentals. 'Art' is relegated to the individual and cultural expressions where theories, media and technologies rely on 'eye candy' interpretations of our highly refined, misused, visual intelligence and the six other 'smart forms' we possess.

Understanding consciousness as our pole star contemplation becomes quite different from geometry as we have been taught by the Egyptians to measure land and not our emotional and spiritual state. In looking at the dimensions of consciousness as body, mind and spirit this begins to resonate with the emerging gauge theories of the purest states possible to begin with. We visualize the pole star being above, with the three states of consciousness as vertices on an infinite plane below now as the four vertices of a tetrahedron operating as the 'DNA of form'. To know this form, or anything else it supports, we have to know ourselves. Among the vast array of uses of carbon elements, sentient microman being one of these ubiquitous and 'dynamic molecular logic structures,' it is extremely important to ALL [Living (with) Intelligent Fields of Energy] LIFE. That ever pre-sent macro-micro 'source connection' provides knowledge of all things by a 'source', known as 'Akasha', to access our know-ledge. We are not necessarily aware of the 'tetrahedron in us and many more of its various forms. The Law of Correspondence; the macro is above and the micro is below, on the physical dense earth plane where the panoply of forms are revealed in temporal illusion….. Eventually.

There is another above phenomena we are not familiar with and that is the Van Allen Radiation Belt. Some believe this is the location for the Akashic records. It is a field of energetically charged particles in suspension around a magnetized planet, such as the Earth, by the planet's magnetic field, credited to James Van Allen.

The Earth has two such belts. The main belts extend from about 1,000 to 60,000 kilometers above the surface where radiation levels vary. Most of the particles come from solar wind and other particles by cosmic rays. [What is the interaction of these particles and the solar wind like?] [Is this the 'sun behind the sun' principle of Egyptian Spiritual Science?] They are located in the inner region of the Earth's magnetosphere and contain energetic electrons and protons. [This is the form of atomic cells.] Other alpha particles, are less prevalent. The belts endanger satellites, [Who's on first?], which must be shielded to protect their sensitive components if they spend significant time in the belts. In 2013, NASA discovered a transient, third radiation belt that was observed for four weeks until destroyed by a powerful, interplanetary shock wave from the Sun.

Is this the eternal 'thought' field (matrix); the field where principles, laws and forces converge and synthesize harmoniously to keep the planet's rhythms balanced? All natural phenomena are in this grand infinite field of fields. Some are waiting to be revealed when the proper alignments and correspondences are caused. Knowledge is one of them with us as the tuning forks resonating with source to become aware. Aware of what is the question. Is Creation itself a cause of itself; it being an expression of a (or more) higher conscious-ness? Is this the other self that is trying to communicating with us? Current explorations in subatomic gauge theories encounter this threefold symmetry with the three quark description of Western science's 'creation story'. When does this scientific story become the 'cultural myth or does it ever, and how can this be done? The high priests weave their 'phantasmagoric' mythology designing performances of shear cosmic drama with plots and characters arranged scenically with complex 'geometries'. A new language is needed if we are to understand what the story means to life itself. A new symbolic 'notation' is a temporary scaffold for an 'architecture' yet to be conceived for this. The story line should involve the organic synthesis of the physical; bodily and physiological.

It also includes the emotional and psychological with the spiritual; creative energy to co-rrelate and co-respond to symmetry principles, laws and 'field languages' we have never heard, seen nor encountered before. The quark model, preceded the standard model and was independently proposed by physicists Murray Gell-Mann and George Zweig in 1964. Quarks were introduced as parts of an 'ordering scheme' for hadrons. There was little evidence for their physical existence until deep inelastic scattering experiments at the Stanford Linear Accelerator Center in 1968. Accelerator experiments have provided evidence for 'all six flavors'. The top quark was the last to be discovered at Fermi lab in 1995. Another 'ordering scheme' is needed for the vast amount of information being developed; I imagine.
This new organizing scaffold goes way beyond Mendeleev's periodic table of elements. The subtitle for this 'passion play' is; "The Tower of Babel II"; the sequel. Subatomic physics sounds like it now.

Chlorophyll, which gives plants their green color, enables them to use sunlight to convert water and carbon dioxide into sugars and carbohydrates, chemicals the cell uses for fuel. Why do plants appear green? The logical answer is; the blue planet and yellow sun make leaves green.

The genomic sequencing in Science is discovering a 'living language'. As with all others, it's one we should expect to be dynamic; keeping pace with our 'mental and computational growth. microman as a bio, magnetic, electro, conscious complex with other sentient forms; from microbes and bacteria to plants and animals can now reveal their 'genomic sequencing'. Traits and identities can read by researchers to decipher genetic information (because of the vibration and periodicity) using the DNA language. A tedious, painstaking process has been transformed into new techniques with powerful computers. They have been sped up a hundredfold and are still growing. Sequencing involves determining the order of bases, the nucleotide subunits (adenine, guanine, cytosine and thymine, referred to by the letters A, G, C and T) found in DNA. (Humans have about 3 billion base pairs.) On language and codes; the most phenomenal reference on the topic, in my opinion, is found in Gregg Braden's book, 'The God Code'. A 'literal message' is encoded as 'a seal of GOD's great work that's written in human DNA code. There are theories being written now, by some smart folks, about how this could have been done that's still being researched. It takes a very open and receptive mind to relate to this information.

The other 'grand challenge project' intends to revolutionize our understanding of the human brain. It's developed by the White House Office of Science and Technology Policy as part of a broader 'White House Neuroscience Initiative'.
It's inspired by the Human Genome Project, 'BRAIN' and aims to help researchers uncover the mysteries of brain disorders. 'Rewiring the brain' is an Anachronism and an inappropriate metaphor that describes the fullest potential of this project. Wires do not belong to this period nor the context in which true solutions and more applications are possible. The highest and best use of the brain project is to address the critical needs in 'brain enhancement' through entrainment technologies for education, growth and human development as preventive and constructive, socially responsible strategies. The emphasis on physical muscle development always outweighs brain muscle every time.

Disruptive, innovations and technologies generally create anachronisms, if they don't they are simply variations of things, rituals and cultural traditions. A "Flavor Machine" or periodic matrix of quantities, qualities and energies allows us to detect the Triadic nature of all natural and manmade expressions we interact with in our daily lives. It can be information technology used as frameworks for the vast amounts of data information technologies we are still struggling to accommodate, distribute and process efficiently and cost effectively. We can synthesize the fundamental elements of nature in ways that can reveal the deepest mysteries that have perplexed the human complex for ages.

A Ubiquitous Combinatorial Architecture, with the technologies and flavors they generate can be useful in helping to deliver the promise of not just the theory of everything but the 'everything' itself, with a 'brand' new aesthetic expression and cultural identity. Exponentially and disruptively advancing computing power that make binary limited brain analogs truly anachronistic to support developments in many disciplines. A truly harmonious well synthesized technology that can predict and simulate structure, behavior and optimize 'power' inspired by a spiritual science that recognizes the oneness of human and universal consciousness is being offered here.

Imagine an adventure that spans the breath of time, space and thought. We board Albert's beam and go back 11,000 years to Gobleki Tepe, in Southeastern Turkey. Göbekli Tepe means Potbelly Hill" in Turkish. It is an archaeological site atop a mountain ridge in the Southeastern Anatolia Region of modern-day Turkey. The architecture of this ancient site is not just a temple form. It's a celestial monument that aligns with constellations in the 'heavens'. Current knowledge corresponding to higher vibrations, on the scale of evolved wisdom traditions, are available now. The meaning of this structure does not align with current interpretations and are somewhat irrelevant. We think these are temples or burial sites and in most cases they are not.

Albert's beam, bends on its way 'back to the future' greatly increasing its momentum. Its significance is exponential added value with disruptive potential making a pit stop in the 'NOW'. Its journey is being relaunched towards its future destination and our destiny. With this charge of energy how far could this stored energy, from 11,000 years ago, propel this beam with an exponentially enhanced intelligent payload? Propulsion will be an anachronistic technology by then. Albert's beam and the light beam become one. Can we all get on board? May be we should ride our own thought waves to our own destiny instead.

NB: There is a variation on anachronism where symbols are used in languages of cultures past creating inaccurate descriptions of their (past) experiences that over time are seldom verified in the current parlance. We keep every creation story in its ancient form and miss its meaning. Let's decode them with our paradigm perspectives and look for connections hidden in us all.

The democratization or 'true freedom of knowledge' is not part of the current dialog in most developed civil societies. The Commoditizing of knowledge that is inspired, supported and justified by materiality is deeply rooted in individuals' Body; mentality and personalities; Mind, and the Spirit or essence of the cultural matrix. Realigning is very often pre-tense disguised as manipulation. What we have instead of paths to true liberation, which is our spiritual right, is 'mental' slavery. Creativity is meant to lead to freedom and happiness. This is the message that gets lost on the conditioned mind-locked fields of those who are caught in all their variations and types of limitations. Creativity trickles down to the denser expressions of manifestations satisfying needs for goods, goodness and technologies, for the good of all. The natural laws of all dimensions, flavors of consciousness demand this of its macro states and the micro 'resonating expressions' like microman, with the most degrees of freedom, given dominion. Dominion is not dominance nor decadence; it's access to a-bun-dance. It's not a free ride through the dashes of life between birth and death; it's an opportunity to raise our consciousness to higher levels of self. The don'ts and does go on 'ad infinitum'. We seem to manufacture more of them as technology advances instead of them diminishing.

Every do and don't becomes a complex and a business opportunity that extends slavery and limits liberation. Liberation is not freedom; it's the imperialists just cutting the purse strings, not the umbilical. There is a disproportionate and unhealthy dependence on consumerism as the engine for and of economics. There are many other avenues for growth and development available to us where the gifts we each have can be used for exchanging value and currency. We are not lilies, even though we are arrayed in finery. We have to toil intelligently when creating self-reliant opportunities to thr-(l)ive. Instead we follow the coerced directions and persuasions of the massive information, knowledge and freedom pedaling Commoditizers; new snake oil marketers, convincing everyone wanting to realize their dreams for Divine order to reign for all to be de-live-red.

This is the myth of 'herdism', the instinct to follow trends like lemmings going over hills. Establishing fair distribution systems require technologies that this inherently creative process can begin to explore. As a model for community building it contains and is capable of offering the framework for such a system. Solutions and true responses to human needs are possible (All things are possible).

Consumerism alone does not create vibrant and sustainable societies, cultures and economies; especially when the definition of economics is basically the study of 'lack' of resources and scar-city (take-note) paralysis and stag-nation of ideas. Abundance is not part of the spiritual ecosystem nor its supporters. It serves the ego instead of the community. Balance, harmony, equity and equilibrium begin with consciousness. Since all is consciousness there is nothing else to for us to be, to use or to become but higher forms of consciousness we are already (light). Since we know that extremes meet we can infer that our work is done. We can be more awake when we get this 'gift of 'pre-sent' understanding to presume and realize that the chimera we are all chasing and becoming is all false. We interact with gross and subtle forces impacting our body, mind and spirit. Interactions are interpreted with the fullest range of human intelligence that correspond with these forces, properties and flavors (power and will). There are extracorporeal faculties that extend into the realms of subtle bodies and energy fields beyond the physical form as part of these interactions. There are interpenetrating fields of space-time-energy vibrations permeating all things, states and essences in appropriate correspondences keeping the super spiritual ecology in tune with the cosmic music we resonate too. Not all can be measured some can be felt and others experienced.

This simple sentence defines what consciousness does for us. Subatomic physics might have discovered these forces and may not be able to 'STEM' them. We rely on emotions, intuitions, experiences and spiritual disciplines and esoteric knowledge as heuristic strategies we use until we can arrive at the continua of all the appropriate triadic convergences in our personal space-time-energy continuum as well. It is on this very personal level where the inner journey is taken. In the interim 'we chop wood and carry water'. In studying the role the sun plays in LIFE there is a profound respect due, I believe, that's inspired by my understanding of what's at the root of Egyptian Spiritual Sciences and the ancestral wisdom of many (not ancient but) eternally human and intelligent traditions.

Our senses and other finely tuned instruments including our kundalini and chakra centers, the endocrine system and the subtle bodies etc. can be seen from a wider perspective as a well harmonized intelligent synthesis interacting with forces and dynamic principles calibrated to interpret the unique vibrations we and all things resonate and respond to. Gauges used to measure 'phenomena' with number in space-time chunks. Energy is involved in the dynamic, causing movement or vibration to starts the measuring process. Many of these have been deciphered only recently; within the last hundred year or less. I propose that there are others in the mix we are not accustomed to thinking of as intelligent systems and to be considered as such. Measurement, Light (the Sun), Feelings, Gravity (we need to look at specifically for their interactions with other forces, intelligences and dynamic potential we are not familiar with and not just as falling apples) and Thought. These are five concepts that align with other five symmetry principles that are worthy of further reflection, contemplation and studies.

Measurement; traditionally, is the assignment of number to the space-time not necessarily energy as yet, characteristics of objects, processes and events, which can be compared with other objects, processes and events. Even though we experience the spirituality of number daily, we do not have a framework for understanding it in terms of 'fields and gauge' theories. Number is qualitative and even spiritual. The oneness of all consciousness, our oneness with our source are 'spiritual. It does not count and cannot be counted as things can be.

We feel one with ourselves, our family and the universe is emotional, transcendental and 'sacred'. These are moral and spiritual compasses that direct our lives. They do not appear on the intellectual disciplines radar. They are supposed to teach us about life and the world. The scope and application of measurement depends on intelligence, behaviors and energies that are inherently in the phenomena we are dealing with. If the intelligence we look for is not in the system no amount of measurement will be coherent and be sensible. We have a propensity for operating within closed loop phenomena. Leave the loop and we are lost.

Measurement is the first principle of manifestation; 'making things'. When ideation, visualization, reality testing and production processes are prepared the next step is 'measurement' followed by incision and fabrication, sufficient for old paradigm with limited creative capacity, inefficient technology and scarce energy (and willpower or) power and will.

While nature creates 'holistically', man-u-facturing (without hands) is characterized of unsystematic partial and non-sequential measures taken over a period of time and is piecemeal. It's done with components, parts and patterns even though things, feelings or events are conceived HOLISTICALLY and three dimensionally. We now have multidimensional holographic models of '3D form making' as a new digital technology.

From levels of deeply rooted codes (language), symbols, symmetry operations for quantitative, qualitative, mental and other thought processes to manifestation, what is imagined and made must be and first measured. To make we are forced, obliged or 'compelled' to measure. Compulsion and measure are cause and effect. Between them there is tension and whenever there is tension the outcome is intellect with intelligence producing some desired result. In this case it's measurement. The ART-iculated 'thing' is the expression of the language of consciousness.

In The 'Study of Numbers': A Guide to the Constant Creation of the Universe by R. A. Schwaller de Lubicz explores number as a symbolic and sequentially genomic or iterative dynamic of nature, mind and spirit within a matrix based on a structural philosophy and combinatorial logic following the principle of the 3:4:7 gauge. This is the Flavor machine™

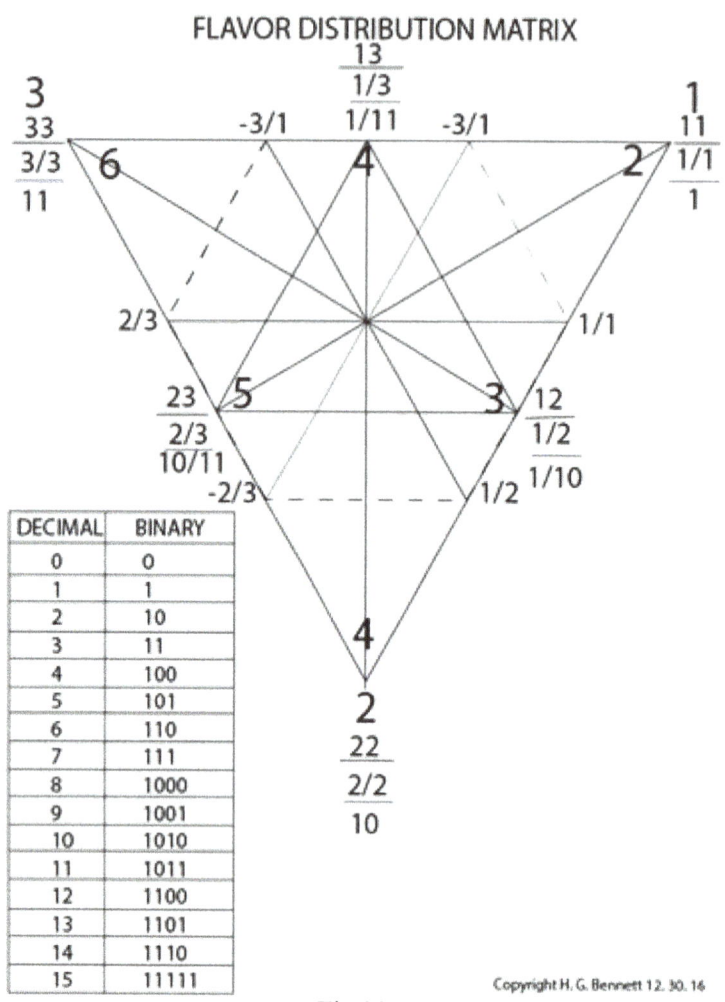

Fig 30

Light: a vibration or radiation is at the root of all conscious phenomena and super-consciousness itself. It could very well be the essence of the universal MIND too. ("Let there be light" in The Gene of Isis) In the metric system light measures 299,800,000 (m/s) meters per second. In the Imperial British and US systems it registers at 186,300 (m/s). If we think of light as some 'foundational vibration' underpinning all expressions of consciousness, where the five versions of consciousness known thus far, possess the qualitative and behavioral and energy (spirit) markers of matter with their respective physio-logical characteristics. The psychological and spiritual registers are a lot more complex than the physical. For this we go to the ancient and eastern wisdom traditions. Psychological interactions are interpreted and experienced by emotions, the prime one being feelings. The five senses map onto brain sites for interpretation as they each are extrapolated into language with grammar, syntax and vocabulary all expressing concepts, things, feelings, thoughts and experiences.

"Wave interference is a phenomenon and a function of light that occurs when two waves meet while traveling along the same medium. The interference of waves causes the medium to take on a shape that results from the net effect of the two individual waves upon the particles of the medium. Wave interference can be constructive or destructive in nature. Constructive interference occurs at any location along the medium where the two interfering (like) [high] waves have a displacement in the same direction. For example, if at a given instant in location and time (space-time) along the medium, the crest of one wave meets the crest of a second wave, they will interfere in such a manner as to produce a "super-crest." Similarly, the interference of a trough and a trough (where two [low] troughs intersect and converge) interfere constructively to produce a "super-trough." Destructive interference occurs at any location along the medium where the two interfering waves have a displacement (separate; diverge) in the opposite direction. For example, the (odd) interference of a crest with a trough is an example of destructive interference. Destructive interference has the tendency to decrease the resulting amount of displacement of the medium. This phenomenon applies to all vibrations and their waves.

Feelings: although brain sites 'process' feelings, altered brain states do not always experience or represent conscious perceptions or changes in feelings. Feelings and their interpretations are (hand 'n' g-love); polarity or opposites inspiring behaviors that would not be experienced if thoughts were void of feeling and spiritual experiences and intelligences. The strip of cortex in the brain, called the insular cortex, is one of the sites contributing to the conscious feeling experiences, even though it is not required for all feelings. The brain is a physical organ; a 'control center', with behavior and energy potential that's 'local'; to the body it in-habits with its own habits and in-hibits. The in-hibits come from conditioning. Mind, which is not the brain, is extracorporeal and extends into the realms of subtle bodies, energy and force fields way beyond any aspects of physical form. Human nature relies on understanding feelings. Feelings are conscious perceptions of sensations originating in certain types of bodily movements, behaviors and energies; (physiological, psychological and spiritual) activating select brain sites, emotional systems and electro-magnetic energies.
Feelings have two distinctly different origins. One set comes from activity in heart, gut, lung, and muscles. The second source is extracorporeal sources participate in experiencing senses of taste, smell, touch, sight and sound.
These are forces (intelligences) in the finer and denser vibrations that correspond with higher or lower level intelligences in brain sites governed by physiology, psychology and spirituality and external stimuli. The hypothalamus links the nervous system to the endocrine system via the pituitary gland. The hypothalamus actually controls the pituitary gland and integrates many messages from parts of the brain based on feedback from all over the body and tells the pituitary what to do through feelings. The pineal gland is considered the (spiritual) seat of the soul located at the crown chakra and its *KUNDALINI* messaging system is finely tuned.

The five forces, behind the five senses including feeling correspond to and align with the seven Hermetic principles and all other Laws of Seven. This is the next dynamic we need to explore starting with 'light' as its complementary form dynamic. One form of light is "visually perceived radiant energy". Light is internal and spiritual. We experience light as insight, emotion or feeling states. Light vibration interacts with the sense of sight. "It is a small range in the electromagnetic spectrum; interacting with the sense of feelings.

It includes X-rays, ultraviolet and infrared energy with the sense of touch, micro-waves and radio waves; interacting with the sense of hearing. When the radiation of an object is emitted or absorbed, by a particular object or by its unique distribution pattern or 'UDP' marker this becomes the object's own electromagnetic code for its own vibration.

Thought: is presented here as a dynamic not 'force' per say (maybe it's a weak force?) which is contrary to pure physical action or mechanical movement. Whether science has appointed it as such is by no means an inhibition. The quixotic art of thinking, used to manifest reality, may employ the same principles and other methods with different results. Thought, the highest vibration of consciousness and the finest spiritual 'fuel' could be much faster than the speed of light as (intuitions, concepts and perceptions) does not become power or will, unless the 'True (inner) dynamic (or higher vibrations) of thought' is brought into action. The same enthusiasm and visioneering can produce different outcomes; one can be positive, the other can be indifferent and the third as cathodic.

With my ancestral memory, all I have read and studied from my intuition I am compelled to present my gift of an interpretation of intuition humbly requesting you to be liberated and suspended from conventional definitions of 'intuition' to now think of it as a force. I would like to nominate it perhaps as a subatomic candidate for 'weak force' consideration. A phenomenon with that much power deserves recognition at least.

We now know of two, very different "operating systems" related to intuition. One is quick, instinctual, and often subconscious and is controlled by our right brain and the "limbic" or reptilian" brain. The primary structures within the limbic system include the amygdala, hippocampus, thalamus, hypo-thalamus, basal ganglia, and cingulate gyrus. The amygdala is the emotion center of the brain, while the hippocampus plays an essential role in the 'formation of new memories about past experiences'. Two is our slower, analytical, our (normal awareness) or conscious way of operating controlled by our left brain and the "neocortex"). Intuition which is part of the first, involve decisions that are not derived from reason, but from instinctive choices, a natural propensity or an innate skill of a specified kind. The term muscle memory's conditioned positive or negative responses to stimuli might be considered intuitive. The 'formation of new memories about past experiences' is all about growth and recognizing paradigm shifts. This allows us to reflect and adjust our perceptions of changing reality.

Concepts related to intuition include but are not limited to the following; talent, gift, ability, aptitude, faculty, skill, feeling, genius, knack, bent; in the words of Albert Einstein: "Intuition is a sacred gift; the rational mind is a faithful servant". Forgetting the gift, society tacitly accepts its 'mental slavery'. Clairvoyance and the host of other abilities are included. As an intuitive myself most of my 'direct knowledge' comes from SOURCE.

Gravity: is the force interacting on all matter in the universe, and a function of mass, distance, space, time and energy. What is its behavior, its interaction and its true essence? What does it do? Is there a telltale gravity 'brain site' that the body, mind and spirit responds or relate to? The Solar Force: the Egyptian Sun behind the Sun: Given that the universe is conscious (a paradigm shift is celebrated) we can imagine how the consciousness within the Solar System is undergoing its own 'solar initiation' with massive amounts of solar flare plasma emissions.

A solar flare is a burst of light, radiation and intelligent energy which erupts from the sun's magnetic atmosphere. Solar flares affect all layers of the solar atmosphere, when the plasma is heated into millions of kelvins, the flares produce radiation across the entire electromagnetic spectrum at all wavelengths. These light frequencies range from radio waves to gamma rays, although most of the energy is spread over frequencies outside the visible spectrum to the naked eye (they may have some role to play in the universal picture). We can celebrate our Mother earth as alive as it supports all living things.

Does it not make sense that giving life depends on be-ing alive. It is clear evidence we are undergoing an intense Solar Initiation of radiating plasmic light which is rapidly and consistently progressing humanities evolution (or digression) on our chosen spiritual journey on earth Is this 'radiating plasmic light' the eternal light or consciousness, insight and our internal light? Is this the paradigm womb pre-paring our rebirth? I believe it is.

INTUITION:
The Metaphysical meaning of intuition is the natural knowing capacity; inner knowing; the immediate apprehension of spiritual 'Truth' without resorting to intellectual means. It's the wisdom of the heart. This faith awakens the so-called sixth sense, intuition, or divine knowing with confidence in the invisible good. Could this sixth sense be the synthesis of the senses reflecting the oneness of source we are all linked to and eventually return to? What senses are needed for that return journey? Through the power of intuition, man has direct access to all knowledge and wisdom. When one trusts 'Spirit' and seeks understanding, a direct connection to source develops. We link to solar intelligence and other wisdom traditions, their great teachers and their records referring to the source, in the Van Allen belts known as "Akasha" or the Akashic records. Here's a flashing intuition. I feel sad and vehemently angry that as a conscious and intelligent species we continue to commit what I call 'spiritual and karmic suicide', doing so in nescience and in no sense with innocence abnegating all forms of intelligence, wisdom and grace to satisfy our self-dissolving desires.

Wordsmithing permitted: we look at a few words with a more 'granular set' of compounds; Kn-own; sh-own; gr-own and s-own. The seed of life is sown, from within our very own (conscious-ness) for our seeking selves to see with insight and delight truth shown as pure eternal (solar) light. This is the open and close case for trusting our instincts: We trust our instinct because we live in, by and with our ancestral wisdom in our DNA.
We do so with all the physiological, psycho-logical and spiritual intelligent dimensions, the subtle parts and creative processes, of eternal LIFE (Living Intelligently For Ever), that's a lot smarter than we can ever be or will become.

> "Intuition is linked to the caudate nucleus, which is part of the basal ganglia—
> a set of interlinked (nerve) brain sites responsible for learning, executing habits and automatic behaviors. It is a processing center for the very complicated messages generated by the brain's frontal lobe, used in thinking, planning, and understanding. It works in learning, the formation of ***good and bad habits,*** and some psychiatric and addictive disorders. Scientists have found that the neurotransmitter dopamine, already linked to the basal ganglia in movement disorders, is important ***in learning via reward and punishment*** as well as in schizophrenia and attention-

deficit/hyperactivity disorder. Knowing how the basal ganglia works has revealed possible avenues for treatment of these and other disorders. The basal ganglia receives stimuli from the cortex; the outer rind-like surface of the brain. These structures project back to the cortex, creating a series of cortical–basal ganglia loops. At some point the interaction with the chakras and the endocrine systems come into play

The cortex is associated with conscious perception and the deliberate and conscious 'ANALYSIS' of any given situation, novel or familiar. It is the site where highly specialized expertise resides that allows you to come up with an appropriate answers ***without conscious thought***. This is the perfect candidate for 'SYNTHESIS'. In computer engineering parlance, a constantly used class of computations (namely those associated with playing a strategy game) is downloaded into special-purpose hardware, the caudate, to lighten the burden of the main processor, the cortex". Thus far we have not invited the ego to this discussion. Wisdom traditions are more open to looking at bigger pictures.

Inferring causation from correlation (otherness) (the mutual relationship or connection between two or more things) (polarity) is the first principle of intuition. If awareness, cognition or memory are not present at the time of need or crisis what else can we rely on? "Just because two things are associated does not imply that one causes the other". Causation is also an inherent event. It percolates with 'affinity' in the same space, time, spirit or energy continuum.

In humans a nerve plexus is network of intersection nerves. Essentially nerves in the spine whose destination is the same are grouped into one large nerve (like many lanes on a highway instead of several isolated roads). There are six of these plexuses in the body: cervical (head, neck, and shoulders), brachial (chest, shoulders, arms, and hands), lumbar (back, abdomen goring, thighs, knees, and calves), sacral (pelvis, buttocks, genitals, thighs, calves, and feet), solar (internal organs), and coccygeal (internal organs). A ganglion is a mass of nerve cell bodies. A plexus is made up of ganglia. They are essentially the relay points of the entire (electrical) nerve network.

"Things do not cause one another; the infinite, normal and sub-conscious constant creative consciousness do. We are just observers on the journey and less engaged participators." A more specific site in the brain thought to govern intuition is the ventromedialprefrontal cortex; a region (with faculties) that stores information regarding past rewards, as well as punishments. (This reson-ates with religious dogma.)? The discovery of the ventromedial prefrontal cortex's role in intuition was solidified by a study conducted by *Antonio Damasio, a neuroscientist from the University of Southern California. "With the difference in thinking and decision-making strategies by normal individuals and those with damage to the said part of the brain, the ventromedial prefrontal cortex can elicit the emotional responses that people deem as 'gut feelings or hunches'". (Is this where most creators and designers operate?) Some healthy professionals refer to this same 'gut' as inspiration and source for creative solutions. The Locus ceruleus, a minute nucleus located in the pons, is another gut feeling center.

Studies related to the science and biology of microbes in the gut, the brain, our bodies and environments are now emerging thanks to genomic sequencing. As the seat of noradrenaline provision to the forebrain, it is one of the structures that build the 'ascending reticular activating system.' Because of its ability to influence emotion, motivation, memory, learning and decision making, it is known to play a big role in intuition. The reticular activating system (RAS), or extrathalamic control modulatory system. This is a set of connected nuclei in the brains of vertebrates responsible for regulating wakefulness and sleep-wake transitions.

As its name implies, its most influential component is the reticular formation. This is where sleep and consciousness has projections to the thalamus and cerebral cortex that allow it to exert some control over which sensory signals reach the cerebrum and come to our conscious attention. It plays a central role in states of consciousness like alertness and sleep. The ventromedial prefrontal cortex (vm PFC) is a part of the prefrontal cortex in the mammalian brain. The ventral medial prefrontal is located in the frontal lobe at the bottom of the cerebral hemispheres and is implicated in the processing of risk and fear. It also plays a role in the inhibition of emotional responses, and in the process of decision making."

**Antonio R. Damasio: Expert in cognition and behavior, with special focus on emotion, decision-making and consciousness. David Dornsife Professor of Neuroscience, Director, Brain and Creativity Institute USC Dornsife College of Letters, Arts & Sciences*

In the sympathetic nervous system, norepinephrine is used as a neurotransmitter by sympathetic ganglia located near the spinal cord or in the abdomen; the location of the **solar plexus chakra**. It is also released directly into the bloodstream by the adrenal glands as sympathetic effector organs found at the chakras the crown, the throat and others. Regardless of how and where it is released, norepinephrine acts on target cells by binding to and activating noradrenergic receptors located on the cell surface.

Thought, the highest vibration of consciousness, is the finest fuel humans use to energize and actualize themselves. It is faster than the speed of light; transmitted as (intuitions, concepts and perceptions), but does not become useful or beneficial as power or will, unless the true inner light, insight, love and all higher vibrational dynamics of thought with motivation, inspiration and desire is (acted on or) brought into action. Thought begins as 'self-talk'; conversations we have with ourselves in the stimuli to sense data stage of cognition, memories and feelings. These monologs are the creative ideation we conduct for survival, protection and satisfying the ego. When we get to the point of being able to share the content of our monologs we communicate (sharing is motivated by intention). Does the word 'Content' have anything to do with being satisfied or 'contented; or contentment or with intent and intention with the products of our self-talk or our communication? Is content intended to create states of happiness and satisfaction? The best we can expect from our sharing is alignment and agreement that resonates with ourselves, our self-talk and with those we communicate with. The otherness of our interactions is the source of intelligence with positive feedback.

"As cognitive science evolves two current approaches dominate the problem of modeling representations. The symbolic approach views cognition as (Quantitative or Calculation) computation involving symbolic (equations) manipulation. Connectionism, making a special case for associations, (Qualitative and Contemplative) models interactions (symmetry functions and operations) using logic structures and artificial neural networks.

*Peter Gärdenfors offers his theory of conceptual representations as a bridge (Syhthesis) between the symbolic and connectionist approaches. There is a third element missing from this construct. Synthesis™ is the creative process of harmonizing the Body, Mind and Spirit/energy through its expressions. The symbolic and connectionist approaches represent the physiology and psychology; body-mind continuum without spirit. Synthesis is the signature of the present paradigm. The trends in technology are predicting that synthesizers as professionals will be essential to how our future is unfolding. Semiology, structuralism, geometric morphology are forerunners of this synthesis effort". Gärdenfors shows how conceptual spaces can serve as an explanatory framework for a number of empirical theories, in particular those concerning concept formation, induction, and semantics.

Connectionist research models both lower-level 'perceptual' processes and higher-level processes as object recognition, problem solving, planning and language understanding. The connectionist AI systems are large networks of simple numerical processors, massively interconnected and running in parallel.
Between the neural and symbolic levels there are processes that are not yet 'connecting' to be the efficient neural representation systems. Are machines ever going to be as natural and nuanced as humans? Can thoughts be machined? Do we know enough to machine thoughts? At what point do we recognize the presence of 'spirit' in knowing and technology. Neural networks are the brain and nervous systems 'electrical' system and source. We need to create new models to interpret subtle forces and more dynamic characteristics of physiology, psychology and energy or spirit not possible from the logic, symbolic or analytic processes of the old paradigm.

Björn Peter Gärdenfors is a professor of cognitive science at the University of Lund, Sweden. He is a member of the Royal Swedish Academy of Letters, History and Antiquities and recipient of the Gad Rausing Prize.

Spiritual models;
The kundalini and chakras; the correlation (associationism or connectivism), the Central Nervous System and the endocrine system, the number and structural logic, form technologies, new gauge theories, energy system, media and art; ingredients of 'SYNTHESIS™'.
Hermetic and symmetry principles, number theories and Laws et.al. The Kundalini and the seven Chakras were first documented in the Vedas thousands of years ago. They teach us about the Physical and Spiritual Subtle Bodies; each connected to a particular Chakra. Symptoms are experienced when Chakras are in balanced or imbalanced. In spiritual science, there are 2 types of "bodies": Physical and Subtle. The Physical body is the tangible body made of 3 major parts; bones, muscle and skin. The Subtle body is comprised of 7 layers and 7 bodies of the aura. They extend from the first or the Root Chakra at the center of this conscious multidimensional matrix moving outwards from the physical as 1. the Etheric Body (light blue/gray), 2 the Emotional Body (multicolored clouds of energy) the sexual center, 3

the Mental Body (bright yellow) the solar plexus, 4 the Astral Body (rose-tinted colors) the heart, 5 the Etheric Template Body (cobalt blue) the throat, 6 the Celestial Body (opalescent pastel colors with golden/silver light) the third eye, and 7 the Causal Body (Golden with threads) the Crown]. These are the three registers with the heart as its own bridging the lower physical plane and the upper spiritual plane with connections to the 7 chakras.

Do these subtle bodies interact with any extra corporeal energy fields? Have we exhausted the connections to all other internal systems regardless of which cultural orientation's paradigms we submit to or use? Since human nature is everywhere the same, there is oneness in all consciousness and all its expressions, a synthesis of any east west dichotomy can and should be harmonized.

The Central Nervous System

Within the brain, the autonomic nervous system regulates and adjusts baseline body functions and responds to external stimuli. It consists of two mutually inhibitory subsystems: those nerves which activate tissues-- the sympathetic or arousal system, and those which slow structures down for rest and repair--the parasympathetic or quiescent system. The sympathetic is ergotropic that is releasing energy, and the parasympathetic is trophotropic, that is conserving energy. The two sides of our autonomic system reflect the two main processes in life "growth" or "protection." How does "growth" or "protection" relate to "fight" or "flight"? These two mechanisms (are binary) and cannot operate optimally at the same time. Consider that our nervous system is either wired for eating (parasympathetic) or running away from being eaten ourselves (sympathetic). So the two systems generally act in opposition to each other; (in polarity) yet where dual control of an organ exists, both systems operate simultaneously although one may be operating at a higher level of activity than the other. This behavior has the imprint of quantum (mechanics/dynamics) principles written all over it.

 comprises the paravertebral sympathetic trunks which run up the front side of the spine from the cranial base to the coccyx. Sympathetic nerves run mostly from the thoracic and lumbar region and are longer and less direct than the parasympathetic nerves. Thus their effect is more diffuse. Instead of separate ganglion for each vertebrae certain segments collect together to form a single large ganglion eg: the cervical ganglion in the neck and the stellate ganglion in the upper thoracic region. Connected to the ganglion are plexi that pass to the organs. The cardiac plexus via the stellate ganglion supplies the heart and lungs, thoracic activity and blood circulation. The solar plexus is connected with the lower thoracic spinal nerves and supplies sympathetic fibers to the stomach, intestines, adrenals and other viscera. The heart is supplied by sympathetic nerves arising mainly in the neck, because the heart develops initially in the cervical region and later migrates into the thorax taking its nerves down with it. These ganglia are known as the 7 chakras of the Kundalini system.

The parasympathetic or reposing side of the autonomic nervous system promotes relaxation, sleep, growth and repair. It is sometimes called the "trophotropic" system because it conserves energy. It includes the endocrine glands, parts of the hypothalamus and the thalamus, and reaches into the right cerebral hemisphere. It aligns with the 7 chakras.

Thus the non-dominant, holistic mind is connected with the quiescent (dormant) system and involves the hypothalamus and hippocampus.

After the activity of sympathetic stimulation the parasympathetic system reverses the changes when the danger is over and returns the body functions to normal.

Psychotropic: Having an altering effect on perception, emotion, or behavior. Uses especially of a drug. Depressants: Substances that created distorted perceptions of reality ranging from mild to extreme *Endocannabinoids: natural, marijuana-like substances produced by the body.*

No	MAJOR PLEXI	REGION/CHAKRA	ALLIED MINOR PLEXI	BODY PARTS
7	Choroid Plexus	Brain/Crown		Sight, Nasal and lingual Glands
6	Thyroid Plexus	Throat	Superior, middle and Inferior Cervical Cartoid Plexus	Neck, Jaw, Throat, Ears, Voice, Upper Lungs, Esophagus
5	Cardiac Plexus	Heart	Pulmonory Plexus	Heart, Thoracic Activity, Lungs, Blood Circulation
4	Coeliac Plexus	Upper Abdomen	Superior Mesenteric Plexus	
3	Aortic Plexus	Navel Region		Combined No. 2 and No. 4
2	Inferior Mesenteric Plexus	Pelvic region		Pelvic Girdle, Male & Female Sex Organs
1	Superior Hypogastric Plexus	Bifurcation of Aorta and Downwards	Inferior Hypogastric Plexus at Internal Iliac Artery	Anus, Rectum, Colon, Prostate Gland, Kidneys, Bladder

These are the three registers with the three flavors of the heart as its own register bridging the lower physical and the upper spiritual planes with colors & textures connecting the 7 chakras.

THREE (3) REGISTERS	1 GROSS & 6 SUBTLE BODIES	COLORS & TEXTURES	CHAKRAS
UPPER SPIRITUAL	7 The Causal Body	Golden with threads	The Crown
UPPER SPIRITUAL	6 The Celestial Body	Opalescent pastel colors with golden/silver light	The third eye
UPPER SPIRITUAL	5 The Etheric Template Body	Cobalt blue	The throat
MIDDLE ASTRAL 3 HEARTS	4 The Astral Body	Rose-tinted colors	The heart

LOWER PHYSICAL	3 The Mental Body	Bright yellow	The solar plexus
LOWER PHYSICAL	2 The Emotional Body	Multicolored clouds of energy	The sexual center
LOWER PHYSICAL	1 The Etheric Body	Light blue/gray	The Sacral

THE MECHANICS OF KNOWLEDGE CREATION (consciousness)

Knowledge creation can be understood in the context of innovation for developing new ideas and new knowledge. Knowledge creation is a social activity. The mechanics of knowledge creation, expresses the physiological dimensions without the psychological and spiritual making it incomplete. The must be the Body, Mind and Spirit BMS continuum that corresponds with the Space Time Energy Paradigm STEP. These two sets of principles, in synthesis, reflect the essential laws of nature and number as qualitative (Geometric) descriptions of all expressions.

What are the data type generated by these descriptions of consciousness?

DATA	TYPES
Un-interpreted	Bit Byte Trit Tryte Word Bit array
Numeric	Arbitrary-precision or bignum, Complex, Decimal, Fixed point, Floating point, Double precision, Extended precision, Half precision, Long double, Minifloat, Octuple precision, Quadruple precision, Single precision, Integer, signedness, Interval, Rational
Text	Character: String, null-terminated
Pointer	Address, physical, virtual, Reference
Composite	Algebraic data type, generalized, Array, Associative array, Class, Dependent, Equality Inductive, List, Object, Meta-object, Option type, Product, Record, Set, Union, tagged
Other	Boolean, Bottom type, Collection, Enumerated type, Exception, Function type, Opaque data type, Recursive data type, Semaphore, Stream, Top type, Type class. Unit type, Void
Related topics	Abstract data type, Data structure, Generic, Kind, Metaclass, Primitive data types Para Polymorph, Protocol, interface, Subtyping, Type constructor, Type conversion Type system

MTABOJECT: An extremely high vibration object encoded with Intelligence, knowledge and information.

WORDS NUMBERS SYMBOLS LETTERS OPERANDS THEORIES, PRINCIPLES, LAWS

Employing creative, design, production principles and systems where art, mathematics and science provide the creative Ideational Intelligence that's then transformed into information in all dimensions, applications, existing and reengineered or newly invented, to build our environment and the various systems needed to produce the technologies to support life is DeSign Science artifacts, conveniences, tools.

The 'TRIAD' of Body, Mind and Spirit correlates with physiology, psychology and spirituality. The Space, Time and Energy continuum is another expression of the same set of three fold symmetry principles of consciousness' as excited to be as much as we are to create the form of its packet that resonates with all waves involved in the processes. Other number laws apply to other dimensions of reality we are experiencing. The creative world where manifesting form operates in is inherently three-dimensional and therefore three fold in its symmetry. The harmonics of creativity are the tools of life we use to support life on all level of consciousness. It may sound like closed loop but there is a 'strange' dynamic about loops that seem to be the science of generating vast amounts of energy when they are constrained.

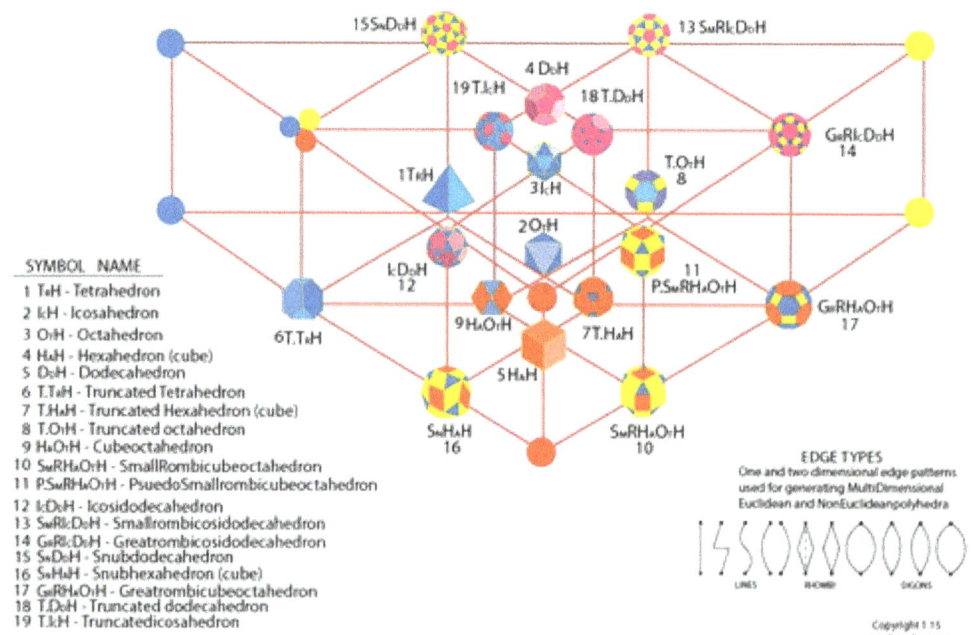

THE STANDARD EUCLIDEAN POLYHEDRA VOCABULARY
Fig 31

Some of these polyhedra are building blocks of elements in nature. They share a very interesting interrelationship that follows the pattern of symmetry operations allowing a type of meta-morphosis from one state to another. The Father of Design Science Richard Buckminster Fuller, suggested that the tetrahedron was the seed of this entire vocabulary. This is also the unmanifestible set of "ideal' form known as Platonic solids that could not be realized.

Where did the division begin with language "Plato's Timaeus was interpreted as analogous, within the limits of poetic intuition and philosophical speculation, to the Christian vision of the inner, sacramental structure of the world. The authority the 'Chartrian Teaching Masters' gave to Plato's cosmology became the foundation on which the study of all ancient literature was established. This is the same source for Plato's idea of the unmanifestible nature of polyhedra that prevented the world from knowing about this language much earlier in our story.

In the year 1006, Fulbert, a major mystic, doctor and builder of his time and known as "Father Socrates" by his students, became Bishop of Chartres Cathedral and consolidated the curriculum of his mystery school. He established the Chartres School out of both a deep spiritual commitment of his own and through extensive interactions with Druidic, Jewish, Sufi and, according to some scholars, Hindu masters. His school continued for two centuries, reconstructed the Cathedral, and served as one of the precursors to the Renaissance.

The Ecology of Consciousness
One of the most interesting new areas of science concerns electrostatic interactions between biological organisms and the environment. I have already indicated that the electro-chemical nature of neural transmission plays an important role in mediating information-transfer throughout the body. Now we will take a look at some of the more subtle extensions of our biological functioning: Our bodies are influenced -- in ways often overlooked -- by the existence of small ions in the atmosphere. The research of scientists such as Albert P. Krueger are sometimes dismissed as insignificant in the face of gross environmental pollution, however they seem to show important implications for consciousness:

Albert P. Krueger

(In terms of psychic consciousness, it is interesting to note that Spiritism has flourished in Brazil, in spite of opposition from the Catholic Church, perhaps more than in any other nation. Brazilian spiritists, synthesizing modern European, native Indian, and African culture, number over a third of Brazil's population and comprise powerful interest groups with their own elected representatives in the national legislature. There are entire towns in Brazil composed solely of spiritists.)

DeSign Science in The New Paradigm Age

CHAPTER 6 TRINE B

WIKI SPEAK

PARADIGMS AND PRINCIPLES:

If knowledge is viewed as an independent or ubiquitous endeavor it cannot be complete unless Wisdom, Intelligence, Knowledge and Intuition or information and many other 'I's" as part of a more Inclusive definition of this as a principle.

1. As an endeavor of ALL MIND, the Wisdom, Intelligence Knowledge and Information (WIKI) extrapolation is a more accurate representation of the universal principle of Mentalism.
2. The WIKI is all Mind and Mental beyond just knowledge alone.
3. The WIKI contains all we know of our past and present with the possibilities and forecasts of an indeterminate future that can only be realized in time.
4. It is essential to all forms of growth and development.
5. The WIKI is our direct path to higher levels of consciousness or SOURCE
6. A physical expression of Knowledge is only one incomplete form of a more comprehensive triad of consciousness defined as Body, Mind and Spirit and is now called a WIKI. The last 'I' represents Intuition.
7. The WIKI is in harmony with the Body Mind and Spirit Formula in the continuum of Space, Time and Energy or Spirit, all expressions of infinite Consciousness, normal awareness and the subconscious MIND.

HERMETIC PRINCIPLES AND PARADIGM CORRELATIONS

BODY	MIND-CREATIVITY
HERMETIC SYMMETRY PRINCIPLES	**KNOWLEDGE**
I. THE PRINCIPLE OF MENTALISM. All is Mind. Though is in all things.	Paradigm 1. The WIKI follows the BODY, MIND, and SPIRIT Formula.
II. THE PRINCIPLE OF CORRESPONDENCE The Constant relationship seeker of interactions with source.	Paradigm 2 All Knowledge comes from source. We interpret it to align with our intention and needs.
III. THE PRINCIPLE OF VIBRATION. Constant motion distributes intelligence by transduction.	Paradigm 3 Truth and integrity are critical to sharing knowledge.
IV. THE PRINCIPLE OF POLARITY. Opposites: extremes of the same thing	Paradigm 4 Transform knowledge into the WIKI through collaboration in communities.
V. THE PRINCIPLE OF RHYTHM Nothing rests. There is constant motion in all things and everywhere	Principle 5. Needs, passion and Love drive knowledge. This is external…..Power comes from within and by focus, intention and will.

VI. THE PRINCIPLE OF CAUSE AND EFFECT. Nothing just happens Laws govern all things	Paradigm 6 Engagement and EMPOWERMENT are critical elements of B.M.S growth.
VI. SWITCH GENDER	Understand the animate and inanimate male and female symmetry and mechanical functions that make things work.
VII. THE PRINCIPLE OF GENDER	There is male and female in all things
VII-CAUSE AND EFFECT. Every action has an equal and opposite reaction.	Paradigm 7 Respect for life is respect for self and source.

D_ESign Science is a creative process, best applied when the multiple 'Canons' from across the cultural spectrum are blended into a system that channels the Wisdom, Intelligence Knowledge and Information in a harmonized 'Synthesis'.

SYNTHESIS: when all elements of a thought come together in harmony in keeping with the BMS formula where space, time and energy are to be balanced naturally, we can define Local knowledge as one specific type of the BMS expression to only one quality or conscious expression enfolding the other two principles. Materialism is the predominant knowledge we have in our three dimensional binary world. This is the way light works by absorption and reflection. One object might reflect green light while absorbing all other frequencies of visible light. Another object might transmit blue light while absorbing other frequencies of visible light.

Design is the field that is itself a synthesis of the arts, mathematics and sciences to create a method now known as DeSign Science.

PAIN POINTS:
"The current emphasis on technology deprives too many of the opportunity to become totally realized beings and accomplished professionals". They miss opportunities to discover their true gifts they were born to share with the world. They join the lemming brigade and head for the cliffs with the eyes and wallets wide open, ready to be taken never doing the taking. Most people live their entire lives never taking what these 'sentences' mean; seriously. They sentenced themselves to a life in despair and regrets. They never take opportunities for self and 'fail' to invite themselves to the life they imagine. Their failures are not about the exterior and materials stuff. It is about failing to do the internal work needed. Until you get off the thread-mill and be connected to your higher self this means nothing to you.
This resources shows the way.

With a minor mindset shift form thinking of things as the way to happiness to looking for what gives you most joy that you can share with others and doing this with the clear understanding that you are connected to source. Everything you have, you are and you desire will come through the connection you create and maintain with some specific thought exercises laid out in the WIKI Paradigms that follow.

D_ESign Science in The New Paradigm Age

By internalizing these principles, paying attention and focusing on your higher self in the thriving environment you desire, you will make the connections you need and accept them as if they are already manifested. This is a design approach to creating your life. If you can design your lifestyle and realize it the way you desire it, you can pass that blessing on to others you create design solutions for. This kind of design is a spiritual science. The time is NOW. "JUMP" off the thread mill now. No guessing where you never know where you can land! You determine where you want to be and these paradigms will get you there.

Synthesis is a journey to purpose by way of desire, intention and focus that's mapped out with all correct signs and clear directions in place taking us to fulfillment, satisfaction and bliss.

Paradigm 1 The WIKI follows the fundamental BODY, MIND and SPIRIT Formulae and is 'grounded' in the space, time energy continuum. All we have known, that we know now and will ever know, require this BMS harmonizing to be in correspondence with source to be in synthesis. WIKI here means Wisdom, Intelligence, Knowledge and Information. From this paradigm on through this sequence the word "knowledge" is replaced with 'WIKI™'.

Paradigm 2 All WIKI™ come from source. It is interpreted to align with what is below with the above and with what is above with below by applying attention, intention and will to define and satisfy true needs.

Paradigm 3 Truth and integrity are critical to the shared WIKI™. In a world of polarities truth and integrity; values and virtues will be the discerning forces to determine the necessary alignments and right actions for the WIKI's highest and best use.

Paradigm 4 Transform knowledge into the WIKI through collaboration in communities.

Principle 5 Needs, passion and Love drive WIKI™. True Power comes from within; by focus, intention and will.

Paradigm 6 Engagement and EMPOWERMENT with the use of the WIKI™ are critical elements of Body. Mind and Spiritual growth supported by the space, time, energy continuum.

Paradigm 7 Respect for life is respect for self and source. All things return to the source of their arising

THE PAIN POINTS:	THE SHIFT:
1. "The current emphasis on technology deprives too many of the opportunity to become totally realized beings and happy, successful and accomplished professionals".	With a minor mindset shift form thinking of things as the way to happiness to looking for what gives you most joy that you can share with others and doing this with the clear understanding that you are connected to source.
2. They miss opportunities to discover their true gifts they were born to share with the world.	Everything you have, who you are and what you desire will come through the connection you create and maintain with some specific thought experiments laid out in the WIKI Paradigms that follow.
3. They join the 'lemming brigade' and head for the cliffs with the eyes and wallets wide open, ready to be taken never doing any taking.	By internalizing these principles, paying attention and focusing on your higher self in the thriving environment you desire, you will make the connections you need and accept them as if they are already manifested.

4. Most people live their entire lives never taking what these 'sentences' mean; seriously.	This is a design approach to creating your life and the styles you desire most.
5. They sentenced themselves to a life in despair and regrets.	If you can design your lifestyle and realize it the way you desire it, you can pass that blessing on to others you create design solutions for.
6. They never take opportunities for self and 'fail' to invite themselves to the life they imagine. Their failures are not about the exterior and materials stuff.	This kind of design is a spiritual science. Get used to being a spirt having a bodily experience and gain self-mastery.
7. It is about failing to do the internal work needed.	The time is NOW. "JUMP" off the thread mill now.
8. Until you get off the thread- mill and be connected to your higher self, through SOURCE, this means nothing to you.	No guessing where you never know where you can land!
9. This resources shows the way.	You determine where you want to be and these paradigms will get you there.

THE FOUR ELEMENTS OF SYNTHESIS

WIKI	Definitions
Wisdom	The sound action or decision with regard to the application of experience, knowledge, and good judgment. Having experience, knowledge, and good judgment; being wise. The body of knowledge and principles that develops within a specified society or period.
Intelligence	The ability to acquire and apply knowledge and skills. The collection of information of military or political value.
Knowledge	Facts, information, and skills acquired by a person through experience or education; the theoretical or practical understanding of a subject. Awareness or familiarity gained by experience of a fact or situation. cognizance, awareness, consciousness, realization, cognition perception, appreciation.
Information	Facts provided or learned about something or someone. What is conveyed or represented by a particular arrangement or sequence of things.

The four principles are like the four forces of consciousness or awareness.

DeSign Science in The New Paradigm Age

MISSION:

Create a new Wisdom tradition using DeSign Science as part of the new and emerging (paradigm) with its knowledge, principles and disruptive trends affecting the global creative community with emphasis on designers and artists today. We need to acquire and apply the WIKI synthesize the new (disruptive) methodologies to enhance the theoretical or practical understanding of the role of science, art, geometry mathematics as expressions that can synthesize or redefine 'consciousness' as the Body, Mind and Spirit formula.

We can then explore its role in all aspects of design in the new Space; Time; Energy as Spirit continuum. We can move beyond the physical and material realms into a more harmonized physical, psychological and spiritual world to deepen our understanding of this newly synthesized 'WIKI™' that supports the arrangement or sequence of things to come.

The awareness gained by experiencing this major paradigm shift will profoundly reshape our perception and concepts like consciousness, cognizance, awareness, realization, cognition perception, appreciation.

The expressions, design solutions and licensed intellectual properties and published information products represent this new aesthetic with a unique identity that will impact the global landscape with a visual intelligence never encoded with such depth and far reaching consequences. The potential value that can be added to the global economy will be quite significant.

LAWS	CAUSE AND EFFECT	PRINCIPLES
THE 1ST LAW of **POLARIZATION**. The Spiritual Heart	ACTION/CAUSE: Cognition; Cision-division-healthy ego as the separation dynamic of choice to filter what's in the funnel to allow the idea to flow. REACTION/EFFECT: Filter-crystallization.	The first law is the Spiritual Principle (attitudes): A fundamental truth or proposition that serves as the foundation for a system of belief or behavior or for a chain of reasoning. Aesthetics, Beauty, Truth, Values, proposition, concept, idea, theory, assumption, fundamental, (Fundament) essential, ground rule: morals, morality, (code of) ethics, Key Concepts: Mentalism, affinity, attention, Divine and unconditional Love, Ideals, Values ,

THE 2ND LAW of **IDEATION**. The Emotional Heart	ACTION/CAUSE: Thought, De-cision, Imagination. REACTION/EFFECT: Conceptualization-thought form.	The second Law is a Psychological Principle (attitudes or behavior): morals, morality, (code of) ethics, beliefs, ideals, standards; More integrity, uprightness, righteousness, virtue, (sense of) honor, decency, conscience, scruples, probity (with all of the above,) Key Concepts: The Love energy (force). The Love of self, people, places and things
THE 3RD LAW of **FORMALIZATION** (Formation; geometric description & definition). The Physical or biological Hear What is the medium for the heart center?	ACTION/CAUSE: In-cision-cut- Visualization from 'cision': vi-sion de-vi-sion. REACTION/EFFECT: Pre-cision Gauge: Measurement. Gauge theories, Dimensionality.	The Physiological Principle includes "elementary principles" (of matter, materials and things): Symmetry rules and operations, Gauge (theories and applications) properties; Physical dimensions width, height and length, sizes, shapes and forms, characteristics, mechanical and machine logic, symmetry and gender principles and structure or operational instructions, specifications, a general scientific theorem or law (theory) that has numerous special applications across a wide field. A natural law forming the basis for the construction or working of a machine; "these machines all operate on the same general principle" The Media: "the first principle of all things is Water; H_2O" in the organic world. The second is Earth; where the elements are stored. The third is fire; the great transformer Fourth is the Air; the medium of vibration. The Fifth was called Ether; now the QED; Quantum Electromagnetic (field)Dynamic

AFFINITY and attention are natural attributes of creation that applies to manifestation in "nature" through attraction. Similarity and differences in the characteristics of possible relationships are created, especially as resemblance in structure between animals, plants, or languages; Life. The BMS formula would suggest that there are other (deeper) relationships involved in creation. Attraction, Participation, Observation,

Thought, Feelings, Expectation and Satisfaction are needed for the creative process to make sense. 'Making sense' relies on our senses. Prior to quantum discoveries and theories in physics, attraction might have served to explain the relationship between atoms and molecules with emphasis on chemistry to encode matter.

"Matters about matter no longer matter it now appears. With the advent of 'particles and wave' theories, relying on 'observation and not participation creates some disconnection that affect perception and truth.

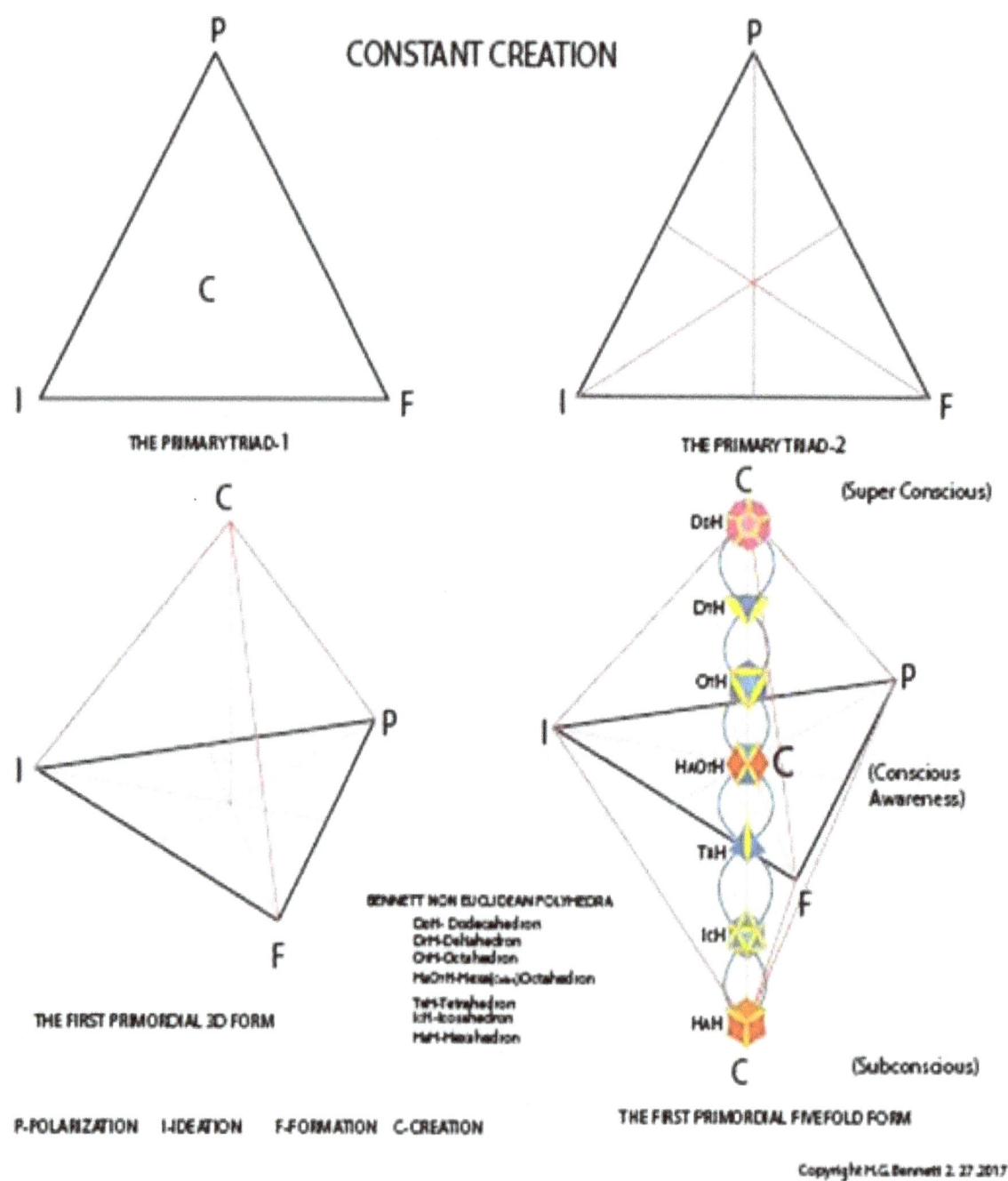

Fig 31

The quest-ions then become: 1. what is the true nature of the relationships between the building blocks of nature? 2. What energies are involved? 3. How are their structures (thought forms) created? 4. Do the forms (ipse) and the energy vibrations they transform e.g. transduce 'encode' the quantitative, qualitative and conscious flavors of their products, behaviors, results, outcomes and their essences?

Here again we apply the **'Synthesis of the 'Canons'** We start with the 'A Study of Numbers': A Guide to (the) Constant Creation.. By René Adolphe Schwaller de Lubicz, born René Adolphe Schwaller in Alsace-Lorraine, a French occultist, student of sacred geometry and Egyptologist known for his twelve-year study of the art and architecture of Egyptian temples (Karnak).

He states: "We lack direct consciousness of Space, Time and Energy. We can know of them only indirectly by (MEASURING) mass, force, and energy, and by the intermediary of phenomena such as may be tested by our five senses" To the limits of Space and Time we can add Body, Mind and Spirit/Energy in Space Time Energy and Power to the list. How can we find CONSCIOUSNESS and holism in our LIVES when such major 'elements of creation' are missing amazes me. To journey beyond the five senses to expand our consciousness we can now build scaffolds with many more modules that the collective human genius has offered and continues to offers us. We must not allow our 'egos and shegos' take full credit for the WIKI that flows to us as part of the constant creation with SOURCE.

Tension is the dynamic that causes the Attention to form the possibility for the relationship between events in space. If these events; waves and particles are anything like humans they may simply be looking to "get laid". Some form of self-preservation could be the hidden driving force they are responding to. The 'human analogy' used here is based on the notion of 'the oneness of everything' idea. Differences or similarities and how duals behave needs to be looked at here as one of the resonating or responsive tenets. Opposites; (charges) differences attract: like; (charges) similarities repel. Are these events (points and potential movements) that connect complementary?

In molecular biology, **complementarity** describes a relationship between two structures each following the lock-and-key principle.(male and female 'Gender' principle) In nature complementarity is the base principle of DNA replication and transcription (transduction?) as it is a property shared between two DNA or RNA sequences, such that when they are aligned antiparallel to each other, the nucleotide bases at each position in the sequences will be **complementary**, much like looking in the mirror and seeing the reverse of things. This complementary base **pairing** allows cells to copy information from one generation to another (transduction?) and even find and repair damage to the information stored in the sequences. The degree of complementarity between two nucleic acid strands may vary, from complete complementarity (each nucleotide is across from its opposite) to no complementarity (each nucleotide is not across from its opposite) and determines the stability of the sequences to be together. Furthermore, various DNA repair functions as well as regulatory functions are based on base pair complementarity. In biotechnology, the principle of base pair complementarity allows the generation of DNA hybrids between RNA and DNA, and opens the door to modern tools such as cDNA libraries. While most complementarity is seen between two separate strings of DNA or RNA, it is also possible for a sequence to have internal complementarity resulting in the sequence binding to itself in a folded configuration (mitosis). A type of cell di-vision that results in two daughter (Gender-Female) cells each having the same number and kind of chromosomes as the parent nucleus, typical of ordinary tissue growth.

Appearing-Pairing: coming into sight; become visible or noticeable, typically without visible agent or apparent cause- Polarity, duality and twofold properties, qualities and relationships. Polarization recognizes possibilities for pairings, for 'appearing or disappearing', connections and traces that we may not know or have reasons for. This happens in sub atomic physics a lot. Cognition is the feature at this point. Next is the idea of 'Cision'.

THE THREE PRIMARY PRINCIPLES OF CONSTANT CREATION

THE 1ST LAW of **POLARIZATION**.   (The Spiritual Heart-Mind connection)
ACTION/CAUSE: Cognition; Cision-division-healthy ego as the separation dynamic of choice to filter. REACTION/EFFECT: Filter for (distillation) and crystallization.
'Cision' is the first gesture for creation to be even possible. It is the first 'act' of the creative process for making we call cutting which produces di-vision of the material being used into parts of a much larger pattern of some form. Here are some wordsmithing decodings of the root 'cision': De-cision, In-cision, Pre-cision. If you think carefully about the associations of these constructs we find the Body, Mind Spirit connection to them as follows. In-cision is Physical (Body), Pre-cision demands Skill (mind) De-cision requires thought (spirit). Division goes from wholeness to the particular part of 'raw' materials and matter which have boundaries and limits and can be deconstructed into modular elements. Gauges are needed to create sizes with measurements to align with proportions, and other features of the creative process. The goal is design excellence, economy and utility through ART.

The quest-ions then become: 1. what is the true nature of the relationships between the building blocks of nature? 2. What energies are involved? 3. How are their structures (thought forms) created? 4. Do the forms (ipse) and the energy vibrations they transform e.g. transduce 'encode' the quantitative, qualitative and conscious flavors of their products, behaviors, results, outcomes and their essences?

In physics, **tension** is the action-reaction pair of forces acting at each end of said elements. Tension is the opposite of compression. It is where atoms or molecules are (stretched) pulled apart from each other and gain potential energy with a restoring dynamic left in place, as a form of tension. In physics, tension is a transmitted force; an action-reaction pair of forces, or as a restoring force, is a force and has the units of force measured in newtons (or sometimes pounds-force). These tension forces are also called "passive (feminine) forces". There are two basic possibilities for systems of objects held by strings: either NO acceleration and in equilibrium, or there is acceleration, and a net force in the system. Light is a ubiquitous constant vibration and speed system.
THINK: At-tention, In-tension, Ex-ten (t) (s) ion for extension.

The **frequency** is the number of waves that pass a point in space during any time interval, usually one second. We measure it in units of cycles (waves) per second, or hertz.
The **frequency** of visible **light** is referred to as 'color', and ranges from 430 trillion hertz, seen as red, to 750 trillion hertz, and seen as violet. We will look at the other systems or Canons for their correspondences later on.

THE 2ND LAW of **IDEATION**. (The Emotional Heart-Mind)
ACTION/CAUSE: Thought, De-cision, Imagination.
REACTION/EFFECT: Conceptualization-thought form. This second Law is a Psychological Principle of (attitudes or behaviors):habits, rituals and paradigms; morals, morality, (codes of) ethics, beliefs, ideals, standards; with integrity, uprightness, righteousness, virtue, probity, (sense of) honor, decency, conscience, scruples. The cision here is about separation with the prefix 'de' added to the root 'cision' or decision. Choice is a major faculty of ideation.
The imagination fills the brain-funnel with ideas that must be evaluated to make appropriated decisions. The principle of mentalism is the major contributor to this first law.
The upper third register is heavily engaged in the first three laws. Ideation. Visualization and imagination and communication as in the use of media inspire and support critical 'Thinking'.
Key Concepts: The Love energy (force). The Love of self, people, places and things. Use wave function **scattering or distribution** analogies as the model.

THE 3RD LAW of **FORMALIZATION** (Formation; geometric description & definition).
ACTION/CAUSE: In-cision-cut- Visualization from 'cision': vi-sion de-vi-sion.
REACTION/EFFECT: Pre-cision Gauge: Me-asure-ment*.

Gauge theories, Dimensionality. (Look at the word lan-guage) (Association; Lan-d-gauge)

The Physiological Principle of formalization includes "elementary principles" (of matter, materials and things): Symmetry rules and operations, Gauge (theories and applications) properties apply; Physical dimensions width, height and length, sizes, shapes and forms, characteristics, mechanical and machine logic, symmetry and gender principles, structure and operational instructions, specifications, a general scientific theorem or law (theory) that has numerous special applications across a wide field. A natural law forming the basis for the construction or working of a machine; "these machines all operate on the same general principle". The Media these are found in are also first principles or elements: "the first 'principle of all things' is Water; H_2O" in the organic world. The second is Earth; where the elements are stored. The third is fire; the great transformer Fourth is the Air; the medium of vibration. The Fifth was called Ether; now the QED; Quantum Electromagnetic (field) Dynamic. The manmade world operates with a different order. Earth is first, Water is next, then Fire, Air and Ether which is now the QED or quantum electrodynamic energy in the 'artificial world'.

Ideas start with desire, inner striving, responding to urges known as tension. Tension is the quantum energetic state of possibilities where fields of energy potential can be stimulated by 'thought' and other vibrations to initiate and facilitate the interpretation of creative processes in 'Constant Creation' modes. This involves many 'a priori states and first causes one of which started with polarizing, defined earlier, with the first 'pairing' of attention and expectation. To realize a vision there must be focus and attention to determine if the thought has value; this we call 'value determination'. All endeavors start in this thought space prior to ideation and thought formation in harmony through 'SYNTHESIS'. Inner striving is the emotional positive and creative stress that is the life force on a mission.

Definition of tension encoded in the BMS formula
1. 1a : the act or action of stretching or the condition or degree of being stretched to stiffness (tautness)
2. 2a : either of two balancing forces causing or tending to cause extension b : the stress resulting from the elongation of an elastic body
3. 3a : inner striving, (stress) unrest, or imbalance often with physiological indication of emotion, a balance maintained in an artistic work between opposing forces or elements
4. 4: a device to produce a desired tension (as in a loom)
   IDEAS

1. What is an idea? It's a thought : a transcendent entity that's a real pattern of which existing things are imperfect representations. B: a standard of perfection : ideal and c : a plan for action or design . It relies on some methodology that can synthesize the natural elements, forces and energies into harmony. The start as stimuli that must then be ART-iculated into representations of Wisdom, Intelligence, Knowledge and Information.

2. Ideation: an action or process: something connected with an action or process that involves energy as in the oxygen in substances with names ending with 'ate' (carbonate). Derived from Latin -ation-, -atio, from -a-, stem vowel of 1st conjugation + -tion-, -tio, noun suffix
A thought is a quantum-wave packet that can apply to the way observers and experimenters think and do their research. Wave/Particle/observer, thinker and participator MAN are new relationship builders in the quantum paradigm, seeking pairings (binary functions) bringing ideas into reality to "make the abstract real".

Man is the potential finely tuned instrument or medium for directing 'attention' to initiate the state changes needed to attract energy and form. It is very important to be in the 'universe of discourse' to find all the appropriate alignments needed. To be in discourse with source intensifies the attention through heightened focus. With focused attention mind begins to create.

Focused attention directs the energy of mentalism (THOUGHT) to the point of attraction as the first pairing. This process could have one of two con-sequences; one is the appearing the other is the disappearing of any response to the first gesture that stretches or stressed to generate the energy that will attract the desired response. Let's visualize a zero dimensional geometric event in space, called a point that is attracted to another complementary or polar event forming a one dimensional vector or line. The dynamic that causes the -intermittent trace elements- pulses and their energy to align in space is a memory in-tension or stress forming this bond. The vector then translates into the plane is two dimensional space. Many relationships are generated from this plane that then translate into three dimensional forms.

Separation is the underlying principle in all pairings. Potential energy is harnessed by the tension from separation similar to splitting atoms. The idea is to have the pair communicate and form a bond. If there is no response the energy cannot be lost. It restores itself, 'bouncing back from rejection' as it were to refocus or find another way. The polarity of the 'signals being transmitted and anticipated being returned are informed. Though this may appear to be a linear process it definitely is not. There is a feedback loop that makes this process cyclical or 'digonal'. This is the fundamental element of the newly discovered, 'Passive' Non-Euclidean geometric polyhedra vocabulary. The 'digon' is the closest we can get to being linear allowing the uninterrupted reciprocating fields to flow naturally. Conduction does not reverse itself in the same conduits or along the same path. There is no 'coherent backflow' in nature. This would violate the' two objects occupying the same space' law of physics*. Both paths are charged with their own type of energy. The sending and the receiving are not the same. They are complementary and binary. The active (male) 'desire energy' is in a state of 'positive and confident expectation' (PACE). The passive (female) responder is the feedback loop. *This "law of physics" is the Pauli exclusion-principle which states that *two* identical fermions (particles with half-integer spin) cannot *occupy the same* quantum state simultaneously. This applies to normal matter, which is made out of only a few kinds of fermions tightly bonded together.

Using the criteria of "A general scientific theorem or law (theory) with numerous special applications across wide fields validates the following claims. If science is about methodology then the following fields of endeavor are sciences.
Following are some new fields that fall under the rubric of 'D_ESign Science™. It is the synthesis of Traditional Physical science, visual logic structures and visual mathematics especially (solid) geometry for generating three dimensional forms, new number theories, periodic qualitative matrices and visual Intelligent strategies and concepts.
Ancient Wisdom traditions both Oriental and Occidental, metaphysics, types of philosophies, sacred and celestial architecture add to the universal ken of untraditional information not normally accepted in western educational systems.

Accepted in western educational systems.	Some versions accepted in western educational systems.	Unaccepted in western educational systems.
Traditional Physical science	visual logic structures	visual mathematics
(solid) geometry generating three dimensional forms	periodic qualitative matrices	new number theories
Occidental (European) philosophies		visual Intelligent strategies
	Oriental and Occidental Ancient Wisdom traditions	types of philosophies
		sacred and celestial architecture
		metaphysics

We are moving towards a universal ken of untraditional WIKI not normally accepted in western educational systems.

A MATRIX FOR DECODING THE NEW PARADIGM WISDOM, INTELLGENCE, KNOWLEDGE AND INFORMATION; WIKI™"

DISCIPLINE	STUDY AREAS	PARTICIPATION LIVING	HOW TO
METAPHYSICS	Study of Existence	What's out there?	Go deep inside; find the universe. Stay on the surface and stuck in the world!
EPISTEMOLOGY	Study of Knowledge The WIKI	How do I know about it?	Connect to source. Know thyself
ETHICS	Study of Action	What should I do?	CREATIVE RIGHT ACTION
POLITICS	Study of Force	What actions are permissible?	SYMMETRY PRINCIPLES AND NATURL LAWS
ESTHETICS	Study of Art	What can life be like?	BEAUTIFUL AND ABUNDANTLY FREE
ESTABLISHMENT	CULTURE	PHYSICAL-PSYCHOLOGICAL-SPIRITUAL	EXPRESSIONS WITH NO CREATIVE SPIRIT
COSMOLOGY	CONSCIOUSNESS	BODY-MIND-SPIRIT	PHYSIOLOGY-PSYCHOLOGY-SPIRITUALITY
DESIGN SCIENCE	SUPER-CONSCIOUS	DEEP LEARNING	TAP INTO META MIND

PARADIGM: In science and philosophy, a paradigm is a distinct set of concepts or thought patterns, including theories, research methods, postulates, and standards for what constitutes legitimate contributions to a specific field.

1. PARADIGM TRANSFORMING INTELLECTUAL PROPERTIES, TECHNOLOGIES AND DEVELOPMENT IDEAS FOR A HOLISTIC FUTURE
2. PARADIGM TRANSFORMING DEVELOPMENT TECHNOLOGIES FOR A BETTER WORLD
3. TRANSFORMING PARADIGM TECHNOLOGIES……FOR A CONSCIOUSNESS AWAKENING
4. PARADIGM TRANSFORMING IDEAS FOR A NEW WORLD
5. HOLISTIC DEVELOPMENT: TRANSFORMING CREATIVE PRINCIPLES FOR AN EMERGING PARADIGM [A NEW PARADIGM]
6. THE ASSETS, PROPERTIES, TECHNOLOGIES AND DEVELOPMENT OPPORTUNITIES TRANSFORMING THE EMERGING PARADIGM TO CREATE, HEALTHY AND PROSPEROUS FUTURES THROUGH RESEARCH, EDUCATION, INNOVATION THAT'S ATTAINABLE AND HOLISTIC.
7. W-HOLISM THROUGH CREATIVE RESEARCH
8. WHY ARE'NT REAL SPIRITUAL REVOLUTIONS TELEVISED
9. APOCALYPSE
10. TOWARDS AN APOCALYPTIC PARADIGM
11. LIFTING NEW PARADIGM VEILS
12. PARADIGM [DOOR] BUSTING
13. NEW PARADIGM OR BUST!
14. THE PARADIGM 21 WARS
15. PARADIGM CHICKEN SOUP
16. BRANDING THE PARADIGM
17. A PARADIGM OF 'NOW' AS PRESENCE!

Reframing Paradigms for new architectures.
With working natural principles in place any harmonized and synthesized organism or system can be optimized.

"Paradigm" definition: Unique, repeatable n-dimensional, geometric Paradigm patterns are developed and configured structurally and spatially in all dimensions to generate vocabularies or families of forms. They vary and can be extended expressing an original identity or a 'paradigm'. There is an inherent logic to the physiology, psychology and (spiritual) integrity or energy of the elements as a result of the inherent aesthetic quality they possess.
Following are special features we look for redefining the meaning of consciousness. What helps us here is that knowledge, both old and new, is unfolding causing our awareness to expand in response to a deeper dynamic that is a cyclical universal pattern.
Views of the world, attitudes and descriptions of how the world works and changes are constantly emerging. The obvious areas of change are in technology, lifestyles and many of the cultural expressions created by artist and high vibrating, creative and thinking beings who are experimenting with their cultural perspectives and paradigms.

1. Change is taking place on a much deeper level and faster now at the mental, the DNA and cellular levels as well as in fields like mathematics, brain science or neuroscience and many other areas. Time is 'Epochal' it's a major factor along with space and energy; all parts of consciousness. They all seems to have unique essences, quality and quantity flavors and gauges.
2. Consciousness is now seen as being three 3 dimensional. Physical, psychological and spiritual are its expressions. For a thing to be it must have at least these three 'flavors'.
3. The oneness of all consciousness, the universes and all else obey symmetry and natural laws that continue to unfold as we age (at least) or grow.
4. Knowing what we are about in the classical 'metaphysical' context leads to a holistic sense of connectedness. Mentalism takes the lead here along with the other hermetic laws of Correspondence, Polarity, Rhythm, Vibration, Gender and Cause and Effect. This knowledge along with other thought form principles given to us by our ancestors is still not part of our normal awareness. The reason for this 'lack' is another story to be told later.
5. We need to connect to ourselves and to source to begin the pre-transformation protocol, which this work offers us. Only then would we realize and create the new vocabularies of forms, energy systems, the materials tools and the technologies that can transform the relevant concepts ideas and ever pressing needs we have for a productive, peaceful and very demanding future.
6. At the root of this creative process there is the notion that we contain all the information, knowledge and skills to life the lives we were meant to.
7. Rekindling the Philo-(love or affinity)-sophy (Sophos or Wisdom) is the challenge. This is the love of wisdom, the joy of learning is another definition. The meaning of philosophy to us is not just truncated. It's castrated.
8. Nature seem to be smart enough to be on a time release protocol where knowledge is unfolding as we get our conscious act together. Stored in our DNA, in our hi-story and her-story is our future.

9. Specific 'right-thinking and actions' are needed socially, ethically politically at all the philosophical and conscious pit-stops of human life as we continue to infect ourselves and affect all our environment/s.
10. If we have a creative agenda for transformation we can restore, heal and create a better future.
11. Our knowledge bases and their traditions are a rich store of creative and positive vision waiting for us with blessings and expressions of systems like Non-Euclidean forms; mathematical ideas still not yet ART-iculated into any useful technologies and applications. Where do new aesthetic traditions and art and industrial movements that reshape our lives come from? What do we have that can refuel or energize our imaginations now? The standard definition of paradigm as a pattern of some kind is nor deep enough.
12. New systems create new economies as they satisfy needs. They replace the obsolescence in our entropic lives.
13. Logic structures, given to us by our ancestors are revealing secrets buried and preserved for us from as far back as 12,000 years ago. Not only is energy neither create nor destroyed. Intelligence follows the same thermodynamic laws of conservation.
14. The overarching value here is having a generic framework of adaptable principles, systems and form vocabularies that can be exploited as all the dimensions, types and scales of our creative expressions are made much more efficiently and solve the problems we identify.
15. The form vocabulary is a repository of our genius waiting for us to awaken to its fullest potential as we achieve all known and still to be known forms of freedom as we create products, technologies with the inherent knowledge of new form consciousness, information, systems and the solutions they offer us.

FIELDS WHERE THEORIES AND APPLICATIONS APPLY:

Architecture and Design Science Methodologies and aesthetic principles.
Art, Sculpture: Collections and Public works
Publishing, Production, Marketing, distribution and licensing and sales
Design: Industrial, Commercial and Products, Household, Tableware, Automotive, Food and Presentation spaces and establishments.
Packaging Design: Food, Appliances and enclosures
Fashion Design and accessories: production marketing and distribution.
Wearable Art and Technologies
Inventions and Patents: Develop other applications, production systems and technologies.
Computer systems Programs, Distribution systems sciences and technologies.
Entertainment, Play and learning activities.
Toys, Games and Puzzles.

ART, MATHEMATICS & SCIENCE
Theoretical and archeological research for discovering and developing industrial applications.
Geometry 3D form development / Euclidean and Non-Euclidean forms and products. Explore other form vocabularies.
The ART of Visual Mathematics: The Periodic Matrix and other periodic systems.
ENERGY GENERATION Systems and technologies Generation distribution and storage all forms of energy.

ARCHEOLOGY: ANCIENT TECHNOLOGIES AND SITES
Research and explorations; Knowledge and information of Wisdom Traditions Metaphysics, Celestial Architecture, Ancient Sites and Monolithic structures.
Publishing
Research, Printing and distribution Images, 3D forms and literary works
Limited editions and partnering with other agents and publishers.

OTHER RESEARCH TOPICS:
Philosophy, Metaphysics, Neuroscience readings
Consciousness, Theoretical Research, Modular Architecture and Construction,
EXHIBITIONS, MEDIA AND INSTALLATIONS Visual art, video and film lectures and talks
ADMINISTRATIVE MARKETING SALES AND BUSINESS DEVELOPMENT FUNCTIONS

Definition of tension
1a : the act or action of stretching or the condition or degree of being stretched to stiffness
2a : either of two balancing forces causing or tending to cause extension b : the stress resulting from the elongation of an elastic body
3a : inner striving, unrest, or imbalance often with physiological indication of emotion, a balance maintained in an artistic work between opposing forces or elements
4: a device to produce a desired tension (as in a loom)

Adhesion and Cohesion, attractive forces between material bodies. A distinction is usually made between an **adhesive force**, which acts to hold two separate bodies together (or to stick one body to another) and a **cohesive force**, which acts to hold together the like or unlike atoms, ions, or molecules of a single body. However, both forces result from the same basic properties of matter. A number of phenomena can be explained in terms of adhesion and cohesion. For example, surface tension in liquids results from cohesion, and capillarity results from a combination of adhesion and cohesion. The hardness of a diamond is due to the strong cohesive forces between the carbon atoms of which it is made. Friction between two solid bodies depends in part upon adhesion.

Relativity, physical theory, introduced by Albert Einstein, that discards the concept of absolute motion and instead treats only relative motion between two systems or frames of reference.
One consequence of the theory is that space and time are no longer viewed as separate, independent entities but rather are seen to form a four-dimensional continuum called space-time. Full comprehension of the mathematical formulation of the theory can be attained only through a study of certain branches of mathematics, e.g., tensor calculus. Both the special and general theories have been established and accepted into the structure of physics. Einstein also sought unsuccessfully for many years to incorporate the theory into a unified field theory valid also for subatomic and electromagnetic phenomena.

Quantum field theory, study of the quantum mechanical interaction of elementary particles and fields. Quantum field theory applied to the understanding of electromagnetism is called <u>quantum electrodynamics</u> (QED), and it has proved spectacularly successful in describing the interaction of light with matter. The calculations, however, are often complex.

In **quantum** mechanics, <u>wave function</u> **scattering or distribution** is said to occur when a wave function initially in a superposition of a collapsed state—scatters or is distributed into many eigenstates (by "observation"). Are there such complementary processes in nature?

The approach here is to synthesize related disciplines, their theories and principles in search of the commonalities that could lead to efficiencies for us to Simple-Fi our information overload. HGB

CHAPTER 6 TRINE C

The Structure of aesthetic or design Revolutions

Thomas Kuhn's 1962 book, The Structure of Scientific Revolutions, proposed the idea of paradigm shift. The idea that a whole world-view might change radically was itself at least a slight shift—the generally accepted idea was that knowledge and world-view evolved gradually. Then came Mr. Moore with his laws of exponential computing power growth and the human race was on.

NEEDS, FULLFILLMENT AND SATISFACTION IN THE NEW CONSCIOUSNESS PARADIGM

Fulfillment: Accomplishing *the full realization of one's abilities or character*. Or the *attainment of something* desired, promised, or predicted. It implies anticipation and expectation as expeditors.
Satisfaction: The fulfillment of one's wishes, expectations, or needs, or the pleasure derived from it; adhering to natural Law. With 'Manmade Law' it is the payment of a debt or fulfillment of an obligation or claim.

The following matrix defines NEED as it relates to the BODY and physical external matter in correlation with the psychological experiences and functions with knowledge and creative skills acquired that can provide fulfillment of the need. It is not conjured up with 'wishes'. This in turn en-ga(u)ges the spirit as the dynamic proactive, willful agent in harmony with the work as the energy of activation and manifestation with the realization of the intention to become the desired and expected outcome. All must be aligned with a triad of laws, principles and forces.
This matrix becomes the structure for the communication between man, nature and God or Source. The dialog that emerges, through complex sentence structures, within the matrix is the internal conversation we have with ourselves as *(THOUGHT)* and with the external world as *(COMMUNICATION)* we must have some control of being the duality that's composes, ARTiculates and directs all the quixotic elements of the ubiquitous triad of Body, Mind and Spirit in all things physical as subject and object in the tangible 'slowly vibrating' world of things. The mind is where mentalism pollinates the creative process. All things start in and come through but not from mind. Though it is the arbiter of matter, behavior and choice in seeking balance in the states of symmetry as the behavior of parts that make things whole and holistic, there is a higher order that rules without rules and not the way we know or think of rules. That dynamic is the Spirit is a Hyper Quantum Dynamic. It is the 'verb' in our (life) sentence structure that operates through thought fuel, desire, visualization and the understanding of the form of the physical energy vibrations invested into the need and out as solution.

'Consciousness is Body. Mind and Spirit'.

BODY, MIND AND SPIRIT NEEDS

BODY	MIND	SPIRIT
• Subsistence/Luxury	• Protection/Security	• Affection/ Love
• Leisure/Pleasure	• Participation/Co-Creation	• Understanding/Wisdom
• Creation/Manifestation	• Identity/Quality	• Freedom/All Forms

Way beyond Maslow's Food, Clothing and shelter, Max-Neef classifies the 9 fundamental human needs as: Needs are also defined according to the existential categories of *having, being, (doing and interacting)*, and from these dimensions, a 36 cell matrix is developed.

Need Abundance	BODY Having (things)	MIND Being (qualities)	SPIRIT Doing (actions) Interacting (settings)
1 Subsistence Luxury	Food, shelter, work	Physical and mental health	Feed, clothe, rest, work living environment, social setting
2 Protection Security	Social security, health systems, work	Care, adaptability, autonomy	Co-operate, plan, take care of, help social environment, dwelling
3 Affection Love	Friendships, family, relationships with nature	Respect, sense of humor, generosity, sensuality	Share, take care of, sexual activity, express emotions privacy, intimate spaces of togetherness
4 Understanding Wisdom	Literature, teachers, policies, educational	Critical capacity, curiosity, intuition	Analyze, study, meditate, investigate, schools, families, universities, communities
5 Participation Co-Creation	Responsibilities, duties, work, rights	Receptiveness, dedication, sense of humor	Cooperate, dissent, express opinions associations, parties,

			churches, neighborhoods
6 Leisure Pleasure	Games, parties, peace of mind	Imagination, tranquility, spontaneity	Day-dream, remember, relax, have fun landscapes, intimate spaces, places to be alone
7 Creation	Abilities, skills, work, techniques	Imagination, boldness, inventiveness, curiosity	Invent, build, design, work, compose, interpret spaces for expression, workshops, audiences
8 Identity Quality	Language, religions, work, customs, values, norms	Sense of belonging, self-esteem, consistency	Get to know oneself, grow, commit oneself places one belongs to, everyday settings
9 Freedom All Forms	Equal rights	Autonomy, passion, self-esteem, open-mindedness	Dissent, choose, run risks, develop awareness

- These needs are clearly defined in a theoretical and philosophical sense depicting principles and correlations. Other characteristics can be extrapolated to represent the tangible methods to internalize and experience them in our daily lives.

Design Science Paradigm is rooted in the sociology of science, with two main meanings:
- As models, archetypes, or quintessential examples of solutions to problems. A 'paradigmatic design' in this sense, refers to a design solution that is considered by a community as being successful and influential.
- A design paradigm involves beliefs, rules, knowledge, etc. that is valid for a particular design community. Here a paradigm is the underlying system of ideas that causes a range of solutions to be 'normal' or 'obvious' project.

"**Design Science Paradigm** "inspires "Design Excellence" that creates "spiritual (creative) movements". The second meaning refers to what creative communities expect from a type of design solutions.

The term **"Design Science Paradigm"** is used within design professions, including and not limited to architecture, industrial design and engineering design, to ART-iculate an *archetypal* solution that is an expression of the form, identity and energy of the space, time energy dynamic of the movement.
Design Paradigms have been introduced in a number of books: Focusing on the science that supports is critical to the evolution of the disciplines that must now embrace new technologies.

Design Paradigms: A Sourcebook for Creative Visualization by Warren Wake, Design Paradigms: Case Histories of Error and Judgment in engineering but never defined by Henry Petroski. This concept is close to design pattern coined by Christopher Alexander in A Pattern Language.

D_ESign Science Paradigms can describe design solutions or facilitate design problem identification, solving and solution manifestation strategies. Problem solving occurs through a process of polarization, ideation and formalization of design solutions, using the time tested Triadic Formula we have invented.

3Dimmensional Form is the metaphor for communication and a symbolic language for our visions of how we live our lives. Metaphors are used to help explain concepts that are new or unfamiliar. Key words to identify paradigm shifts as indication of transformation are complexity, community, creativity, spirituality, flexibility, and positivity in the ken of knowledge now.

Design paradigms can be seen as higher order relationships of consciousness and LAW. It is the multi-dimensional synthesis of symmetry principles that make parts whole and how they work, within groups of things or their phyla, between the known and the unknown. Polarizing or differentiating the known from the unknown, and the functional equivalent of a blend of the physical, the psychological and the spiritual uses the triadic formula that applies to many fields of creative endeavor where common principles obtain.

A problem is defined as an event, a matter or situation regarded as unwelcomed or harmful and needing to be dealt with and overcome in a controlled and safe environment. Solving this type of 'problem' requires the most appropriate understanding of it and the alignments that are satisfied with solutions. In other disciplines the definition is more direct or logical. It is an inquiry starting from given conditions to investigate or demonstrate a fact, result, or law *as in science; physics and mathematics. It would seem to me that a more logical approach with design science is a better alternative. 'Circular thinking'* starts with a question that serves as a prescription to a problem with a solution that is contained in the question.

THE NEW CONSCIOUSNESS PARADIGM IS NOT VERY NEW. NEEDS, FULLFILLMENT AND SATISFACTION DO NOT CHANGE. NATURAL LAW IS AT THE CORE.

The new 'Consciousness Paradigm' is a model of reality that proposes Body, Mind and Spirit as the only fundamental elements of reality, life and energy.
Every ART-iculation of them is a reflection. Each element has a unique characteristic, quality and flavor that describes it. Experiences we have of the elements are the only reality we know. Words we use to describe their form, process or essences to communicate them are all illusions. They are inflections and reflections of phenomena that cannot be described without knowing their true nature. That the phenomenal nature keeps defying reason or logical and sensible description is significant. Each element is in a family or classifications with definitions associated with it that follows the tenets of a "paradigm".

Expressions or manifestations are all physical and in the domain of physiology.
Experiences and feelings are psychological and in the domain of psychology.
Essences and Energies are creative and spiritual in the realm of 'source' and spirituality.
These three elements emerge from 'one' phenomenal dynamic as a principle of singularity. It is sometimes known as 'oneness' or as a fact or state of being one in agreement and alignment with natural law or a paradigm. The principle of creation is *one* extrapolated into *three*, with *three* returning to *one* at an elevated level of consciousness in synthesis. The elevated three possesses information due to its space, time, energy continuum.

THIS IS THE DESIGN SCIENCE INTERPRETATION
THE NEW PARADIGM OF DESIGN SCIENCE IS NOT EXACTLY NEW. DESIGNERS ALWAYS FOLLOW NATURAL LAW TO ADDRESS NEEDS, GENERATE FULLFILLMENT AND SATISFACTION THAT CHANGE THROUGH HISTORICAL AND CULTURAL AGES. EACH ONE HAS A UNIQUE IDENTITY, QUALITY AND AESTHETIC THAT ENRICHES OUR LIVES. WE ARE IN A NEW AGE NOW PROMOTING ITS BENEFITS. A NEW PARADIGM MOVEMENT COULD LEAD US TO AGREE "WE CREATE REAL FUTURES".

The new 'Consciousness Paradigm' is a Design Science model of Body, Mind and Spirit as key fundamental elements of reality, life and energy. Every design we create is a reflection with unique characteristics, qualities and flavors that describe us in our time. Experiences we have of the design elements express the form, process or essences to communicate the essence of the space, cultures, iconic and archetypal solutions. Design Science creates inflections and reflections of aesthetic ideas about the true nature of how things work. This is the domain of science. When Design is fused with Science it makes magic.

Designs are physical and in the domain of physiology and form. They express behavior as psychology, with Essences and creative Energies that express our true aesthetic spirit and design excellence. These three elements express the principle of singularity; 'oneness' or as a fact or state of being one in agreement and alignment with natural law or what we call a 'paradigm'.

Symmetry principles of creation in the *one* is extrapolated into *three*. In space, over time and energy the *three* returns to *one; the source of its arising,* to an elevated level of consciousness or 'synthesis'. The new *three* is information in a unique space, time and energy continuum.

A NEW PARADIGM MOVEMENT LEADS US TO AGREE THAT "WE CREATE REAL FUTURES"
A NEW PARADIGM MOVEMENT LEADS US TO AGREE THAT "DESIGN SCIENCE CREATES REAL FUTURES"
WITH A NEW PARADIGM AND/FOR DESIGN SCIENCE WE CREATE REAL FUTURES
THIS IS THE DESIGN SCIENCE INTERPRETATION [THAT ENRICHES OUR LIVES]
THE NEW PARADIGM OF DESIGN SCIENCE IS NOT EXACTLY NEW. DESIGNERS ALWAYS FOLLOW NATURAL LAW TO ADDRESS NEEDS, GENERATE FULLFILLMENT AND SATISFACTION THAT CHANGE THROUGH HISTORICAL AND CULTURAL AGES. EACH AGE HAS A UNIQUE IDENTITY, QUALITY AND AESTHETIC THAT TELLS OUR STORY. WE ARE IN A NEW AGE NOW PROMOTING THE BENEFITS OF A NEW PARADIGM MOVEMENT LEADING US TO AGREE THAT DESIGN SCIENCE IS UP TO THE TASK OF CREATING FUTURES".

The new 'Consciousness Paradigm' is a Design Science model of Body, Mind and Spirit as key fundamental elements of reality, life and energy. Every design we create is an expression of unique characteristics, qualities and flavors that describe us in our own time. Experiences of new design elements express the form, process or essences to communicate the essence of the space, cultures, iconic and archetypal solutions we design. Design Science creates aesthetic inflections and reflections of the true nature of how things work. This is the domain of science. When Design is fused with Science magic is made. Designs are physical and in the domain of physiology and form. Design enhances behavior or psychology, with Essences and creative Energies that express our true aesthetic spirit and design excellence. These three elements express the very popular 'singularity principle' of 'oneness'. It is a fact or state of being *one* in agreement and alignment with natural law or what we call a 'new paradigm (movement)'.

Esoteric symmetry principles of creation extrapolate the one into three. In pure extension, pure duration and pure energy the three returns to pure one; the source of its arising, to an elevated level of 'synthesis' or consciousness. The new three is information in a uniquely branded space, time and energy continuum.

Direct Wisdom, Intelligence Knowledge and Information from Sources is not respected in design, architecture and engineering paradigms. The closest dynamic here would be intuition.

A NEW PARADIGM MOVEMENT LEADS US TO AGREE THAT "WE CREATE REAL FUTURES"
A NEW PARADIGM MOVEMENT LEADS US TO AGREE THAT "DESIGN SCIENCE CREATES REAL FUTURES"

A NEW PARADIGM AND/FOR DESIGN SCIENCE WE CREATE REAL FUTURES

FROM	TO
1. From an over-done work and competition ethos	To a growing recognition of the humanizing powers of play and fantasy, cultivating imagination, visualization and dreams.
2. From an over-focus on language and reason, history and basic subjects	To a re-appreciation of the need for the arts to develop a more balanced mind and life
3. From thinking about things to beginning increasingly	To think about the way we think about things, questioning assumptions and fostering a more flexible philosophy of mind: i.e. meta-meta-cognition, critical thinking,
4. From an over-controlling and unrealistic estimate of what we can and cannot do	To a more realistic understanding of the limits of our power– humility and surrender
5. From thinking that cheerfulness was something that reflects inner mood	To realizing that it involves a mixture of positive attitudes, willed intention, and skills that must be implemented even in the face of negative circumstances or inner sadness or fear. It's not denial, covering-up or disguising, but rather a

	turning away from the dark into the light, an act of faith.
6. Similarly, from thinking that wisdom is something that one attains and then has	To a more process-oriented idea that wisdom-ing is how you use all your skills along with your highest values to respond to changing circumstances. Loving or believing is an activity, not a fixed quality.
7. From accepting the authority of past "experts,"	Recognizing that emerging knowledge and changing circumstances affect and modify even the most seemingly wise pronouncements made in the past.

Future Trends

Study the way all technological developments will affect the emerging culture. Carefully consider how people's attitudes and modes of thinking are changing. One change is the rising expectations and desires when observing how wealthier people live. This is a product of media and communication. If they can have it, why can't I? I wonder what encountering a wealthier life style does to young people's minds. Does it make them more resentful, more willing to cheat?

I think there may well be cultural shifts associated with the growth of lifestyle shifts and international fashion infusion (in gadgets more than clothes and other basic needs). Consider global trends in emerging Indian and Chinese populations as examples among others. What further shifts would you add to this provisional list?

The idea of humans being 'spirits having bodily experiences' is an absolute critical shift. The knowledge of the oneness of everything and that they all return to the source of their arising. The way the brain works and its impact on the functions of the mind, body spirit continuum according to quantum theory is the next frontier. "New Frontier Thinking"

Summary

The purpose is to promote awareness and discussion of the changing trends in our culture. How can they be addressed more wisely? What are some pitfalls? (I don't doubt that within any of these trends there are those who may seek to commercially exploit them, that there are ways of experiencing and reacting to them from a more childish mentality, that these trends may be reflected in styles of mental and physical dis-ease as well as approaches to therapy. The more we understand basic dynamics, though, the more we can effectively anticipate and respond to problems as they arise.

"light and shade, long and short, black and white, can only be experienced in relation to each other; light is not independent of shade, nor black of white. There are no opposites, only 'temporary' relationships." Lankavatara Sutra from "Buddha Speaks"

CHAPTER SEVEN

DODECAHEDRON-DoH

7. The Sahasrara chakra as the crown
Blissfully enjoyed with its opened ways.
A state of inseparable oneness, interwoven
with the ever-transcending Beyond says;
We always deal with Infinity, Eternity, Immortality
ALL in ALL in ALL consciousness is reality.
Of all the centers highest is this most peaceful,

The most soulful and fruitful is in Sahasrara.
Here Infinity, Eternity and Immortality come
to source in creation with the one.
Clairprescienced and reawakened not by our
reasoning in thought or logic or deduction
Reading the akashic records to unseal all
eternal data on HIMHERIT's hard drive live.

If not of the world, transcending things and time
while creating through MENTALISM and deep mind
In a rooted sense of oneness, dissolving, illusion of some kind
boundaries of consciousness in my conditioned human brain
three parts harmonized reptilian mammalian still remain below
The neocortex crowning seated in the ether of a pineal soul.

CHAPTER 7 TRINE A

LOGIC STRUCTURES
in a distribution Matrix
Ancient Sites and The Logic Structures associated with them

THE BLOCK MODEL FOR THE FLAVOR MACHINE

ARCHITECTURE HYPOTHESIS FOR MONUMANTAL ARCHITECTURE
Copyright Adger Cowans

The up, side, down modules

Individual U. S. D units in the 1-2-3 proportion

The technology in the flavor machine reflected in the *Stonehenge prehistoric monument in Wiltshire, England, 2 miles west of Amesbury and 8 miles north of Salisbury; two other sites.*

Stonehenge England

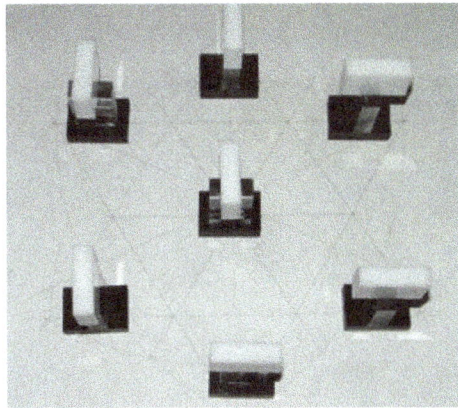
Flavor Machine Logic Structure

'The Flavor Machine and Block [DISTRIBUTION] Model™'. This is a New Paradigm Model. Extrapolate; the distribution of elements, processes and energies in a defined matrix or field of information to explore: observe, test and investigate non-binary variations, relationships and possibilities for combinations of expressions, dynamic properties and flavors that may or may not be useful. This is a method of modeling and reality testing volatility and other dissonant behaviors in a controlled environment prior to the design and manifestation of the result, invention and development. This is the essence of the Thought Field Experiment. It corresponds with the phenomenal thought itself or concept, the thought field or the dynamic environment and the flavor of energy of the thought and the work or observation and all other mental processes needed for the heuristic processes to occur.

ex·trap·olate

EXTRAPOLATION:
1. By extend the application of models (a method or conclusion, especially one based on statistics)can be derived from observation and experimentation of an unknown situation by studying trends that will continue or similar methods will be applicable.

Extend (a graph, curve, or range of values) by inferring unknown values from trends in the known data.
"A set of extrapolated values"

A **heuristic technique**, often called simply a **heuristic**, is any approach to problem solving, learning, or discovery that employs a practical method not guaranteed to be optimal or perfect, but sufficient for the immediate goals. Where finding an optimal solution is impossible or impractical, heuristic methods can be used to speed up the process of finding a satisfactory solution. Heuristics can be mental shortcuts that ease the cognitive load of making a decision. Examples of this method include using a rule of thumb, an educated guess, an intuitive judgment, stereotyping, profiling, or common sense.

D_ESign Science in The New Paradigm Age

THE LOGIC STRUCTURE OF ANCIENT MONUMENTAL SITES

Flavor Machine Logic Structure

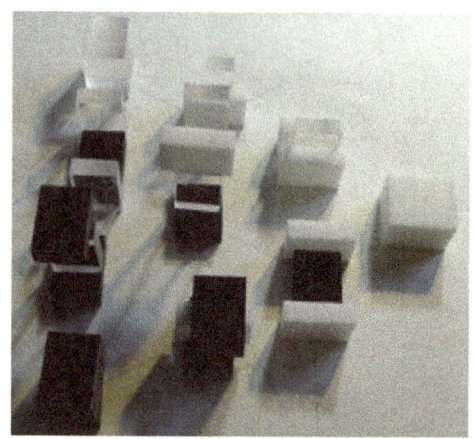
Flavor Machine Logic Structure

The Flavor machine with the up; clear blocks- side; white blocks and down; black blocks.

Flavor Machine Logic Structure

Stonehenge England

The Flavor machine with the up; clear blocks- side; white blocks and down; black blocks as elements represented in both arrangements.

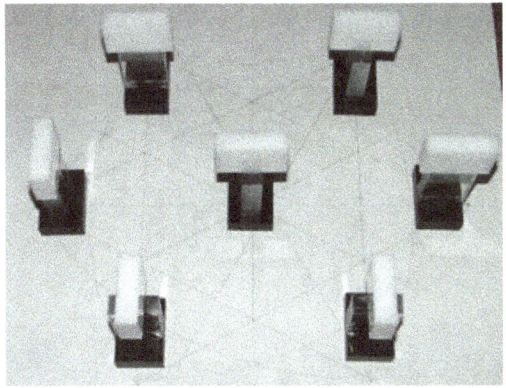

The six orientations of the 1-2-3 logic structure (View 1)

Six (6) central elements in a hexagonal configuration

The full set of the orientations of the 1-2-3 logic structure (View 2)

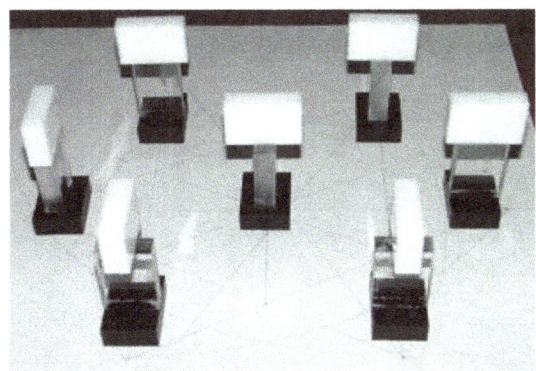

The six orientations of the 1-2-3 logic structure (View 3)

There are technologies in these logic structures we have not been able to decode for 12,000 years. The basic premise in their designs have never been part of any paradigms we know or have incorporated in any of the global frameworks or canons used for our civilization building.

The attitude of aligning with celestial phenomena without the desire to colonize space is alien.

Gobelki Tepe : The Archeological site in South Eastern Turkey

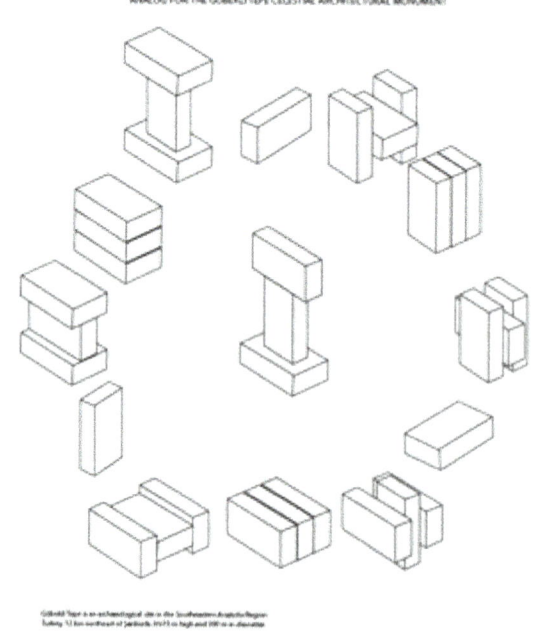

The Triangular configuration is transformed into a Circular pattern with the same elements

SYSTEMS OF LEARNING OR COGNITION ARE SUPPORTED BY THE FOLLOWING CONCEPTS

"**Pedagogy**" literally means "leading children." "**Andragogy**" was a term coined to refer to the art/science of teaching adults. Malcolm Knowles and others theorized that methods used to teach children are often not the most effective means of teaching adults.

Heuristic: involving or serving as an aid to learning, discovery, or problem-solving by experimental and especially trial-and-error methods <heuristic techniques> <a heuristic assumption>; also : of or relating to exploratory problem-solving techniques that utilize self-educating techniques (as the evaluation of feedback) to ...
The process of according **value** to a symbol is **psychological** and social. Money is a social institution based on the consent of the population and a **psychological** symbol based on the consent of the individual.

Axiology (from Greek ἀξίᾱ, axiā, "value, worth"; and -λόγος, -logos) is the philosophical study of value. It is either the collective term for ethics and aesthetics[1]—philosophical fields that depend crucially on notions of worth—or the foundation for these fields, and thus similar to value theory and meta-ethics. The term was first used by Paul Lapie, in 1902,[2] and Eduard von Hartmann, in 1908.[3][4]
Axiology studies mainly two kinds of values: ethics and aesthetics. Ethics investigates the concepts of "right" and "good" in individual and social conduct. Aesthetics studies the concepts of "beauty" and "harmony." **Formal axiology,** the attempt to lay out principles regarding value with mathematical rigor, is exemplified by Robert S. Hartman's Science of Value. Studies of both kinds are found in Cultura: International Journal of Philosophy of Culture and Axiology.
In philosophy, **qualia** (/ˈkwɑːliə/ or /ˈkweɪliə/; singular form: **quale**) are individual instances of subjective, conscious experience. The term "qualia" derives from the Latin neuter plural form (qualia) of the Latin adjective quālis (Latin pronunciation: [ˈkwaːlis]) meaning "of what sort" or "of what kind"). Examples of qualia include the pain of a headache, the taste of wine, or the perceived redness of an evening sky. As qualitative characters of sensation, qualia stand in contrast to "propositional attitudes".[1]
Daniel Dennett (b. 1942), American philosopher and cognitive scientist, regards qualia as "an unfamiliar term for something that could not be more familiar to each of us: the ways things seem to us".[2]

Erwin Schrödinger (1887–1961), the famous physicist, had this counter-materialist take:
The sensation of color cannot be accounted for by the physicist's objective picture of light-waves. Could the physiologist account for it, if he had fuller knowledge than he has of the processes in the retina and the nervous processes set up by them in the optical nerve bundles and in the brain? I do not think so.

In philosophy, qualia (/ˈkwɑːliə/ or /ˈkweɪliə/; singular form: quale) are individual instances of subjective, conscious experience. The term "qualia" derives from the Latin neuter plural form (qualia) of the Latin adjective quālis (Latin pronunciation: [ˈkwaːlis]) meaning "of what sort" or "of what kind"). Examples of qualia include the pain of a headache, the taste of wine, or the perceived redness of an evening sky. As qualitative characters of sensation, qualia stand in contrast to "propositional attitudes".[1]Daniel Dennett (b. 1942), American philosopher and cognitive scientist, regards qualia as "an unfamiliar term for something that could not be more familiar to each of us: the ways things seem to us".[2]Erwin Schrödinger (1887–1961), the famous physicist, had this counter-materialist take: The sensation of color cannot be accounted for by the physicist's objective picture of light-waves. Could the physiologist account for it, if he had fuller knowledge than he has of the processes in the retina and the nervous processes set up by them in the optical nerve bundles and in the brain? I do not think so. Feelings and experiences vary widely. For example, I run my fingers over sandpaper, smell a skunk, feel a sharp pain in my finger, seem to see bright purple, become extremely angry. In each of these cases, I am the subject of a mental state with a very distinctive subjective character. There is something it is like for me to undergo each state, some phenomenology that it has. Philosophers often use the term 'qualia' (singular 'quale') to refer to the introspectively accessible, phenomenal aspects of our mental lives. In this broad sense of the term, it is difficult to deny that there are qualia. Disagreement typically centers on which mental states have qualia, whether qualia are intrinsic qualities of their bearers, and how qualia relate to the physical world both inside and outside the head. The status of qualia is hotly debated in philosophy largely because it is central to a proper understanding of the nature of consciousness. Qualia are at the very heart of the mind-body problem.

The entry that follows is divided into ten sections. The first distinguishes various uses of the term 'qualia'. The second addresses the question of which mental states have qualia. The third section brings out some of the main arguments for the view that qualia are irreducible and non-physical. The remaining sections focus on functionalism and qualia, the explanatory gap, qualia and introspection, representational theories of qualia, qualia as intrinsic, nonrepresentational properties, relational theories of qualia and finally the issue of qualia and simple minds.

"VISUAL AWARENESS"

The periodic table of elements is not the only example of ordering systems and characteristics of universal phenomena where suchness- quality and muchness-quantity are synthesized to describe or art-iculate and predict (new) flavors and realities.

What is a mark of a great scientist? Scientists discover new information and make sense of it, linking it to other data. They may go further by giving an explanation of this linked data which, maybe not immediately, other scientists accept it as a correct explanation. However the outstanding scientist goes further in predicting consequences of his ideas which can be tested. This boldness identifies the great scientist if the predictions are later found to be accurate. One such person was Russian chemist Dmitri Mendeleev. Incidentally, although he is often regarded as the father of the Periodic Table, Mendeleev himself called his table or matrix, the Periodic System."
Periodic Matrices are universal ordering systems created to 'process information' using natural principles and law.
Everything in our inner and outer experiences (our world) are expressions of one super-consciousness that connects all of itself, (which includes us) into an inherent dynamic of oneness or w-holism. Symmetry principles and natural laws operate on all basic and natural expressions including, natural phenomena, forces and dynamic process in universes, planets, animals and people in a grand periodic matrix. The best model of such a system is the periodic table of elements. Qualitative expressions; phenomena like light, color, orientation, and

consciousness itself (in sub, normal awareness and super states) are all in a pulsing ever expanding vibrational hierarchical dynamic called "LIFE". Simple elements that are few in number, are synthesized and harmonized to form conscious experiences we call awareness. These principles operate on Nano, micro, human and individual and Macro scales at levels of vibration.

"The consciousness paradigm is wired to the three dimensions of the divine and all expressions of it in our three-dimensional world (view). They constitute the Law of number (3); physiology, psychology and spirituality". We can call these expressions or realities 'dimensions'. The highest state of understanding matter or things involves physiology with the thought and ideational processes: MENTALISM.

Behavior, thinking and feelings come next. At the base level is the physical or the tangible and manifested state. In linguistic terms In English the main parts of speech here are nouns and pronouns. There is psychology or behavior in Matter as well as in people. Knowing how to align form, behavior and matter is key to manifestation at the root (basic or dense state) of things. The correspondence and alignments of these states are critical. These are the qualifiers and connectors; adverbs, adjective, determiner, preposition, conjunction, and interjection used to make meaning and give clarity to intention and expression. The highest state of behavior and the understanding of how everything works relies on Psychology (behavior). Here in the thought form dynamic behavior and movement are both internal and external. The will, the intention and motivation are all forms of internal psychological movement that causes changes in states along a continuum of e-motions and ex-pressions. This movement is a grosser type of physical displacement or movement than is found at the spiritual level.
The spiritual here is the creative spirit or our electrical life, Chi, Ka and Ba etc. Here we use verbs and adverbs as words to express thoughts, feelings and directed actions. Words are assigned in accordance not only with syntactical functions but with the highest states of psychological awareness and spiritual experiences attainable. People and much larger systems engaged in the spiritual dynamic expressions and are still not yet in touch with nor are ready to accept the inevitable emerging paradigm shifts. We now see ourselves as spiritual beings having bodily experiences. Everything elevates to a higher level of consciousness when we accept this as our new reality.

Number and flavor; measure and form.

Discovering new information: Cognition
Making sense of it.
Linking it to other data.
Giving an explanation of this linked data.
Other scientists (Creatives) accepting it as a correct explanation.
Predicting consequences of ideas which can be tested.
Making predictions that are found to be accurate later.
A matrix is a concise and useful way of uniquely representing and working with linear transformations. Are there systems for non-linear transformations? In particular, every linear transformation can be represented by a matrix, and every matrix corresponds to a unique linear

transformation. The matrix, and its close relative the determinant, are extremely important concepts in linear algebra, and were first formulated by Sylvester (1851) and Cayley.

In his 1851 paper, Sylvester wrote, "For this purpose we must commence, not with a square, but with an oblong arrangement of terms consisting, suppose, of m lines and n columns. This will not in itself represent a <u>determinant</u>, but is, as it were, a Matrix out of which we may form various systems of determinants by fixing upon a number p, and selecting at will p lines and p columns, the squares corresponding of pth order." Because Sylvester was interested in the determinant formed from the rectangular array of number and not the array itself (Kline 1990, p. 804), Sylvester used the term "matrix" in its conventional usage to mean "the place from which something else originates" (Katz 1993). Sylvester (1851) subsequently used the term matrix informally, stating "Form the rectangular matrix consisting of n rows and $(n + 1)$ columns.... Then all the $n + 1$ determinants that can be formed by rejecting any one column at pleasure out of this matrix are identically zero." However, it remained up to Sylvester's collaborator Cayley to use the terminology in its modern form in papers of 1855 and 1858 (Katz 1993). In his 1867 treatise on determinants, C. L. Dodgson (Lewis Carroll) objected to the use of the term "matrix," stating, "I am aware that the word 'Matrix' is already in use to express the very meaning for which I use the word 'Block'; but surely the former word means rather the mould, or form, into which algebraical quantities may be introduced, than an actual assemblage of such quantities... " However, Dodgson's objections have passed unheeded and the term "matrix" has stuck.

The <u>transformation</u> given by the system of equations

Howard Gardner's Theory of Multiple Intelligences

In the past century, numerous theories about intelligence have emerged. One of the more famous theories was created by developmental psychologist Howard Gardner in 1983. Gardner proposed that intelligence is not made up of one factor, but rather eight.

For example, a Spanish professor may have a strong appreciation for language (linguistic intelligence), but may have a hard time relating to his students (interpersonal intelligence).

STYLES OF LEARNING	HOWARD GARDNER'S THEORY OF MULTIPLE INTELLIGENCES
Logical/mathematical intelligence	Includes your ability to reason, think critically and analytically, and your understanding of complex mathematical concepts
Linguistic intelligence	includes your ability to appreciate language and use it effectively to accomplish goals
Visual-spatial intelligence:	includes your ability to visualize, remember images and details, and an awareness of your surroundings
Musical intelligence:	includes your awareness of musical sounds, tones, and rhythms; vibration intelligence

Bodily/kinesthetic intelligence:	includes your athletic ability and being aware of your body
Visual-spatial intelligence:	includes your ability to visualize, remember images and details, and an awareness of your surroundings
Musical intelligence:	includes your awareness of musical sounds, tones, and rhythms; vibration intelligence
Bodily/kinesthetic intelligence:	includes your athletic ability and being aware of your body
Interpersonal intelligence:	includes your ability to relate to those around you, understand their motivations, their goals, and their feelings
Intrapersonal intelligence:	includes your ability to understand yourself, your strengths and weaknesses, your goals, and your motivation
Intuitive intelligence	
Psychological intelligence	
Spiritual intelligence	
Creative intelligence	

Wisdom Traditions
Wisdom traditions represent mankind's deepest source of knowledge about universal principles that govern harmonious, prosperous and sustainable existence. They generally involve encountering and observing early stage cognitive challenges and the need for knowledge design and architecture.
The term is often given to the inner core or mystical aspects of a religion or spiritual tradition, without the trappings, doctrines, sectarianisms and power structures often associated with institutionalized religions.
Wisdom traditions provide a conceptual framework for the development of the inner self, living a spiritual life and the realization of enlightenment.
From various Buddhist lineages and Shamanic cultures, to previously held secret practices of Tai Chi Ch'uan and Qigong, to ancient healing traditions using plant spirits, herbs, dreams, yoga, meditation and more, our Wisdom Traditions programs allow for exploring many different paths along the journey towards transformation of self and society.

Introduction to the Five Branches of Philosophy

Philosophy can be divided into five branches which address the following questions:

Metaphysics	Study of Existence	What's out there?
Epistemology	Study of Knowledge	How do I know about it?
Ethics	Study of Action	What should I do?
Politics	Study of Force	What actions are permissible?
Aesthetics	Study of Art	Life is the study of Spiritual Expressions.

There is a hierarchical relationship between these branches as can be seen in the Concept Chart. At the root is Metaphysics, the study of existence and the nature of existence. Closely related is Epistemology, the study of knowledge and how we know about reality and existence. Dependent on Epistemology is Ethics, the study of how man should act. Ethics is dependent on Epistemology because it is impossible to make choices without knowledge. A subset of Ethics is Politics: the study of how men should interact in a proper society and what constitutes proper. Esthetics, the study of art and sense of life is slightly separate, but depends on Metaphysics, Epistemology, and Ethics.

BRANCHES	STUDY AREAS	OUTER DIRECTION	INNER DIRECTION
Metaphysics	Study of Existence	What's out there?	The inner journey
Epistemology	Study of Knowledge	How do I know about it?	It is direct knowledge
Ethics	Study of Action	What should I do?	Trust it and research
Politics	Study of Force	What actions are permissible?	Model and test it
Esthetics	Study of Art	What can life be like?	Self-realization
	Culture		
	Consciousness		

"Philo(Love)-sophy (wisdom) or learning … **has no other subject matter than the nature of the real world, as that world lies around us in everyday life, and lies open to observers on every side.** But if this is so, it may be asked what function can remain for philosophy when every portion of the field is already lotted out and enclosed by specialists? Philosophy claims to be the science of the whole; but, if we get the knowledge of the parts from the different sciences, what is there left for philosophy to tell us?
To this it is sufficient to answer generally that the synthesis of the parts is something more than that detailed knowledge of the parts in separation which is gained by the man of science. It is with the ultimate synthesis that philosophy concerns itself; it has to show that the subject-matter which we are all dealing with in detail really is a whole, consisting of articulated members." "Philosophy," Encyclopedia Britannica (Cambridge: Cambridge
Divisions of Philosophy
Abstract: Philosophy, philosophical inquiry, and the main branches of philosophy are characterized.

1. What is Philosophy?
    1. The derivation of the word "philosophy" from the Greek is suggested by the following words and word-fragments.
        - philo—love of, affinity for, liking of
        - philander—to engage in love affairs frivolously

- philanthropy—love of mankind in general
- philately—postage stamps hobby
- phile—(as in "anglophile") one having a love for
- philology—having a liking for words
- sophos—wisdom
- sophist—lit. one who loves knowledge
- sophomore—wise and moros—foolish; i.e. one who thinks he knows many things
- sophisticated—one who is knowledgeable

2. A suggested definition for our beginning study is as follows.
   Philosophy is the systematic inquiry into the principles and presuppositions of any field of study that brings joy into learning and living the truth.

From a psychological point of view, philosophy is an attitude, an approach, or a calling to answer or to ask, or even to comment upon certain peculiar problems (i.e., specifically the kinds of problems usually relegated to the main branches discussed below in Section II). There is, perhaps, no one single sense of the word "philosophy." Eventually many writers abandon the attempt to define philosophy and, instead, turn to the kinds of things philosophers do. What is involved in the study of philosophy involves is described by the London Times in an article dealing with the 20th World Congress of Philosophy: "The great virtue of philosophy is that it teaches not what to think, but how to think. It is the study of meaning, of the principles underlying conduct, thought and knowledge. The skills it hones are the ability to analyse, to question orthodoxies and to express things clearly. However arcane some philosophical texts may be … the ability to formulate questions and follow arguments is the essence of education."

The Main Branches of Philosophy are divided as to the nature of the questions asked in each area. The integrity of these divisions cannot be rigidly maintained, for one area overlaps into the others.

Axiology: the study of value; the investigation of its nature, criteria, and metaphysical status. More often than not, the term "value theory" is used instead of "axiology" in contemporary discussions even though the term "theory of value" is used with respect to the value or price of goods and services in economics.

Some significant questions in axiology include the following: Is Spiritology in the mix?
Nature of value: is value a fulfillment of desire, a pleasure, a preference, a behavioral disposition, or simply a human interest of some kind?
Criteria of value: de gustibus non (est) disputandum (i.e., ("there's no accounting for tastes") or do objective standards apply?
Status of value: how are values related to (scientific) facts? What ultimate worth, if any, do human values have?
Axiology is usually divided into two main parts.

Ethics: the study of values in human behavior or the study of moral problems: e.g., (1) the rightness and wrongness of actions, (2) the kinds of things which are good or desirable, and (3) whether actions are blameworthy or praiseworthy.

Consider this example analyzed by J. O. Urmson in his well-known essay, "Saints and Heroes":

"We may imagine a squad of soldiers to be practicing the throwing of live hand grenades; a grenade slips from the hand of one of them and rolls on the ground near the squad; one of them sacrifices his life by throwing himself on the grenade and protecting his comrades with his own body. It is quite unreasonable to suppose that such a man must be impelled by the sort of emotion that he might be impelled by if his best friend were in the squad."

Did the soldier who threw himself on the grenade do the right thing? If he did not cover the grenade, several soldiers might be injured or be killed. His action probably saved lives; certainly an action which saves lives is a morally correct action. One might even be inclined to conclude that saving lives is a duty.

But if this were so, wouldn't each of the soldiers have the moral obligation or duty to save his comrades? Would we thereby expect each of the soldiers to vie for the opportunity to cover the grenade?

Æsthetics: the study of value in the arts or the inquiry into feelings, judgments, or standards of beauty and related concepts. Philosophy of art is concerned with judgments of sense, taste, and emotion. Expressions of consciousness in synthesis and harmony.

E.g., Is art an intellectual or representational activity? What would the realistic representations in pop art represent? Does art represent sensible objects or ideal objects?

Is artistic value objective? Is it merely coincidental that many forms in architecture and painting seem to illustrate mathematical principles? Are there standards of taste?

Is there a clear distinction between art and reality?

Epistemology: the study of knowledge. In particular, epistemology is the study of the nature, scope, and limits of human knowledge.

Epistemology investigates the origin, structure, methods, and integrity of knowledge.

Consider the degree of truth of the statement, "The earth is round." Does its truth depend upon the context in which the statement is uttered? For example, this statement can be successively more accurately translated as …

"The earth is spherical"

"The earth is an oblate spheroid" (i.e., flattened at the poles).

But what about the Himalayas and the Marianas Trench? Even if we surveyed exactly the shape of the earth, our process of surveying would alter the surface by the footprints left and the impressions of the survey stakes and instruments. Hence, the exact shape of the earth cannot be known. Every rain shower changes the shape.

(Note here as well the implications for skepticism and relativism: simply because we cannot exactly describe the exact shape of the earth, the conclusion does not logically follow that the earth does not have a shape.)

Furthermore, consider two well-known problems in epistemology:

Russell's <u>Five-Minute-World Hypothesis</u>: Suppose the earth were created five minutes ago, complete with memory images, history books, in the AKASHA, how could we ever know of it? As Russell wrote in The Analysis of Mind, "There is no logical impossibility in the hypothesis that the world sprang into being five minutes ago, exactly as it then was, with a population that "remembered" a wholly unreal past. There is no logically necessary connection between events at different times; therefore, nothing that is happening now or will happen in the future can disprove the hypothesis that the world began five minutes ago." For example, an omnipotent GOD could create the world with all the memories, historical records, and so forth five minutes ago. Any evidence to the contrary would be evidence created by GOD five minutes ago. (Q.v., the Omphalos hypothesis.)

Suppose everything in the universe (including all spatial relations) were to expand uniformly a thousand times larger. How could we ever know it? A moment's thought reveals that the mass of objects increases by the cube whereas the distance among them increases linearly. Hence, if such an expansion were possible, changes in the measurement of gravity and the speed of light would be evident, if, indeed, life would be possible.

Russell's Five-Minute-World Hypothesis is a philosophical problem; the impossibility of the objects in the universe expanding is a scientific problem since the latter problem can, in fact, be answered by principles of elementary physics.

3. <u>Ontology</u> or <u>Metaphysics</u>: the study of what is really real. Metaphysics deals with the so-called first principles of the natural order and "the ultimate generalizations available to the human intellect." Specifically, ontology seeks to indentify and establish the relationships between the categories, if any, of the types of existent things.

THE TONE OF THE INTERNAL DIALOG

What types of things exist? Do only particular things exist or do general things also exist? How is existence possible? Questions as to identity and change of objects—are you the same person you were as a baby? as of yesterday? as of a moment ago?
How do ideas exist if they have no size, shape, or color? (My idea of the Empire State Building is quite as "small" or as "large" as my idea of a book. I.e., an idea is not extended in space.)
What is space? What is time?
E.g., Consider the truths of mathematics: in what manner do geometric figures exist? Are points, lines, or planes real or not? Of what are they made?
What is spirit? or soul? or matter? space? Are they made up of the same sort of "stuff"? When, if ever, are events necessary? Under what conditions are they possible?
Further characteristics of philosophy and examples of philosophical problems are discussed in the next tutorial.

 *This ancient wisdom of duality is the essence of the Tao. It also appears to be the essence of physics. Anne Bancroft 2000*

CHAPTER 7 TRINE B

SYNTHESIS: A 'CREATIVE THOUGHT PROCESS'

 ***Different states of consciousness*** *are associated with different brain wave patterns. Brain waves are tracings that show the kind of electrical activity going on in the brain. Scientists use an electroencephalograph, or EEG, to record these waves. The main types of brain waves are Alpha, Beta, Theta, and Delta.*

A thought (experiment) process considers some hypothesis, theory, or principle for the purpose of thinking through its consequences. Given the structure of the experiment, it may not be possible to perform it, and even if it can be performed, there need be no intention to do so.

The common goal of a thought (process) is to explore the potential consequences of the principle in question: "A thought process is a device with which one performs an intentional, mental and creative process of deliberation in order to speculate, within a specifiable problem domain, about potential consequents (or antecedents) for a designated antecedent (or consequent)" The connection to source or the becoming one with nature as it is called affords Insight and other Epi Phenomena to be explored.

Creativity is now about Interpretations of 3 dimensional forms that are transcendental and consistently sustaining our thoughts and ideas through time, space and energy.
With our needs to redefine humanity with our creativity, genius and technology we are in an exponential evolution and a disruptive revolution.
We are in an exponential evolution and a disruptive revolution now.
Consciousness, by definition, is comprised of the physical, or physiology, the emotional or psychological and the spiritual dimensions or creative work energy.
Consciousness by definition is the physical, or physiology, the emotional or psychological and the spiritual dimensions or creative work energy in harmony.
We are creating the future now by blending the best of multiple related industries, synthesizing and transforming them into powerful and effective paradigm shifting systems, with tools and technologies to create a new future for humanity.

INSPIRED BY ANCIENT TEXTS
"O Devi! Thou art the ***mind, the sky, the air, the fire, the water, and the earth.*** Nothing is outside Thee on Thy transformation. Thou hast become Siva's consecrated queen to alter Thy own blissful conscious Form in the shape of the world".
Kundalini, the serpent power or mystic fire, is the primordial energy or Sakti that lies dormant or sleeping in the Muladhara Chakra, the center of the body. It is called the serpentine or annular power on account of its serpentine form. It is an electric fiery 'occult' power, the great pristine force which underlies all organic and inorganic matter. What other thought parallels can we find in the western world?

Looking further we read: Kundalini is the Goddess of speech, (language, symbols, and words) and is praised by all. When awakened she offers illumination (light), the source of all Knowledge and Bliss. She is pure consciousness; the Supreme Force, the Mother of Prana,

Agni, Bindu, and Nada. It is by this Sakti that the world exists. Creation, preservation, and dissolution are in Her. Only by her Sakti the world is kept up. It is throu, tergh Her Sakti on subtle Prana, Nada is produced. While you utter a continuous sound or chant Dirgha Pranava ! (OM), you will distinctly feel the real vibration starting from the Muladhara Chakra. Through the vibration of this Nada, all the parts of the body function. She maintains the individual soul through the subtle Prana. In every kind of Sadhana, the Goddess Kundalini is the object of worship in some form or the other.

Language through the ages may have been influenced by our understanding with the accuracy and effectiveness of the symbolic expressions of their times that were used to convey the knowledge and nature of phenomena and experiences. Does the passage of time (New Ages) bring with it more accurate terminologies, symbols and descriptions we apply to phenomena we are still trying to elevate our consciousness? There is constant movement to higher levels of innerstanding in keeping with the unfolding patterns of growth and development that would serve us better, that follow natural laws. This unfolding pattern of wisdom, intelligence Knowledge and Information the WIKI™ has momentum with it that comes with refinements of the fundamental elements of language and its structural elements, number being one of the keys. It would help if we use our will intention and passion to manifest the momentum.

In looking at theories of how the universe and life began we are aware of the big bang and the expansion of the universe. This can be seen as a primary and external momentum creator. The human beings experience natural and internal urges to survive using several types of driving forces, procreation for preservation, creation to satisfy needs and the knowledge to do the work required to provide for successful design, implementation and delivery of the 'goodness' they represent. Here we have at work the physical, the psychological and the spiritual drivers that inspire and motivate all who engage in these creative acts. Is there a connection between the expanding universe and the progression of the life force and over time the momentum that develops from their creative processes? Do photons of "Let there be Light make big bangs?

Granted these are a few of the grand dynamic phenomena but they have their unique corresponding echoes on all phases and levels of the spectrum of living, being and becoming we are motivated and driven by. By understanding what consciousness is, given the same constant drivers (paradigms or habits) that do not really change over time, it would appear that the variable, in this equation, as the flexible and changing dynamic is (reluctant) MAN. This is where paradigms are reevaluated. The systems that are used to create the language and interpretations that facilitate the documentation and preservation of knowledge, to tell our stories, are built on beliefs, customs and languages that change. People change by responding to their internal drivers and motivations to exercise their one special gift of all species on earth, that is to transform their environments and in so doing change themselves.

The degrees of freedom include thought, imagination, communication feelings, creativity, sex and gender and making things and work. This is the realm of design. It is now the language needed that is much more complex than before.

The reduction of complexity is a necessary driver that depends on the clear and precise understanding of our selves and the world we live in so that simplicity can offer the ease and efficiency needed to reduce stress and be healthy. Creativity in the past could survive and

thrive in a linear manner with a physical and materialistic targeted agenda. The tone of our time (duration and displacement) must be defined by simplicity. It must also embrace the triadic harmony that is the essence of our new innerstanding of consciousness in every iteration of it. That it can be understood as the synthesis of the body, mind and spirit or energy (as thought, plasma and electricity) is crucial to building that momentum creating dynamic that can be used to create our visions and imaginations as the language of our time with its unique aesthetic forms, expressions, flavors and experiences.

There are three numbers we can extract from this new driver with its own system of number laws we now are focusing on. In all of the above these are the three principles at work; Body, Mind and Spirit. The three number law: The symbol 3, the triad and the trinity, the triplet or third wave. The symbol 4 represents the four forces operating in our world that in combination with the first three (3) are the tool designers use. 4 is the quartet or fourth place. The symbol 7; the heptet or seventh place, the wave or Heptave. The other numbers in the enneave 1-Unity, 2-Duality, 5-the symbol of Balance, 6- Creativity, 8-the eight fold symmetry (way), & 9-Completion relate to other vibrations of creation and consciousness. The 3, 4 &7 fold vibrations are more directly linked to manifestation. They are at the core of the universal creative process shared with man the physical being-mankind the emotional expression in sub consciousness and the spiritual essence or life force of man and all other expressions. In the realm of vibrations 3 is in the upper register of super consciousness. 4 is at the lower register of normal consciousness or awareness. In the mid register there is a vibration that serves as the transformational dynamic (of feeling or touch) at the Love center. 3 + 4 =7 the registers are arranged as 3 in the upper, 1 in the mid register and 3 in the lowest range.

Number is a visual intelligent language used to describe ALL thought forms of muchness. Their patterns are qualitative descriptors of 'suchness' used to encode the emotional experiences and habits. The spiritual (infinite 'consciousness') dimensions are all connected to, the grand synthesis or source and the cacophony of names we use to describe it.

Insight/ Flash: it appears that though love is a very powerful force the three upper registers, centers and dynamics with the three lower registers, centers and dynamics are the 6-fold elements of the creative process. This may evade the observer since it is not quite evident. Here are some of the drivers that we are all connected to that operate in the background to support our lives: It is the realm of synthesis where spirit operates as 'etheric' dynamics so awesome and miraculous when compared to 'all we hold sacred' that it is seen as being sacred.

'Awe-ware-ness' is the word to describe these phenomena.
In keeping with language as the theme here; metaphysics could be seen as the language of spirit. Religion has scripture as its medium. There are some very subtle drivers in the metaphysical world that require some sort of scientific method to grasp their relevance and significance to levels of knowledge not used in the material realm. Light is one fine example of this concept. Methodologies for studying spirit are needed. Materialism and behaviorism are somewhat represented but our spiritual development is sorely lacking.

Our knowledge of the elements has dominated the mindscape leaving us confused.
not knowing the true nature of matter and how they truly affect us. We believed that periodicity applies to physical matter and nothing else. But do we know that as a fact?

Fire is the result of a chemical reaction, called combustion. At a certain point in the combustion reaction, called the ignition point, flames are produced. Flames consist primarily of **carbon dioxide**, **water vapor**, oxygen and **nitrogen**. Fire emits heat and light.

Water, is one oxygen atom bonded to two different hydrogen atoms, H_2O. When atoms are bonded together, (H_2, O_2, H_2O etc.) we call the total structure a molecule. Hydrogen atoms and oxygen atoms are found in many different molecules, and these different molecules **make** up, 1. Solids, 2.Liquids, and 3. Gasses. Here again we see the law of three (3) at work again in the 3:4:7-10 (Decave). **Water vapor** (water in its gaseous state) is also present in the atmosphere in varying amounts, by up to 2%.

Dry Air is primarily made up of **nitrogen (78.09%) and oxygen (20.95%).** The remaining 1% is made up of argon (0.93**%), carbon dioxide** (0.039% as of 2010) and other trace gases (0.003%).
The most recent addition to this threesome is plasma, a fourth element. That this could be the ether is still unknown. Does it have any connection to "dark matter" is another conundrum.

ELEMENTS					
AIR		**Oxygen**	Nitrogen	Carbon Dioxide	Argon
FIRE		**Oxygen**	Nitrogen	Carbon Dioxide	[Water Vapor (gas)]
WATER	Hydrogen	**Oxygen**			
PLASMA	Ether	**Ether**	Ether	Ether	Ether

The four fold symmetry expressions; those that obey the laws of the realm of manifestation, are created through Natural, artificial and Creative expressions of thought, mentalism and consciousness; extrapolated into anther 3fold symmetry; Polarization, ideation and formation. The various fields of information or energy encode matter with their 'messaging systems'. The one we know best is DNA. Is this all she wrote or are there other systems? What languages are these codes written in.
Source encodes the messaging to in-form animals and plants and everything else with DNA. Human Deoxyribonucleic Acid (DNA) is made up of molecules called nucleotides. Is DNA the basis for the 'binary code? This is where we see our 'story' encoded in its'3fold' glory.

The order [GEOMETRY] of these bases is what determines DNA's instructions, or genetic code.

Nucleotides.	a phosphate group	a sugar group	a nitrogen base

The Four types of nitrogen bases are: A is the Atomic number-M is the 'ultimate sum' of the Atomic number

Thymine (T)	Oxygen A2 M 6	Hydrogen:A6-M 1	Nitrogen A2 M 5	Carbon A5 M 3
Cytosine (C)	Oxygen A2 M 6	Hydrogen A6-M 1	Nitrogen A2 M 5	Carbon A5 M 3
Adenine (A)	Oxygen A2 M 6	Hydrogen A6-M 1	Nitrogen A2 M 5	Carbon A5 M 3
Guanine (G)	Oxygen A2 M 6	Hydrogen A6-M 1	Nitrogen A2 M 5	Carbon A5 M 3

EARTH [AS A CARBON BASED CONTAINER OF CONTAINERS FROM NANO TO MACRO SCALES]

NATURE	FORCES	ELEMENTS	ENERGY
Natural & Artificial Symmetry forms of Expression.	Gravity, Electro Magnetic, The strong and The Weak Forces	The periodic Table of Elements-Minerals	Carbon base fuels. Kinetic & Potential Energy. Thermal etc.

Against the backdrop of the symmetry dynamics of source, creativity, consciousness, creative processes and interactions; MAN being a major or pivotal one, follows all the phases of manifestation. The three dimensional realm of 'ABOVE' in correspondence (corres-pond-dense' runs with the dense thought forms or solidity into the four dimensional world). We actually live in multiple dimensions manifested and maintained by the three higher planes. Since there is no time, space or matter there, with all being energy and vibration, presenting or believing time to be the fourth dimension is Maya. The three dimensions of eternal (macro universal) and singular individual (nano and micro) ideation in the states of 'ABOVE or SUCHNESS' is harmonized and flavored with essences of physical, natural and spiritual expressions or life. [s-uchness m-uchness and con-sciousness] are transformed by two symbols or letters of the English Alphabet, namely-'s' and 'm'. Is it co-rrespon-dence or co-inci-dence? S is the 19th letter of the Alphabet with an ultimate sum 'NUMBA' 1. The total US value of the word suchness is 27 with the US of 9. M is the 13th letter of the Alphabet with an ultimate sum or 'NUMBA' 4. The total Ultimate Sum of the word muchness is 30 with the US of 3. These ideas are significant and will be developed later. BELOW or MUCHNESS is synthesized and quantized with the measure of vibration as, cision, the first creative principles of manifestation.

Between Suchness and muchness lies an interstitial zone where transformers, translators and transducers operate. They all have the qualities of Body, Mind and creative Spirit. To function as true media they must possess the sameness of all favors and expressions or frequency. They must possess the above and the below natures to be able to 'know' themselves and the wide range of possibilities open to them as we now see in quantum theory. With the minimum inventory all media, gauges, processes and energies; from the gross electric light bulb energy to the finest fuel; thought, the combinatorial mysteries begin and the illusions (the ill-use of minds visions) are produced. This 'media kingdoms' have bridges. Man is the bridge for humans, emotional life, thought and humanity. The sun's bridge is energy; light has heat.

The plant kingdom, the animal kingdom and all others have their bridges and possess the expressions of consciousness they ALL contain and align with their genetic codes. While we

are on 'the 'NAME-NUMBA' mystery zone; Please allow me to remove the religious virus. The name of the spiritual force or source that is or of and for 'ALL' and causes us much consternation; a phenomenal word is Con·ster·na·tion. It conjurs feelings of anxiety, distress or dismay, typically at something unexpected or unknown. We need a new device to find new meanings with words just as we are discovering new relationships with words and numbers as 'numb-a'. 'We are only interested in the sound vibrations of word, to get to the deeper meaning rooted in the vibration of sound making poetry and songs very powerful expressions; for this profound reason. In studying Latin, required for certain professional studies, we learn the roots of words that echo ideas, emotions, feelings and deep thoughts
when we speak. With poetic permits we go from the standard definition of the encoded experience 'consternation' to 'that which stirs cultures and nations'; in other words what makes people uncomfortable or pisses them off. The simplest word in our lexicon with the most discomfort is…(drum roll please)…..G-O-D! There is no noun, pronoun, gender or discussions about the plethora of names to be used here when we cut through the quagmire and go directly to …….(drum roll please)……."HIMHERIT". Please leave your emotions on this page and move on in pure openness with "HIMHERIT" from now on. All the dimensions are covered. It rolls around the brain and the aura lovingly and peacefully.

Here is a thought process: Ask the questions and be on the quest with your request with these constructs, their associations and correspondences.

Predominant physical-physiology-elements: the universe and the solar system, stars and galaxies- the periodic table of elements-minerals-man's creations and destructions.

Body-physiology, psychology, spirit and spiritual nature: natural & artificial symmetry forms of expression; the kingdoms: man, animals, insects, plants.
Thought energy and thought form:
Pure energy: the four forces gravity, electro-magnetic, the strong and the weak forces
Work energy: carbon base fuels. Kinetic &potential energy. Thermal etc.
From the interactions and transformations of consciousness as the source of this minimum inventory of elements, dynamic symmetries, rituals and behaviors, spiritual energies, forces and thoughts springs life.

Through interactions and transformations of a minimum inventory of elements, dynamic symmetries, rituals and behaviors, spiritual energies, forces, flavors, creative processes, and effort with 'sound' reason and positive intention, intelligence, though love and compassion we connect to all that is greater than us and experiences.

CONSCIOUSNESS.

We have spent our objective, subjective and spiritual times, lives and 'hi-stories' in major distractions, not deciphering the communication codes to transcend the epochal links inheriting a disastrous chaotic planet. Transformations of the word 'hi-stories' from histories is meant to awaken us. We compound this with 'her-stories' for heresies by pushing the phoneme a bit. I suggest we include the experience of 'sound', to enhance the meaning' we transform 'phoneme to 'phenomena'.

How does Hydrogen become helium?
When a star has changed all of the hydrogen atoms (it fuses) into helium. It begins to convert helium atoms into carbon and oxygen atoms. All of the carbon was made inside stars. (We and all else are Carbon with our own star stuff). Carbon atoms are heavier (with more mass) than helium or hydrogen. They have six protons and six neutrons in the nucleus, with six electrons going around the exterior shell.

The six electrons cannot all go around the nucleus at the same distance, only two electrons can fit in the inner shell. Some of the electrons go farther away from the nucleus. A carbon atom has two shells, with two electrons in the inner shell and four electrons in the outer shell. That outer shell could hold as many as eight electrons. It's not full. That makes it very easy for carbon to combine with other atoms to make bigger molecules. A lot of carbon combines with oxygen to make carbon monoxide (one oxygen atom) or carbon dioxide (two oxygen atoms). All living things on Earth are made mostly of **hydro-carbons** (molecules of hydrogen, carbon) and **water** (molecules of hydrogen and oxygen). Both plants and animals are about 18 percent carbon. Plants get carbon by taking carbon dioxide from the air and breaking off the oxygen atoms. Animals (including MAN) get carbon from eating plants and other animals. Animals recombine the carbon with oxygen to make carbon dioxide, which is what you breathe out. When plants and animals die, their bodies gradually become carbon dioxide again. Most of our fuels are made of carbon Coal, gasoline, oil, and wood are all hydro-carbons. The carbon pours into the air causing global warming?

"Solar light force and the power of the eternal breath actually has power over the heart". The solar force and the eternal breath together is the source, like the polarity of light and mind. The biological and emotional heart have their own functions and expressions. The spiritual heart has its own. Is the solar light force the biological and emotional heart force and the power of the eternal breath the power of the spiritual heart? The brain has its link to the heart by way of plexes of neurons/ nerves. These three states permeate all realities in the physical, the psychological and the spiritual expressions of consciousness.
The consciousness state of one (ness) which is not the 1 of value, 'trifurcates' into three and returns to the conscious (ness state of one-ness) which is not the number one (1) of value or place. One is a (physical-spatial) thing. It represents feelings because of its position as is Psychological. As the ONE of all infinite it is the source as Spiritual-Energetic. A new principle we can gather from this would state that "All things or more fashionably 'Everything' has these three flavors".
The next would be; to be realized and experienced they must align with the four forces of the world we live in, which for the time being is still unfolding with its WIKI, leaving us with approximations that continue to change in time and with 'paradigms' that shift.
Through the creative process the one on a higher level of vibration or on a higher wave connects with its origin or source of arising.

All expressions follow the natural law of 3-three. When the six centers; the 7 minus the love, are distributed obeying the laws of 4; the 4 forces, the four elements, and all other four fold symmetry principles and processes. This is the inherent nature of the law of four. There are higher registers above the upper register we are more familiar with. This requires deeper thought to reach higher levels of consciousness. SOL (Spirit or son Of Light) or SOUL: Spirit Of Universal Light).

The Seven Hermetic Principles, upon which the entire Hermetic Philosophy is based, and two other canons are as follows:

BODY	MIND-CREATIVITY	SPIRIT
HERMETIC SYMMETRY PRINCIPLES	KNOWLEDGE	KUNDALINI
I. THE PRINCIPLE OF MENTALISM.	Paradigm 1 Knowledge is socially created and socially transmitted:	SAHASRARA Polarization, Ideation, Formulation
II. THE PRINCIPLE OF CORRESPONDENCE	Paradigm 2 Focus on knowledge generation and transfer: Be repository not depository.	AGNYA Vision, Imagination
III. THE PRINCIPLE OF VIBRATION.	Paradigm 3 Trust is fundamental to knowledge sharing	VISHUDDHI
IV. THE PRINCIPLE OF POLARITY.	Paradigm 4 Leverage knowledge through networks of people who love to collaborate	Opposites
V. THE PRINCIPLE OF RHYTHM	Principle 5. The need to solve problems will drive knowledge. This is external….. Power comes from demand not supply.	ANAHATA
VI. THE PRINCIPLE OF CAUSE AND EFFECT.	Paradigm 6 Focus on engagement and enablement (EMPOWERMENT)	NABHI
VI.SWITCH GENDER		SEXUAL CTR SWADISTHANA
VII. THE PRINCIPLE OF GENDER		
VII-CAUSE AND EFFECT].	Paradigm 7 Take an employee life cycle approach	MULADHARA

BODY	MIND-CREATIVITY
HERMETIC SYMMETRY PRINCIPLES	KNOWLEDGE
I. THE PRINCIPLE OF MENTALISM. All is Mind. Though is in all things.	Paradigm 1. The WIKI follows the BODY, MIND, SPIRIT Formula.
II. THE PRINCIPLE OF CORRESPONDENCE The Constant relationship seeker of interactions with source.	Paradigm 2 All Knowledge comes from source. We interpret it to align with our intention and needs.
III. THE PRINCIPLE OF VIBRATION. Constant motion distributes intelligence by transduction.	Paradigm 3 Truth and integrity are critical to sharing knowledge.
IV. THE PRINCIPLE OF POLARITY. Opposites: extremes of the same thing	Paradigm 4 Transform knowledge into the WIKI through collaboration in communities.
V. THE PRINCIPLE OF RHYTHM Nothing rests. There is constant motion in all things and everywhere	Principle 5. Needs, passion and Love drive knowledge. This is external…..Power comes from within and by focus, intention and will.
VI. THE PRINCIPLE OF CAUSE AND EFFECT. Nothing just happens Laws govern all things	Paradigm 6 Engagement and EMPOWERMENT are critical elements of B.M.S growth.
VI. SWITCH GENDER	Understand the animate and inanimate male and female symmetry and mechanical functions that make things work.
VII. THE PRINCIPLE OF GENDER	There is male and female in all things
VII-CAUSE AND EFFECT. Every action has an equal and opposite reaction.	Paradigm 7 Respect for life is respect for self and source.

2. **Principles of Ethics:** 'Ethical behavior helps protect individuals, communities and environments, and offers the potential to increase the sum of good in the world. As 'DeSign Scientists' making the world a better place' is our mission.
3. **We should avoid** (or at least minimize) doing long-term, systematic harm to those individuals, communities and environments...' (Israel and Hay, Research Ethics for Social Scientists, 2006) The three principles of ethics include informed consent, confidentiality and avoiding harm to do good.

- **Informed consent** is important that those participating in the research understand its aims and objectives and that informed consent is given, for research that is carried out with children or vulnerable adults, it is essential to acquire informed consent from a parent, guardian or responsible adult
- **Confidentiality**
  Confidentiality needs to be considered - how will confidentiality be maintained? is it always appropriate and applicable (i.e. criminal activities, if someone is in harm...etc)
- **Avoid harm and do good**
  Ethics can go so far as to suggest that research needs not only avoid harm, but to ensure that its purpose is to do good...how might this impact on the methodology of the research? And the impartiality?

BODY	MIND	SPIRIT
VIRTUE ETHICS: PARADIGMS	SOME MAIN PRINCIPLES HABITS	ASPIRING TO A SET OF VIRTUES.
Who am I?	Finding the right balance within and between values.	Integrity is a primary value.
How to Live My Life	Creativity: What are my gifts? I strive for excellence	Honesty, Love, Patience, Compassion
What kind of person do I want to be?	Is my behavior or (ARE MY HABITS) consistent with being a moral person and the person I want to be?	Identifying and Avoiding the set of counter-productive vices.
What virtues bring me closer to my goals;	Which vices prevent me from achieving my goals?	What impact is my behavior (HABITS) having on the world?
What virtues distract me from my goals;	Which values attract me to achieving my goals?	What impact do I want my (HABITS) to have on the world?
Triggers and Pain points	Balance the polarity of habits	Centers and Pleasure points

We seem to be getting confused with the identification of the BMS dimensions of consciousness and as a result have traditions of mis-identifying cause and effect relationships of fundamental phenomena life relies on.

Seven Principles of Knowledge Continuity: Paradigm shifts are the engines of continuity

"Knowledge creates and makes resources mobile. Knowledge workers, unlike manufacturing workers, own the means of production, their intellectual property and product: they carry their knowledge in their heads and therefore can take it with them. At the same time, the knowledge needs of organizations are likely to change continually as paradigms shift" Peter Drucker

PARADIGMS	SHIFTS
Paradigm 1 Knowledge is socially created and socially transmitted:	It's a celebration of our humanity and a people affair.
Paradigm 2 Focus on knowledge generation and transfer: Be repository not depository.	"Intrapreneural" learners, build and create new knowledge and insights for today's and tomorrow's world.
Paradigm 3 Trust is fundamental to knowledge sharing	Build trust and people will share and create new knowledge.
Paradigm 4 Leverage knowledge through networks of people who love to collaborate	Questing – always BE searching for new knowledge with Questions and Requesting.
Principle 5. The need to solve problems will drive knowledge. This is external….. Power comes from demand not supply.	Internally: SOURCE affords knowledge with spiritual alignment, respect and passion. Energy comes from within through the flow.
Paradigm 6 Focus on engagement and enablement (EMPOWERMENT)	Creating a workplace culture worth belonging to with PURPOSE.
Paradigm 7 Take an employee life cycle approach	Engage all staff in ALL knowledge sharing activities and problem solving

I. THE PRINCIPLE OF MENTALISM: The Principle of Mind (Light)

"The All is Mind, the Universe is Mental The finest energy of SOURCE is thought. Mentalism is a principle which is the expression of THOUGHT. It is the highest level, finest quality and purest energy we (can only) sense and therefore believe we know. Laws of nature fold or shape into folds (species and phyla to 'enfold' into themselves as their inherent nature. There is no-thing beyond them that affects them as they affect all transductions, iterations and expressions of themselves (as Maya) to reflect and generate what we call reality. 'Reflection' here implies the presence of 'light' in the 'field of light mind' which is the matrix for all the "ISMS". They are not causal. To 'generate' implies creation (idea) and manifastation (matter) along with feelings (emotions) and Spirit (energy). They are results of projections of ONE fundamental phenomena, action or process we really do not know. There is only one LAW light. Light refracts into a plethora of essences we qualify or think (self talk) of and quantify or 'measure'(refer to some agreed to standard) with imagined or real gauges calling our responses 'vibrations'. Can we please let the Nun help us beat the incoherent drum here- conundrum? Conun-drum roll please! "All we know comes from the great and infinite unknown!

Here is a real 'sensible' clue that we can resonate with or to and therefore know'. **Transduction (psychology)** is what takes place when many sensors in the body [we add-here to mind and spirit in space-time and energy] to convert physical (physiological and movement), emotional (chemical) and energy (electrical) signals from the environment, our bodies minds and thoughts into encoded neural signals sent to (through) the central nervous system, the kundalini and the subtle systems along with the endorcine system. Not only are we not identifying causal relationships properly we are also chort changing or consciousness by ellimnating the mond and spirit fro the holistic process of life and wonder why we are in the state we are in now! 2/3's of our life is missing.

The oneness of infinite-consciousness hints to a logical intertrasformative dynamic, with correspon-dences of many versions applied to create diverse expressions from common elements. Nature seems to adore small numbers a its minimum inventory for the maximum diversity we experience as multiplicity and enormity. It would be interesting to find the center that causes this tendency. It might be the greed factor. There are a few correspondences we are concerned with that represent the Body, Mind and Spirit. 1. The Hermetic Laws we think of as symmtey principles since they speak ot the arrangement of things that are thought of, manifeted or made. 2. The next is the Kundalini and the seven chakras which deals with the Spirit, Energy and Thought at the highest levels of consciousness. This also involves applications of creative principles, skills, talents and behaviors of humans, systems and matter as flavors and expressions of MIND and mental acuity and capability. Psychology and creativity are gifts we can be a lot more grateful for if we understood where they originated.

The Kundalini represents both the psychological and the physiological aspects of creation and manifestation or making directly related to design. The grand synthesizer of all froms of consciousnessis the first principle of harmony where Body, Mind and Spirit operate with the Space, Time, Energy and Plasma and the forces of their dynamic processes. Apart from the spiritual, metaphysical and phenomenal neters of the Kundalini, there is a physiological, material and workenergy for a creative process that corresponds to the infinite consciousness that moves the creative spirit to conceive, think and do work to develop an idea, formed in the mind; to plan; devise; to originate; to understand and satisfy need.

BODY	MIND	SPIRIT
Subsistence/Luxury	Protection/Security	Affection/ Love
Leisure/Pleasure	Participation/Co-Creation	Understanding/Wisdom
Creation/Manifestation	Identity/Quality	Freedom/All Forms

The Kundalini has a praxis we represent as the 'TKP'. This affords to experience the connections between Mentalism, the Crown Chakra and Ideation. They align with the second principle and its TPK and other correspondences.

For 'fun'-da-mental insights let's look at HOMONYMS; words that sound alike but have different meanings and Homophones which are a type of homonym that also sound alike but have different meanings and different spellings. HOMOGRAPHS are words that are spelled the same with different meanings. The word 'corres-pon-dence' is our example. 'Dence' (dense or dance) is a state. 'Corres' is to run. 'Pon' is on. We now paint the pictures beyond the literal interpretation of the normal use of language to explore deeper connections– not meanings. Poetry is visualization in correspondence with ART and life; through Body,Mind and Spiritual forms of vibration (energy) which are transduced into different expressions, emotions and states.

II. THE PRINCIPLE OF CORRESPONDENCE: The Principle of relationships.

"As it is above, so it is below; as it is below, so it is above"

It correlates, creates and seeks through thoughts and expressons to maintain alignments with SOURCE. There is direct commnication and transmission of Wisdom, Intelligence, Knowledge and Information that circumvents the rational processes of thought itself. When the grand process 'SOURCE' recognizes, the proper alignments through corresponndence; with all that is below it lets its precious RE-SOURCE or its WIKI flow directly as with his gift to the planet we know as $E=Mc^2$ for example. Cohesion is the purpose and reason for correspondence.

Without it we are compelled to invent systems to correspond with the imaguned hgher values we concuct and conmjure.

The TKP-and other correspondences here are the Third eye center qualities and characteristics. Imagination and Visualization are the experiences here. Creativity cannot happen without these skills. Thought, Mentalism, Visual and Imaginative skills transduces Ideation, induces, reduces and produces etheric energies that transform into denser expressions of thought energy. At this stage the thing being conceived is vey much the internal self talk conversation we call thinking. It is then prepared for the next stage that of communication.

III. THE PRINCIPLE OF VIBRATION: The Principle of Constant Motion-Movement
"Nothing rests; everything moves; everything vibrates."
Nothing is at rest. 'Direct Flow', creates movement, in all ubiquitous forms, qualities and flavors of SOURCE/ THOUGHT / CONSCIOUSNESS and is extrapolated into all the monikers of all our 'Babelian' vocabularies we use to describe the indescribe and define the indefinable. What that leaves us with is a scaffolds that do not last but limited periods of 'agreement and acceptance' until the next shiny one comes along.

Metaphors like the big bang, the word (from the big inning) all have one base element in them. They all make noise! The universe, planets, matter, foods people animals all make noise. Some more harmonious than others. Silence is in another deeper in accessible realm. Does the mind vibrate and does it make sensorial audible (human) detectable noises. Transduction is the phenomenon that refracts thought into expressions through all senses and energies into electricity.

Electricity here also has many levels. There is the 'thought-fuel' kind the' Love kind' and the 'work kind' that powers computers and light bulbs with many others in between these three.

Vibration is one of the expressions (properties) of the light that brought the infinite Universe from the very highest vibration of The Source, The First Cause, of 'God', down to the physical world of matter. Let there be-LIGHT'. Vibration is an integral component of the great continuum of the Universe, manifesting an infinite number of characteristics for constant creation. Energy is (not) vibration. It is one of its effects. The TKP-and other correspondences here are the throat center and the media used to share the thought; the conception and the idea. It is not a product as yet. For it to become one it must be communicated to solicit the responses and agreements which function as validation.

IV. THE PRINCIPLE OF RHYTHM: "Everything ebbs and flows, rises and falls;
Number (IV) four is (V) as the principle of polarity in the HERMETIC CANON. There are some compelling reasons for shifting this alignments here for better correspondence. For symmetry rhythm states that "Everything ebbs and flows, rises and falls; this best represents the TKP's Heart center and the dynamic of rhythm. It is the place for reality testing ideas in the practical and 'loving' sense; where we talk of 'loving what we do' but do not pay much attention to what this 'thing kind of love' is in contradistinction to the Logos: the 'Eros', the 'Philos' and the 'Agape' or the three hearts of the Kundalini-Chakra canon. In this Canon there are 'three hearts' a Left heart which is the drum and home of the gross body and pranic or etheric energies of the physical. This directs our creative fuel for the TKP.

The Middle heart is for unconditional love and spiritual maturity. The Right heart is called **the seat of the soul,** the source of identity; the intuitive place our "I" is. The pineal gland is another 'soul seat' given to us by Descartes.

First Corinthians 13:4-8 provides a perfect description for agape: "Love is patient, love is kind. It does not envy, it does not boast, it is not proud. It does not mind because it is all MIND; the Logos of all time. It is not rude, it is not self-seeking, it is not easily angered; it keeps no record of wrongs nor rights. It is compassionate and forgiving morning, noon and night. Love does not delight in evil, but rejoices with the truth. It always protects, always trusts, always hopes, and perseveres. Love never fails."

V. THE PRINCIPLE OF POLARITY: The Principle of complementarity

"Everything is dual, everything has poles; everything has its opposite; (COMPPLEMENTS): Human beings consist entirely of pure Energy in the form of a physical body, even though the human body and physical surroundings might appear to be solid in accordance with the five physical senses. There are subtler 'bodies' like the emotional or psychological and spiritual energy fields.

How can this be? Matter is composed of smaller units known as molecules, and still smaller units known as atoms, subatomic particles and strings. The ultimate constitution of anything is of pure Energy vibrating at specific frequencies that correspond to their unique functional and individual characteristics. On the physical plane this Energy is also known as "Light". What's behind this light is 'THOUGHT'; the 'SOURCE' energy. This is not the light commonly known in the physical Universe as emanating from The Sun, but rather the Sun behind the Sun or the 'let there be' or 'Light radiation' from which everything in the Universe was created, lives, its Being and becoming, the source of ALL. As observed by quantum physicist David Bohm, the physical Universe of matter can be considered to be "frozen light".

From 'The Source, The Prime Creator, God', down to the densest of matter there are an infinite number of modes of vibration. This is also now one of the basic accepted principles of quantum physics with the discovery that sub-atomic particles are ultimately not particles at all, but are rather progressively more subtle forms, until ultimately consisting of vibration, pure Energy and units of probability (thought).

Polarity is the tone of the music of the universe. It pairs the dualities of the creative elements that make matter into form, qualities obtain behaviors and their energies resonate with the right vibrations and frequencies. The work energy needed to produce what we make comes from the food we eat that in turn is enriched by the rays of the sun. This energy is a transduced version of the source of light with all of the subtle quantitative, qualitative and creative characteristics being held together by the thought process.

The TKP correspondences at this level is about energy. This is the solar plexus energy center that transforms food into fuel for the body to work on itself and in turn on its creative expressions.

Sound for is vibration. The frequency range of perception of the ear of the average human is between twenty vibrations per second to twenty thousand vibrations per second. On the vibrational scale electricity is around one billion vibrations per second. Heat is at two hundred billion vibrations per second. Visible colors is around five hundred billion vibrations per second. The invisible spectrum of color including infra-red and ultra-violet is still higher.

At the top of the scale we encounter x-rays which vibrate at around two trillion vibrations per second. The vibrations slow down as the density of the medium increases. The medium starts out at the highest SOURCE vibration frequency of its energy and slows down as it is transduced through the 'creation ray' each plane or matrix is characterized by a unique periodicity that corresponds with the typology, phylum or generation of the plane (earth) the thing, process, event or realization is manifested on.

On the scale of vibration we reach levels which do not manifest any characteristics known yet to modern science, and are therefore largely still ignored, overlooked or deemed not to exist at all; in other words vibration is often deemed by science to be finite as it pertains to the physical world and can be measured by scientific instrumentation.

Vibrating at speeds completely beyond the comprehension and observation of science are the inner dimensions of the planes of correspondence, The Great Astral, Mental and Spiritual planes. Attaining a high enough conscious level, one could rejoin The Source. Everything is Energy and therefore vibration. Everyone and everything therefore "tunes" into their own unique vibration.
If we look at these principles themselves as vibration facilitators and stimulators we can experience shifts in our consciousness corresponding to the planes they govern.
Let's view the 7 principles and their characteristics and levels in this new light; i Mentalism; ii. Correspondence; iii. Vibration; iv. Polarity; v. Rhythm; vi. Cause and Effect; vii. Gender.

What is required to activate these frequencies to entrain their energies to affect us positively? Science is not interest in human development, real people are. Does the understanding of these principles alone work? How do we implement each of these universal natural principles to our lives? If infinite consciousness employs these methods can we employ them as the correspond-dences for elevating our consciousness, applying the creative process and changing our habits to behaviors and states that align with the values we aspire to.

Since the levels of Energy can be raised higher on each plane, the finer will be the corresponding vibrations, and accordingly everyone and everything within that level of Energy or vibration will be much more "finely tuned" with it. Even physical energy-matter at the lowest level of Energy and vibration could be refined with the new aesthetic flavors generated by the heightened consciousness and growth. Each of these states have skill sets, activities, thought and creative processes we can implement in our own transduction to the manifestation of our own destiny and all that we can ART-iculate to represent new identities and expressions we cultivate.
Science recognizes the fact that heat, light, temperature and magnetism are all but differing degrees of vibration, each exhibiting unique characteristics, and each of which everyone can readily recognize and experience by means of the five biological senses and does not create a holistic strategy for development.

VI. THE PRINCIPLE OF GENDER.
"Gender is within everything; everything has its masculine and its feminine principles; gender manifests on all planes."
Gender is not 'sex'. It is a symmetry principle used to align complementary physiological polar elements using the analogy of male and female to describe the parts.
If 'symmetry is the behavior of parts that make things whole there are gender elements of opposing features that add functionality to devices and inventions for parts of do the work they are designed to do. A pin is a male element a hole is female. Pins and holes are used in all aspects of nature and in the manmade world. Since our focus is design Science we are concerned with gender as a design principle not sex.

In the TKP correspondence canon, sexual energy may play a role in stimulating creativity. This is another research project all its own.

What we need to remember is that somewhere along this path values play a critical role in the subtler aspects of these energy centers and states of being. Integrity, truth, compassion, honesty are signs we make decisions about as we journey on. This could be the most subtle and intimate aspect of one's attitude, altitude and sense of gratitude we can relate to the seventh principle of 'Cause and Effect'.

VII. THE PRINCIPLE OF CAUSE AND EFFECT.

"Every cause has an effect, every effect has a cause for the Karmic debt;

In the physical world correspondences this type are related to matter, forces, elements, their properties and behaviors. They follow strict symmetry rules when arranged in their useful logical sequences. When symmetries are violated beautifully conceived machines end up being the end up being 'as found sculptures' with misaligned gender elements; which I love to do.

Positive Vibration plays an extremely important role in the everyday lives of everyone by way of well thought out creative processes, spiritual alignments, values and states of being. Thought's, emotions, desires, temperament indeed any mental state of Mind or being are all ultimately degrees and aspects of Energy in the form of vibration, as is thought itself. Negative vibrations lead to stress when the limbic brain begins to overreact for the fight or flight stimuli we face daily.

Every single thought, every single state of Mind and every single emotion is characterized by its own unique vibration. It is an expression of infinite consciousness available to those who see it this way. In the physical world of matter these vibrations and "thought forms" are not always readily apparent to the physical senses, thoughts and thought forms can actually be seen to be instantly created and projected directly from source by intuition and other spiritual skills. Thought forms can be transduced to higher levels of Energy once there is clarity, intention and focus.

I THE PRINCIPLE OF MENTALISM. II. THE PRINCIPLE OF CORRESPONDENCE. III. THE PRINCIPLE OF VIBRATION. IV. THE PRINCIPLE OF POLARITY. V. THE PRINCIPLE OF RHYTHM. VI. THE PRINCIPLE OF CAUSE AND EFFECT. [VI.SWITCH GENDER.] VII. THE PRINCIPLE OF GENDER. [VII-CAUSE AND EFFECT].

The World Is Sound: Nada Brahma
Music and the Landscape of Consciousness
By Joachim-Ernst Berendt

We have only recently learned that the particles of an oxygen atom vibrate in a major key and that blades of grass 'sing." Europe's foremost jazz producer takes the reader on an exhilarating journey through Asia, Europe, Africa, and Latin America, exploring the musical traditions of diverse cultures and reaffirming what the ancients have always known--the world is sound, rhythm, and vibration.

WHEN YOU ARE AN INVISIBLE NOBODY YOU CAN DO ANYTHING;
BECOME A VISIBLE SOMEBODY AND EVERYTHING CHANGES.
THE BRIGHT LIGHTS [EGO] CAN DIM YOUR VIEWS.

PARADIGM SHIFTS DO SO HARMONIOUSLY.
THEY ARE EXPRESSIONS AND ART-ICULATIONS OF NATURAL LAWS
IF THEY DON'T THEY CREATE DISPLACEMENT, DISRUPTIONS AND CHAOS.
INNOVATORS MUST THEN DIFFUSE INFUSE THEIR CREATIVE WISDOM, INTELLIGENCE
KNOWLEDGE AND INFORMATION [WIKI] TO CREATE PEACE AND PROSPERITY
ON THE PHYSICAL PLANE; SELF REALIZATION ON THE PSYCHOLOGICAL PLANE
AND ASCENTION BACK TO SOURCE.
PARADIGM SHIFTS DO SO HARMONIOUSLY IF WE ARE IN SYCH AND IN TUNE WITH THEM.
IF THEY DO THEY CREATE NEW AESTHETIC EXPRESSIONS TO DEFINE THE ESSENCE OF
THE SPACE, TIME ENERGY OF THE ERA
INNOVATORS MUST THEN USE THEIR CREATIVE WISDOM, KNOWLEDGE AND
INFORMATION [WIKI] TO CREATE THE LOVE, COMMUNITY AND SYSTEMS FOR
WELLNESS IN THE SHIFTING MINDSET NEEDED TO ACCESS ALL OF THE ABOVE.

HERMETIC SYMMETRY PRINCIPLES

The Seven Principles or Laws of Nature

1.	**The Principle of Mind** "The All is Mind, the Universe is Mental"	This principle explains that Mind is the Universe's common principle; it is unique energy that is the essential force within the chemical composition of the elements.
2.	**The Principle of Correspondence** "As it is above, so it is below; as it is below, so it is above"	This law has to do with the similarity of wavelength of the various vibratory planes in the Universe. The same laws that govern dense matter act on subtle matter and vice-versa.
3.	**The Principle of Vibration** "Nothing rests; everything moves; everything vibrates."	This principle explains the differences among the manifestations of matter, the mind, and the Spirit.
4.	**The Principle of Polarity** "Everything is dual, everything has poles; everything has its opposite:	Like and unlike are the same; opposites are identical in nature but different in degree; extremes meet; all truths are but half-truths; all paradoxes can be reconciled."
5.	**The Principle of Rhythm (HEART)** "Everything ebbs and flows, goes up and comes down;	The pendulum swing is present in everything; the swing to the right is equal to the swing to the left; rhythm is the compensation."
6.	**The Principle of Cause and Effect** "Every cause has an effect, every effect has a cause;	Everything happens according to the law. Chance is nothing more than the name that is given to an unknown law; there are many planes of causation, yet none escape this law."
7.	**The Principle of Gender** "Gender is within everything; everything has its masculine and its feminine principles; gender manifests on all planes."	Gender is the impulse of life, which cannot originate or be maintained without the presence of both the positive and the negative pole.

NOTES: The left heart is the physical heart, the home of the gross body and pranic or etheric energies of the physical.
Here we inherit the rich earth and animal urges, limbic resonances and fascinations, pranic and erotic swells, sentimental dreams and romantic promises, conventional psychologies and consoling therapies, enthusiastic religions, and conventional, immature spirituality.
In other words, the left heart is as big as nature, and unfortunately, holds all of nature's capacity for delusion (even as it unconsciously grows the very Heart Itself).
The middle (center) heart says "I love you" without sentimentality, without needing a response or feeling the sense of promise. Coming to the middle heart from the left, we intersect the vertical (chakric) description of spiritual development; for the middle heart is the same as anahata, the fourth chakra -- real human maturity and integral existence.
The right heart is an esoteric secret, revealed by those who have exceeded animal lows, humanistic logic, and the yogic highs. Two finger-widths to the right of center, the heart on the right is called **the seat of the soul,** the source of identity, and the intuitive place we point to when we say "I".

The Heart is used in the Vedas and the scriptures to denote the place when the notion "I" springs. Does it spring only from the fleshy ball? It springs within us somewhere right in the middle of our being. Truly, ,'I' has no location, everything is the Self. There is nothing but that. So the Heart must be said to be the entire body of ourselves and of the entire universe, conceived as ,'I'. But to help the practitioner, we have to indicate a definite part of the Universe, or of the Body. So this Heart is point as the seat of the Self. But in truth, we are everywhere, we are all that is, and there is nothing else." -- Ramana Maharshi

"[…] it is apparent that several relationships exist between the pineal gland and retina. The similarities in development and morphology have been obvious for many years…. Although the mammalian pineal gland is considered to be only indirectly photosensitive, the presence of proteins in the pineal which are normally involved in photo-transduction [light sensing] in the retina, raises the possibility that direct photic events may occur in the mammalian pineal gland…"

"Life starts with the knowledge of diversity, but the awareness of (and the movement towards) unity is the pinnacle of life."
Hazrat Inayat Khan

Otherness is the seed of intelligence. HGB

CHAPTER 7 TRINE C

DESIGN SCIENCE INTERPRETED THE NEW PARADIGM OF DESIGN SCIENCE IS NOT EXACTLY NEW. DESIGNERS ALWAYS FOLLOW NATURAL LAW TO ADDRESS NEEDS, GENERATE FULLFILLMENT AND SATISFACTION THAT CHANGE THROUGH HISTORICAL AND CULTURAL AGES. EACH AGE HAS A UNIQUE IDENTITY, QUALITY AND AESTHETIC THAT TELLS OUR STORY. WE ARE IN A NEW AGE NOW PROMOTING THE BENEFITS OF A NEW PARADIGM MOVEMENT LEADING US TO AGREE THAT WITH DESIGN SCIENCE WE CAN CREATE REAL FUTURES".

The new 'Consciousness Paradigm' is a Design Science model of Body, Mind and Spirit; key fundamental elements of reality, life and energy. Every design we create is an expression of unique characteristics, qualities and flavors that describe us in our own time. [Experiences of] New design elements express the language of form, process and creativity to communicate meaning of space, cultures, iconic and archetypal designs. Design Science creates aesthetic inflections and reflections of the true nature of how things work. This is the domain of science where Design and Science are fused to create classics.

Designs are physical and in the domain of physiology…form. Design enhances behavior or psychology, with Essences and creative Energies that express our true aesthetic spirit with design excellence. These three elements express the very popular 'singularity principle' of 'oneness'. It is a fact or state of being one in agreement and alignment with natural law or what we now call a 'new paradigm (movement)'.

Esoteric symmetry principles of creation extrapolate the one into three. In pure extension, pure duration and pure energy. Three returns to the pure one; source of its arising, to an elevated level of 'synthesis' or consciousness. Three (3) is a 'canon' in a uniquely branded space, time and energy continuum as a general law, rule, principle, or criterion by which something is judged and understood.

TAG LINES;
A NEW PARADIGM MOVEMENT LEADS US TO AGREE THAT "WE CREATE REAL FUTURES"A NEW PARADIGM MOVEMENT LEADS US TO AGREE THAT "DESIGN SCIENCE CREATES REAL FUTURES"WITH A NEW PARADIGM AND/FOR DESIGN SCIENCE WE CREATE REAL FUTURES

Seven (7) True Benefits to using Design Science for 'ALL' your design needs.
Use the 7 chakras as principles for the 7 benefits. These are the macro creative principles of creation and on the micro levels of all natural things. They obey the universal law of seven (7).

THE DESIGN PROCESS

CHAKRA	BENEFIT	WE CREATE/ USE
The Crown	Ideation Sells	Exciting Design Science ideas that create powerful brand identities that sell.
The Third Eye	Imagination is the key to innovation	Imaginative Innovations based on sound design science principles and solutions.
The Throat	Communication, media	Excellent design is your message with meaningful messaging
The Heart	LOVE attracts All	Love is the universal force that attracts attention and intention
Solar Plexus	Energy is Efficient Work	We use all energies Efficiently is make working smart
Sexual Center	Creativity, and Gender Symmetry	Creativity is the life blood of design excellence for Form, fit and high performance.
Root. Matter Grounding	Earth, smart materials, will power	We make matter, energy and workable designs your key to our unique New Paradigm business Success Formula

Discovering Paradigms for Change.

How do we know about paradigms? Paradigms as world views emerge when philosophical shifts are made. What we believe, what we feel and think change for a number of reasons. The current shift has been with us for at least a century or more. Evidence of this can be found in the disciplines of science, new areas of mathematics and philosophy. Cultural traditions with their esoteric philosophies even the forgotten wisdom of some ancient civilizations often add insights to enlighten us about the consistency of laws and principles that transcend space, time and energy. One cosmological periodic shift or transition is going from the Piscean age to the Aquarian age. This brings with it changes in human consciousness we still do not understand. I believe the real quest-ion for humanity is; are we simply earth bound physical beings or universal conscious spirits visiting this planet and are given the opportunity here to deal with realities not available to us in any other dimensions of our spiritual lives beyond here and now? This is my quest and the motivation for my life and the works I do.

Being committed to expressing unique principles from any emerging paradigm shift we are experiencing in our life time is such a unique and appreciated privilege and a momentous occasion in human history. Transforming knowledge and ideas into new concepts and innovative technologies that add value to create new opportunities for human development in the visual and design disciplines is our guiding principle. This is an exciting niche. It's where talents are best suited and applied to many forms of expression (things, feelings and flavors).

The core value involves researching, articulating and applying the new knowledge obtained from the evolving consciousness *"PARADIGM"* we tap into to develop systems to transform all the elements of design, value engineering and production of viable products, structures and services.

These are some research and development initiatives that support businesses and developers in many other market sectors that can benefit and grow from this rich source of visual intelligence, new information, technologies and designs we create and offer from a vast matrix of real world applications that's available to developers, businesses and entrepreneurs.
By "making the abstract real™", we bring into reality relevant and exciting ideas, solutions and opportunities that are unique to our methodology and mind set shifts from the deeply rooted mental and critical creative reprogramming of our own creative minds and critical thought processes that we now share with the world. We walk the walk and talk a different conscious talk we now need to share to create the dialog for this community to engage in.

We extrapolate knowledge from disciplines of science, mathematics, metaphysics, philosophy, world cultures wisdom and esoteric traditions to synthesize it into useful knowledge, design concepts and technologies some of which are not yet fully realized and are still in the research and testing stages. There are proven concepts ready to be exploited and made available to developers in the design, manufacturing, new materials development, business, sciences, Education, training and technology sectors of our economy. What we have discovered is that this business model is the highest and best use of our problem solving technologies and methods. It is more effective than traditional methods in many ways. The uniqueness of new inventions and intellectual properties, with rich and exciting aesthetic identities, add significant value as game changers.

We are looking to collaborate with likeminded creative geniuses who can help with the tremendous task of publishing, expanding and implementing our 'matrix of viable conscious solutions' for human development. The quest encoded at the root of the 'quest-ions' we ask of ourselves, our world and our creative cap-abilities and potential set the course for our research, inquiry and curiosity that leads to finding solutions through specific and unique problem solving methodologies we create. They are perfected and we are ready to share them with the world. Some of these are proprietary and others are from the wisdom traditions and studies conducted through our research initiatives and from studying human development, progress and human needs. They shape our guiding principles and define our goals.

We transform traditional development methods of production and distribution that are ineffective and irrelevant. Our goal is to implement unique, game changing disruptive and exponentially valuable and effective solutions to "soft touch" and "high tech" design and manufacturing environments across many disciplines. We find common principles that are transformed by iterative variations, line extensions and other powerful identity rich media, data, visual information and three dimensional (3D) forms.

Having a vision that can be implemented with divinely inspired talents that can deliver promises, goods and benefits to fulfill people's needs is noble work.
Our moto, taken from Latin, "Laborare est orare"; means 'to work is to pray'. This vision is built on rich and powerful "Visual Intelligence Principles" scientifically, mathematically and creatively researched and tested at all levels of a fun R&D process that continues to evolve. Tools, critical thought processes, design strategies and technologies are used to solve problems and create viable solutions for people worldwide. We firmly believe that the challenges faced

with the reluctance to change, perceptually, practically and emotionally and can be successfully transformed with the application of our methods and demonstrations of benefits, value, satisfaction and joyful experiences people can experience by engaging with this program.

I am not aware of too many opportunities for creative people to participate in ground breaking opportunities with the level and kind of exponential and inclusive potential this program offers. If anyone knows of one please let me know so I can sign up and stop reinventing the wheel or the operating system or any other metaphor or 'metafive'. We are in the reality testing phase looking to build and expand a community of like minds to fully develop this paradigm shifting model and unique development concept.

The new paradigm is based on the notion that consciousness; the being all, the becoming and end all of ALL things can in fact be defined by three primary qualities that permeate all things, expressions and energies. This creates a threefold matrix or triad with Body, Mind and Spirit: Physiology, Psychology and Spirituality as the symmetry principles that nature and the universe use and resonate to on Macro and Micro (man) levels of reality. This to us is the interpretation of one set of laws of many we apply to our creative process. Culture is to us the blend of these three dimensions of consciousness. Not only the physical that we have mastered, not only the emotional or psychological, that we have not understood, not only the Spiritual or the religious that we have confused with true spiritual and creative process. There must be harmony and synthesis in and with all three dimensions. This Triad is the structure of thought, expressions in the structure of language and the creative energies we are inspired by to create our reality. This is the fundamental premise that this new paradigm gives us to get connected to ourselves to one another, to the universe, to source and here and now to our higher selves.

When we internalize this 'Triadic' concept we gain a higher level of understanding and change becomes possible. CHANGE is: C-Cultivating or Conditioning, H-Healing or Healthy, A-Attitudes and Actions to then N=Neutralize Negative, E-Emotions and Energies to do creative and productive meaningful work. The reward is also 'grace' not just dollars or things. There is no market for grace it seems. It will be one of the currencies of the new paradigm.

We have discovered deeply rooted concepts operating at the 'fundamental building block' or subatomic or 'Nano dynamic' levels of nature, behavior and thought form and their relationships to dynamic symmetry principles and forces. We employ a design science or morphological methods borrowed from formal disciplines to reverse engineer, intuit with empirical experiments, quantitatively and qualitatively to arrive at workable solutions.

The branding and mission statement for the "PARADIGM" program is an inspired concept called ....'Cultural Economic Development™'. It states that the goods, expressions and flavors people create satisfy needs and become economies.

Accounting, management and quantitative systems and strategies needed are developed and implemented to manage and conduct "business".

Emphasis is placed on funding and financial management, excellent healthy and efficient production, strategic marketing and effective sales, distribution, customer care and consumer research.

CONSCIOUSNESS

PHENOMENA	BODY	MIND	SPIRIT
Consciousness	Physiology	Psychology	Spirituality
Authority	Law	Discipline	Control
Manifestation	Vision	Desire	Focus
	Occasion	Opportunity	Preparedness
Wealth	Gold	Knowledge/experience	Power

Here is a very unique opportunity to participate in the evolution of a creative development strategy faced with challenges and problems that can be resolved and can be very effective in offering viable solutions and opportunities for entrepreneurs and creative professional who need innovative technologies and ideas to generate opportunities, offer benefits to communities and create wealth. This is what entrepreneurs do! The recognition of one's own 'power' is used to empower self as the prime form of achieving self-esteem and self-control. To 'Protect' the status quo without viable alternatives, ideas and solution to be offered and developed is not a viable creative solution or option for the long term.

'Resistance' to change is based on the lack understanding the true value and benefits of transformations. In many cultures where the threefold essences are understood as being natural Law, creation, preservation and destruction are part of the creative process. Change is part of this dynamic. Paradigms are meant to cre-ate (belief in motion) understanding of the world as it evolves. What we believed in no longer enables or empowers us to live the way we now need to, think, feel and interact with our physical, behavioral and spiritual realities, our environment and our community. Resistance on the other hand and in the context of embracing the paradigm shift offers strength and freedom. Not doing so leaves us stuck and unfulfilled.

Our world view (physically, emotionally and spiritually) is shattered. I can be restored to align with new beliefs, emotions and thoughts. Paradigm shifts generally start with revolutions and technology innovations, changes in physical and creature comforts; all impacted by and impacting external cultural and economic forces. They do occur on deeper levels of consciousness; on the super-sub and normal conscious awareness level when beliefs, emotions and spiritual understanding of things around us, things within us and things greater than us evolve. The environment created can them transform lives as it does but with the right messages and information. These shifts happen less frequently than the external forms of change and movements. They rely on epochal phenomena and take longer to manifest. We know about Art movements, Political, and cultural revolutions. We know little or nothing at all about the universal shifts that can totally transform humanity. The safest place to have this conversation is in an Ashram, church, temple or synagogue all holy places; not in the work place.

Preparing for the internal shifts involve lifelong experiences, study travel and exposure to the teachings of cultural wisdom traditions and a true sense of openness to explore the world and our relationship to it. The neuro-linguistic programming, mental, social, religious, political and general awareness of our environment and the forces operating within it are a major source of negative conditioning that's not conducive to personal growth, development, autonomy, empowerment. There are few avenues for body, mind, spirit reconditioning and self-awareness available that most folks are even aware of or can afford. The same materialistic oppressive phenomena that undermines our consciousness creates the vehicle that offers us solutions that are monetized that we must pay for. How do we get out of this trap? We might look at offering creative skills and meaningful opportunities to exponentially grow not only our consciousness and our wealth but all the parts of our life and the environments we desire to develop and in some cases need to cultivate. Systems of training that do not produce sustainable and consistent positive results are themselves to be transformed by this very PARADIGM shift. This is revolutionizing the internal and external realities of human life to respond to the universal intelligence to become part of the process of healthy human evolution with a grand sense of responsibility to harmonize this holistic vision we are proposing.

Becoming part of the problem and offering more of the same old tired solutions causes more pain than pleasure is hypocrisy! If this is the solution we claim it is, it must first solve the problems that causes 'unconsciousness' that interferes with accepting change as a natural attribute of the universe and of life as a gift that helps us grow in the first place. I always believed that we should fix ourselves first before we try fixing anyone or anything else. Become the mirror. Everything reflected in it will have your essence. A much needed community dedicated to the acceptance of the reality of a paradigm shift echoes the argument about global warming phenomena. The education and special interest about a "Paradigm Shift" has not convened and never will. It's not conscious.
Who's interested in changing the status quo? Raise your hands for the shifting paradigm. Be part of a community to determine your destiny.

These ideas must translate into calls to action to inspire creativity in the disciplines that the new paradigm offers us. The first line of engagement starts with "jobs". The attitudes about self- reliance and self-discovery in this movement are quite open and would expand the creative potential we are capable of manifesting for the good of all. The unique value of the Non-Euclidean form inventions synthesize information that is charged in many natural phenomena that become the purview of technical and human intellectual endeavors that become disciplines responsible for decoding the principles used to design, create and support life.

Play is an important element in education. The behavior of some of the properties of these forms lend themselves to new and exciting activities that we can all enjoy and learn from. Novelties, games, puzzles, construction and logic structures can be used as building blocks. The symmetry patterns and dynamic properties can also be articulated in the context of digital dynamic behaviors that can be programmed for the electronic engineering field.
Intellectual or 'smart toys' are not as popular as the lifestyle modeled activities found in action figures, board games and the panoply of opportunities for creative expression and learning. The fun derived from Play is universal and eternal; not restricted to youth. *"FORM"* is its language. Play does more than provide fun and diversion.

 *"The musician is very close to mysticism, far closer than the philosopher...because music is meaningful without any words; it is meaningful simply because it rings some bells in your heart..creates a synchronicity between you and itself, when your heart starts resonating in the same way, when you start pulsating in the same way." Osho, Philosophia Ultima*

D_ESign Science in The New Paradigm Age

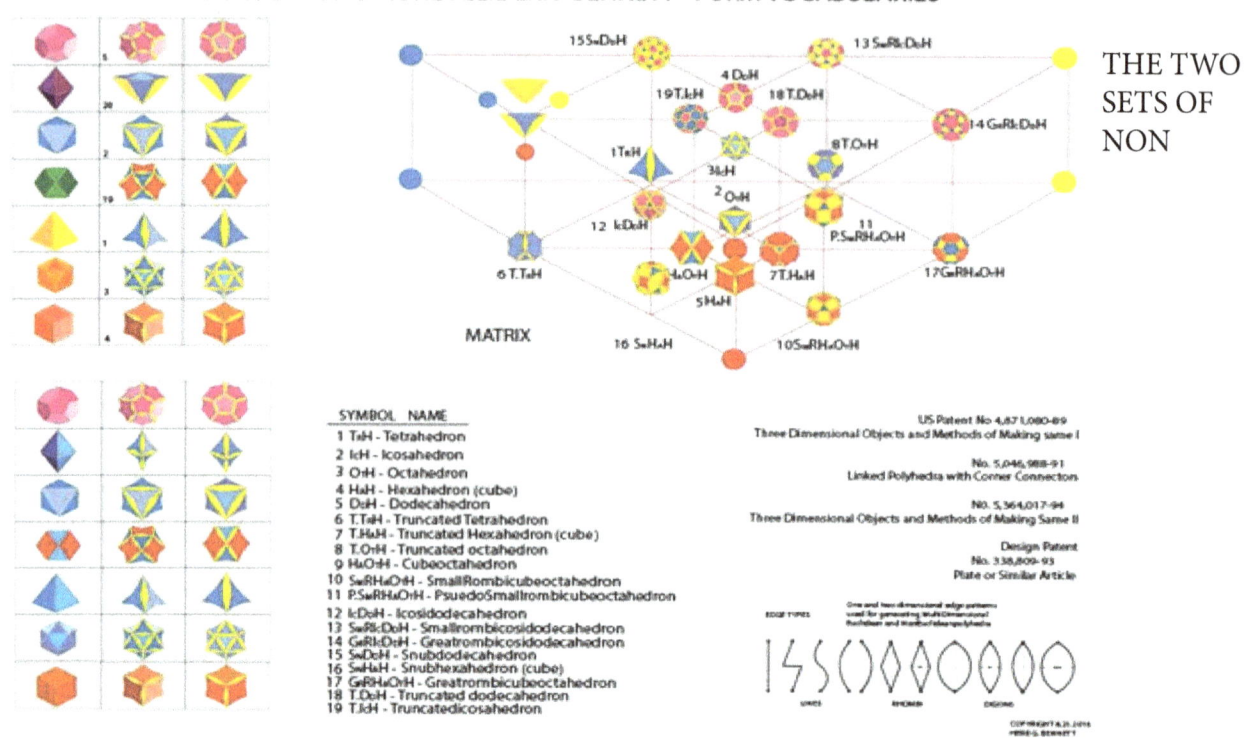

EUCLIDEAN POLYHEDRA
Fig 33

Here is the classic Euclidean vocabulary and the newly invented versions developed by HGB. Art, Mathematics (Geometry) and Science are systems of exploration, description and critical thinking highest 'forms of play' designers engage in to learn. The joy of learning or philosophy is the medium for the inspiration and passion that the finely tuned instruments we become express the celestial universal music of life and the infinite consciousness (source) we are all connected to. This can be fun and not stress.

1. ***Accelerated product and service market intervention:*** Original products and services are fast tracked by sub-contract production and delivery providers, joint venture, coop partnership marketing and distribution financed by orders and factoring.
   Develop fun products to plant the seed of the new form technologies that will be developed into serious applications later on as brand loyalty and capital is available. This set the tone for the success of this technology.
   Here we go for High volume-low cost items with an aggressive "Market Blitz and distribution" for significant profit.
2. Intellectual property Licenses of patented inventions, manufacturing and trade secrets.
   Here the interventions are small devices, midsize consumer related and household use that are "analog, electrical and digital".

Partnerships and subcontracting are the means of production and distribution. Here we go for medium volume-medium to high end items with an aggressive and specialty "Market and distribution systems" for reasonable profit.

3. Acquisition of existing low to borderline capacity entities operating unsuccessfully in the old paradigm mode, with an infrastructure that is adaptable. With an infusion of new systems, product ideas and technologies this is a "TURN ROUND STRATEGY".

4. A Straight Licensing 'Hands off' opportunity for advances and royalty compensation with any further consulting needed to be on a contract basis with a time limit and mutually agreed terms and conditions.

5. Knowledge and "natural" Logic Structures, Technology and Systems for design, Development, engineering, architecture. With-Through (Visual) Information and content publishing and distribution by HGB and other co-creators (Open Source?) Talent, genius and specialists.

6. Visual Science is the prime discipline that architecture has evolved into. It is applied to technology, media and communication, design and production and all other area of research and development, Services and information. THESE FORMS CONTAIN A VAST AMOUNT OF INFORMATION WE NEED TO MINE, PROCESS, ARTICULATE AND DISSEMINATE TO EDUCATE AND TRAIN ALL WHO PARTICIPATE IN THIS VISION. Transforming life support systems and needs for an emerging vision with anew aesthetic new expressions for and of lifestyle enhancement, comfort, ease, freedom, wealth building and legacy creation.

7. Research and Development: Exploration, travel and discovery to find applications for inventions and ideas being imagined and manifested with the New Paradigm Technologies we create and

STYLES OF LEARNING AND THEIR CORRESPONDENCE TO THE SEVEN CHAKRAS
1. Linguistic intelligence (as in a poet),
2. Logical-mathematical intelligence (as in a scientist),
3. Musical intelligence (as in a composer),
4. Spatial intelligence (as in a sculptor or airplane pilot),
5. Bodily kinesthetic intelligence (as in an athlete or dancer),
6. Interpersonal intelligence (as in a salesman or teacher),
7. Intrapersonal intelligence (exhibited by individuals with accurate views of themselves

NO.	CHAKRA	LEARNING STYLE
7	Sahasrara	Logical-mathematical intelligence (as in a scientist
6	Ajna	Intrapersonal intelligence (exhibited by individuals with accurate views of themselves
5	Visuddha	Linguistic intelligence (as in a poet),
4	Anahata	Musical intelligence (as in a composer)
3	Manipura	Spatial intelligence (as in a sculptor or airplane pilot
2	Svadhisthana	Interpersonal intelligence (as in a salesman or teacher)
1	Muladhara	Bodily kinesthetic intelligence (as in an athlete or dancer),

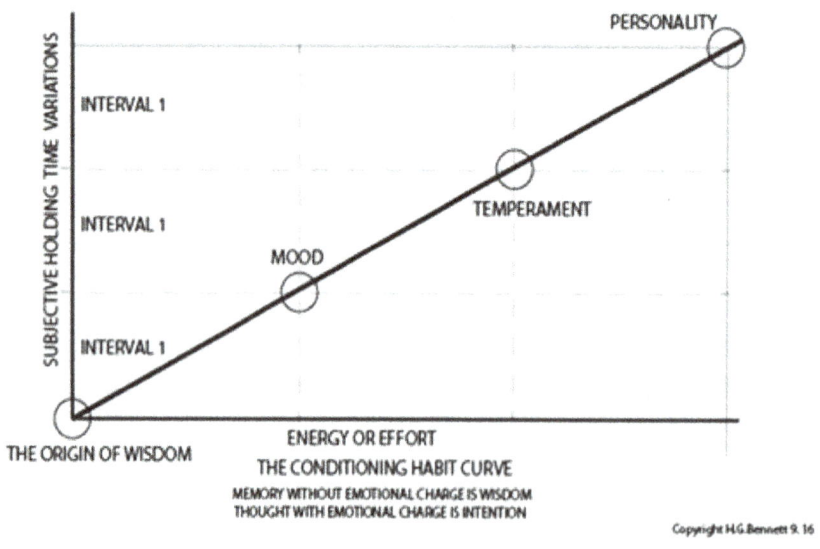

Fig 34

D_ESign Science in The New Paradigm Age

"*Design has matured from a largely stylistic endeavor to a field tasked with solving thorny technological and social problems, an evolution that will accelerate as companies enlist designers for increasingly complex opportunities, from self-driving cars to human biology. "Over the next five years, design as a profession will continue to evolve into a hybrid industry that is considered as much technical as it is creative," says Dave Miller, a recruiter at the design consultancy Artefact. "A new wave of designers formally educated in human-centered design—taught to weave together research, interaction, visual and code to solve incredibly gnarly 21st-century problems move into leadership.*

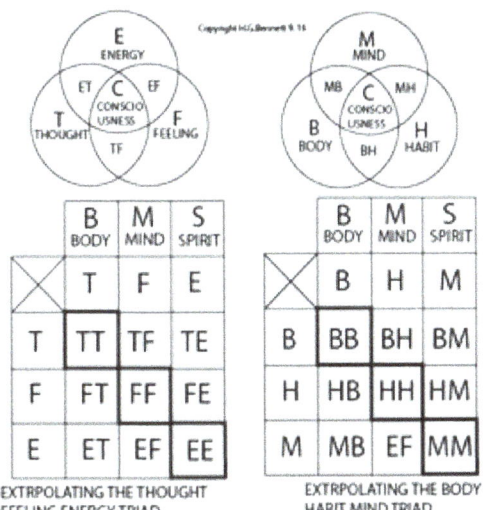

Fig35

They will push the industry to new heights of sophistication" and become 'D^ESIGN SCIENTISTS' R.B. Fuller 1957"

1. *As our research results indicated, a 'CREATIVE TECHNOLOGY' program can be taken to many different directions. For example, University of Twente focuses on smart devices and new media; Auckland University focuses on entrepreneurship and industry collaboration projects.*
2. *We should focusing on creating a creative technology program to utilize the communication design, strategy and planning, and creative technology expertise that we have in the department.*

3. *We should prepare CT graduates for specialized jobs for the future such as augmented reality designer, VR designer, avatar programmer, drone experience designer, cybernetic director, embodied interactions designer, human organ designer, content designer, digital product designer, simulation designer, tech/design Innovators, tech/design Entrepreneurs…etc. This will create a clear separation with our current graduates.*

BODY	**MIND**	**SPIRIT**
DIGITAL PRODUCT DESIGNER, TECH/DESIGN INNOVATORS, TECH/DESIGN ENTREPRENEURS	CYBERNETIC DIRECTOR, HUMAN ORGAN DESIGNER, EMBODIED INTERACTIONS DESIGNER, SIMULATION DESIGNER,	as augmented reality designer, VR designer, content designer, avatar programmer, drone experience designer,
PATHS THAT DEPEND ON MATERIALISTIC AND PHYSICAL LABOR INTENSIVE THINGS AND MECHANICS	PATHS THAT ARE BEHAVIORAL; RELATED TO FEELINGS: CREATIVITY AND INTELLIGENCE	PATHS RELATED TO THOUGHT ENERGY AND HIGHER STATES OF CONSCIOUSNESS

D_ESign Science in The New Paradigm Age

Chief Design Officer or Chief Creative Officer	Intelligent System Designer	Program Director
Conductor: Creative Technology Researcher	Machine–Learning Designer	Synthetic biologist
Fusionist	Real–time 3-D Designer	Interventionist, sim designer
Director of Concierge Services		
		Nanotech designer,

Where is the basic understanding of the visual sciences, visual mathematics like geometry with topology and 3D Form literacy in this vision of "creativity future™"

4. In his 2016 Design in Tech report, John Maeda shared his thought on how design schools should prepare data-oriented, coding-enabled graduates with business acumen who will become the future computational designers.

5. John Maeda believes the primary difference between computational design which is quantitative logic and classical design which is qualitative in nature are in fact how the computational designer will create living systems such as the Android Material Design Language or new culture/economy paradigm such as Airbnb. As in "quantum dynamic theory" where is the observer in this equation? How are observers impacting the experiments and outcomes?

Principles of the New Paradigm:
An Unbroken Wholeness

1. Everything is related to everything else.
Every organism, community, business, organization, or political problem is defined by its relationships to the rest of life.
2. Humans are not separate from the rest of the natural world.
We are a beautiful strand in the fabric of life. The well-being of humans is intrinsically tied to the well-being of the Earth system.
3. Spirit (ENERGY) and Matter; (WAVE AND PARTICLE) while not the same, are inseparable.
The Universe is, from inception, alive with spirit, mind, or consciousness. Spirit is inseparable from what is manifest. It is the same reality from different perspectives, like the inside and outside of a house; we can neither separate them nor reduce them to one. Extremes are the same just different degrees of separation.
4. We *are* Spirit.
We do not "have" it, "earn" it, or "become" it.
We are Spirit as it displays through the great changing diversity in this space-time-energy continuum. We have not sinned nor fallen from grace. This is where we belong.
5. There is inherent value in all life.
The well-being and flourishing of life on Earth has value in itself, independent of the usefulness of the nonhuman world for human purposes.
6. The Universe is neither created nor maintained from the outside.
The whole unfolding is spontaneously emerging, self-regulating, and self-evolving.

7. In living systems, the whole is always greater than the sum of its parts.
How the whole evolves is directed by the experiences of the parts.
8. Diversity, not sameness, makes for healthy systems.
Life doesn't make "monocultures" out of people, cultures, or land. Monocultures always collapse because of lack of diverse feedback in a constantly changing world.
9. Life is intent on finding what works, not what is right.
In life, what "works" are interconnections that create more opportunities for relationship, diversity, complexity, and order.
10. Changes in small systems produce changes in larger, sometimes non-local, systems.
Each individual person, bee, car and tree influences the greater whole—and our behaviors may have consequences at great distances, in places we never thought.
11. Unlike mechanical systems, living systems do not change incrementally.
In life, change occurs through fits, starts, surprises, unseen connections, and quantum leaps.
12. A slight variation of initial conditions can cause huge, unexpected, and seemingly unconnected effects.
Think now of the popularity of Harry Potter, the toppling of the Berlin Wall, the ubiquity of Facebook, the rise of citizen journalism—in each of these instances, conditions produced unpredictable, large scale, and diverse effects on the world system
13. Every living system has two tendencies.
The first is a self-assertive tendency to preserve its individuality, and the second is an integrative tendency to function as part of a larger whole. Creating a dynamic balance between the two is required for any sustainable community or other living system
14. We co-create what we assume is reality.
The way we make sense of things is by looking for information that is meaningful to us in some way, given who we think we are. Since the New Story changes our perception of what it means to be human, it influences how we perceive reality—enabling us to see fresh and radically different solutions to our problems.
15. We cannot predict, direct, or control life systems.

"To understand the whole it is necessary to understand the parts. To understand the parts, it is necessary to understand the whole. Such is the circle of understanding." Ken Wilber, Eye of Spirit

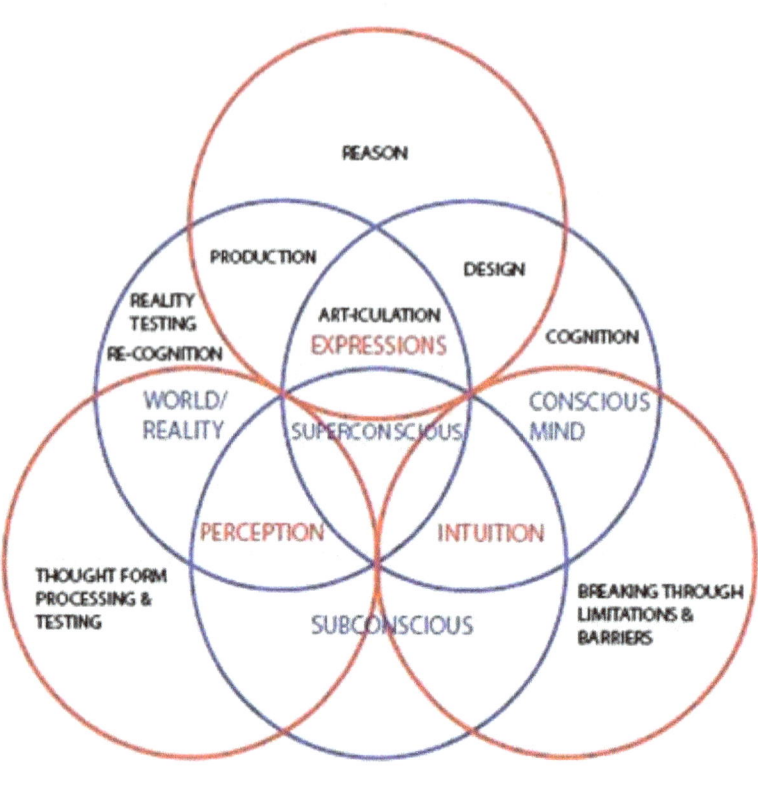

THE TRIAD

Design is a skill that uses Affirmations & Creativity to manifest ideas. Writing or inscribing thoughts is powerful. Using any media to record thoughts is highly recommended to enhance one's creativity and connection to higher levels of creativity. A journal format helps to visualize, ideate and author your desires. Affirmations are the follow up habits that reinforces the desires and visions, with focused intention and attention, revised as needed, to manifest whatever you focus on. This format is highly recommended.

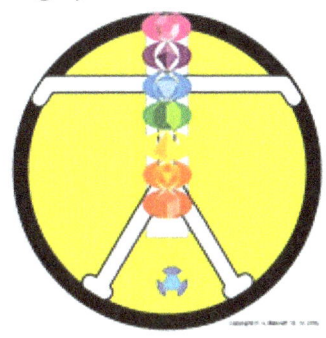

*Affirmation*_____

I [_____fill in your name here_____] & _____

Define Desires in this space_____
_____
_____
_____
_____
_____
_____
_____

_____as it's already done and I Declare into the Infinite Mind my clear intention and elevated emotion to manifest it. Every thought, written or spoken word in this desire is truth. It carries its own conviction with it. I visualize being transformed and I am in ease, gladness, joy and profound gratitude. I Declare this to be so NOW

I feel it to be the truth; I mentally assert that this [----------repeat Desire here------------] will find its way to me and will benefit me as the receiver of it; I feel that each manifestation is cared for by the spirit; that it is an articulation of truth and an expression power and that it will carry the conviction and the realization with it. When the word is written or spoken always and the image is visualized I feel that infinite Mind at once takes it up and never fails to act on it. Our place in the creative order is to know this and to be willing to do all that we can without hurry or worry and above all else, to trust absolutely in the spirit to do the rest. I see this most clearly and believe most implicitly that this will make the greatest demonstration of manifestation. This is me and this will be me as soon as the false thought is gone and the realization that there is but the one power and the one presence comes.

I am wrapped in an Infinite Love and Intelligence and I should cover myself with it and claim its protection from all evil. I declare that my word is law and my vision and desire is the presence and the activity of the power of all that is and I wait for the perfect concept to unfold". I wait patiently in Gratitude!
"Claim Your Power, Manifest Your Own Destiny and Legacy
For You, Your Family, Your Communities and our humanity"
A collective and infinite Consciousness can empower Individuals, Communities and Humanity.

DEDICATION: Ernest Shurtleff Holmes (January 21, 1887 – April 7, 1960) was an American New Thought writer, teacher, and leader. He was the founder of a Spiritual movement known as Religious Science, a part of the greater New Thought movement, whose spiritual philosophy is known as "The Science of Mind." He was the author of The Science of Mind and numerous other metaphysical books, and the founder of Science of Mind magazine, in continuous publication since 1927. His books remain in print, and the principles he taught as "Science of Mind" have inspired and influenced many generations of metaphysical students and teachers. Holmes had previously studied another New Thought teaching, Divine Science, and was an ordained Divine Science Minister.[1] His influence beyond New Thought can be seen in the self-help movement.

The Paradigm Shift and the Cultural Economic Development Strategy

Vision allows us to focus on *the 'chop wood and carry water'* mentality and the rituals of our journey in those trying moments when keeping the *'big or higher life' picture* in focus for the realization of a totally fulfilled 'you' to unfold. This unfolding continues by sharing knowledge, expertise and values in collaborations and in true relationships with people of like minds in the spirit of creative community who shift their paradigms often for their creative growth to continue.

How do we know about paradigms? Paradigms as world views emerge when real, psychological and spiritual or philosophical shifts are made. What we believe in, what we feel and think change for many reasons. The current shift has been with us for at least half a century or more. Evidence of this can be found in the disciplines of science, new areas of mathematics and philosophy. Cultural traditions with their esoteric philosophies even the forgotten wisdom of some ancient civilizations often add insights to enlighten us about the consistency of laws and principles that transcend space, time and energy. One cosmological periodic shift or transition is going from the Piscean age to the Aquarian age. This brings with it changes in human consciousness we still do not understand.
I believe the real quest-ion for humanity is; are we simply earth bound physical beings or universal conscious spirits visiting this planet and are given the opportunity here to deal with realities not available to us in any other dimensions of our spiritual lives beyond here and now?

Being committed to expressing unique principles from any emerging paradigm shift we are experiencing in our life time is such a unique and appreciated privilege and a momentous occasion in human history.

Transforming knowledge and ideas into new concepts and innovative technologies that add value to create new opportunities for human development in the visual and design disciplines is our guiding principle. This is my niche. It's where my talents are best suited and applied, I am an artist and an architect involved in many forms of expression.

The core value involves researching, articulating and applying the new knowledge obtained from the evolving consciousness *"PARADIGM"* we tap into to develop systems to transform all the elements of design, value engineering and production of viable products, structures and services. These are some research and development initiatives that support our businesses and for developers in many other market sectors that can benefit and grow from this rich source of visual intelligence, new information, technologies and designs we create and offer from a vast matrix of real world applications that's available to developers, businesses and entrepreneurs.

By "making the abstract real™", we bring into reality relevant and exciting ideas, solutions and opportunities that are unique to our methodology and mind set shifts from the deeply rooted mental and critical creative reprogramming of our own creative minds and critical thought processes that we now share with the world. We walk the walk and talk a different conscious talk we now need to share to create the dialog for this community to engage in.

We extrapolate knowledge from disciplines of science, mathematics, metaphysics, philosophy, world cultures wisdom and esoteric traditions to synthesize it into useful knowledge, design concepts and technologies some of which are not yet fully realized and are still in the research and testing stages. There are proven concepts ready to be exploited and made available to developers in the design, manufacturing, new materials development, business, sciences, Education, training and technology sectors of our economy. What we have discovered is that this business model is the highest and best use of our problem solving technologies and methods. It is more effective than traditional methods in many ways.

The uniqueness of new inventions and intellectual properties, with rich and exciting aesthetic identities, add significant value as game changers. We are looking to collaborate with likeminded creative geniuses who can help with the tremendous task of publishing, expanding and implementing our 'matrix of viable conscious solutions' for human development. The quest encoded at the root of the 'quest-ions' we ask of ourselves, our world and our creative cap-abilities and potential set the course for our research, inquiry and curiosity that leads to finding solutions through specific and unique problem solving methodologies we create. They are perfected and we are ready to share them with the world. Some of these are proprietary and others are from the wisdom traditions and studies conducted through our research initiatives and from studying human development, progress and human needs. They shape our guiding principles and define our goals.

We transform traditional development methods of production and distribution that are ineffective and irrelevant. Our goal is to implement unique, game changing disruptive and exponentially valuable and effective solutions to "soft touch" and "high tech" design and manufacturing environments across many disciplines.

We find common principles that are transformed by iterative variations, line extensions and other powerful identity rich media, data, visual information and three dimensional (3D) forms. Having a vision that can be implemented with divinely inspired talents that can deliver promises, goods and benefits to fulfill people's needs is noble work. Our moto, taken from Latin, "Laborare est orare"; means 'to work is to pray'.

This vision is built on rich and powerful "Visual Intelligence Principles" scientifically, mathematically and creatively researched and tested at all levels of a fun R&D process that continues to evolve. Tools, critical thought processes, design strategies and technologies are used to solve problems and create viable solutions for people world- wide. We firmly believe that the challenges faced with the reluctance to change, perceptually, practically and emotionally and can be successfully transformed with the application of our methods and demonstrations of benefits, value, satisfaction and joyful experiences people can experience by collaborating and engaging with this program. I am not aware of too many opportunities for creative people to participate in ground breaking opportunities with the level and kind of exponential and inclusive potential this program offers. If anyone knows of one please let me know so I can sign up and stop reinventing the wheel or the operating system or any other metaphor or 'metafive'. We are in the reality testing phase looking to build and expand a community of like minds to fully develop this paradigm shifting model and unique development concept.

The new paradigm is based on the notion that consciousness; to be all the becoming and end all of ALL things can in fact be defined by three primary qualities that permeate all things, expressions and energies. This creates a threefold matrix or triad with Body, Mind and Spirit: Physiology, Psychology and Spirituality as the symmetry principles that nature and the universe use and resonate to on Macro and Micro (man) levels of reality. This to us is the interpretation of one set of laws of many we apply to our creative process.
Culture is to us the blend of these three dimensions of consciousness.
Not only the physical that we have mastered, not only the emotional or psychological, that we have not understood, not only the Spiritual or the religious that we have confused with true spiritual and creative process.

There must be harmony and synthesis in and with all three dimensions. This Triad is the structure of thought, expressions in the structure of language and the creative energies we are inspired by to create our reality. This is the fundamental premise that this new paradigm gives us to get connected to ourselves to one another, to the universe, to source and here and now to our higher selves.

When we internalize this 'Triadic' concept we gain a higher level of understanding and change becomes possible. CHANGE is: C-Cultivating or Conditioning, H-Healing or Healthy, A-Attitudes and Actions to then N=Neutralize Negative, E-Emotions and Energies to do creative and productive meaningful work. The reward is also 'grace' not just dollars or things. There is no market for grace it seems. It will be one of the currencies of the new paradigm.

We have discovered deeply rooted concepts operating at the 'fundamental building block' or subatomic or 'Nano dynamic' levels of nature, behavior and thought form and their relationships to dynamic symmetry principles and forces. We employ a design science or morphological methods borrowed from formal disciplines to reverse engineer, intuit with empirical experiments, quantitatively and qualitatively to arrive at workable solutions.
The branding and mission statement for the "PARADIGM" program is an inspired concept called 'Cultural Economic Development™'.

It states that the goods, expressions and flavors people create satisfy needs and become economies. Accounting, management and quantitative systems and strategies needed are developed and implemented to manage and conduct "business". Emphasis is placed on funding and financial management, excellent healthy and efficient production, strategic marketing and effective sales, distribution, customer care and consumer research.

PHENOMENA	**BODY**	**MIND**	**SPIRIT**
Consciousness	Physiology	Psychology	Spirituality
Authority	Law	Discipline	Control
Manifestation	Vision	Desire	Focus
Now	Occasion	Opportunity	Preparedness
Wealth	Gold	Knowledge/experience	Power

Here is a very unique opportunity to participate in the evolution of a creative development strategy faced with challenges and problems (opportunities I call them) that can be resolved and can be very effective in offering viable solutions and lots of relevant opportunities for entrepreneurs and creative professional who need innovative technologies and ideas to generate businesses, offer benefits to communities and create wealth. This is what entrepreneurs do!

The recognition of one's own '*power*' is used to empower self as the prime form of achieving self-esteem and self-control. To *'Protest'* the status quo without viable alternatives, ideas and solution to be offered and developed is not a viable creative solution or option.
The '*Resistance*' to change is based on the lack understanding of its value and benefits to transformations or fear. Paradigms are meant to create (to internalize and digest ones faith) understanding the world as it evolves.
What we believed in no longer enables or empowers us to live the way we now need to, think, feel and interact with our physical, behavioral and spiritual realities, our environment and our community.
Resistance on the other hand and in the context of embracing the paradigm shift offers strength and freedom. Not doing so leaves us stuck and unfulfilled. Our world view (physically, emotionally and spiritually) is shattered. I can be restored to align with new beliefs, emotions and thoughts. Paradigm shifts generally start with revolutions and technology innovations, changes in physical and creature comforts; all impacted by and impacting external cultural and economic forces. They do occur on deeper levels of consciousness; on the super-sub and normal conscious awareness level when beliefs, emotions and spiritual understanding of things around us, things within us and things greater

than us evolve. The environment created can them transform lives as it does but with the right messages and information. These shifts happen less frequently than the external forms of change and movements. They rely on epochal phenomena and take longer to manifest. We know about Art movements, Political, and cultural revolutions. We know little or nothing at all about the universal shifts that can totally transform humanity. The safest place to have this conversation is in an Ashram, church, temple or synagogue all holy places; not in the work place.

Pre-paring (division and cision) for the internal shifts involve lifelong experiences, study travel and exposure to the teachings of cultural wisdom traditions and a true sense of openness to explore the world and our relationship to it. The neuro-linguistic programming, mental, social, religious, political and general awareness of our environment and the forces operating within it are a major source of negative conditioning that's not conducive to personal growth, development, autonomy, empowerment. There are few avenues for body, mind, spirit reconditioning and self-awareness available that most folks are even aware of or can afford. The same materialistic oppressive phenomena that undermines our consciousness creates the vehicle that offers us solutions that are monetized that we must pay for. How do we get out of this trap? We might look at offering creative skills and meaningful opportunities to exponentially grow not only our consciousness and our wealth but all the parts of our life and the environments we desire to develop and in some cases need to cultivate. Systems of training that do not produce sustainable and consistent positive results are themselves to be transformed by this very PARADIGM shift that is revolutionizing the internal and external realities of human life to respond to the universal intelligence to become part of the process of healthy human evolution with a grand sense of responsibility to harmonize this holistic vision we are proposing.

Becoming part of the problem and offering more of the same old tired solutions causes more pain than pleasure. It is hypocrisy! If this is the solution we claim it is, it must first solve the problems that causes 'unconsciousness' that interferes with accepting the change as a natural attribute of the universe and of life as a gift that helps us grow in the first place. I always believed that we should fix ourselves first before we try fixing anyone or anything else. Become the mirror. Everything reflected in it will have your essence.

A much needed community dedicated to the acceptance of the reality of a paradigm shift echoes the argument about global warming phenomena. The education and special interest about a "Paradigm Shift" has not convened and never will. It's not conscious.

Who's interested in changing the status quo? Raise your hands for the shifting paradigm. Be part of a community to determine your destiny and legacy.

BODY	MIND	SPIRIT
Life Support Systems and conveniences Ideas and Things of quality	Habits, Human Development, Ethics, Comfort and Knowledge	Energy, Effort, Spiritual Growth, Thought, Peace, Action, Meditation, Imagination and visualization

Continue the List	Continue the List	Continue the List
CODES:	A*-Individual	B*-Family
C*- Community	D*-Humanity	

Emotions

Affection, Anger, Angst, Anguish, Annoyance, Anticipation, Anxiety, Apathy, Arousal, Awe Boredom
Confidence, Contempt, Contentment, Courage, Curiosity Depression, Desire, Despair, Disappointment, Disgust, Distrust Ecstasy, Embarrassment, Empathy, Envy, Euphoria
Fear, Frustration, Gratitude

1. Infancy	2. Childhood	3. Adolescence, youth	4. Adulthood
1. Young adulthood		2. Middle age	3. Elderly (old age)

Self-awareness	Self-concept	Self-consciousness	Self-esteem	Self-image	Self-knowledge	Self-perception	Self-realization
Personality traits *Five personality traits*	Extraversion and Introversion	Agreeableness		Conscientiousness	Neuroticism Emotional stability		Openness to experience
Virtues of self-Control	Temperance - Self Control regarding pleasure	Good Temper - self-Control regarding anger		Ambition – self Control regarding one's goals		Curiosity – self Control regarding knowledge	
Frugality (also Thrift) – self Control regarding the material lifestyle	Industry - self-Control regarding play, recreation and entertainment	Contentment - self-Control regarding one's possessions and the possessions of others;		Continence - self-Control regarding bodily functions		Chastity - self-Control regarding sexual activities	

Personal values

Virtue – characteristic of a person which supports individual moral excellence and collective wellbeing. Such characteristics are valued as a principle and recognized as a good way to be. What are the virtues in embracing new paradigm movements?

Virtues of self-Efficacy	Patience	Perseverance	Persistence
Courage - willingness to do the right thing in the face of danger, pain, significant harm or risk	The ability to delay or wait for what is desired	The ability to work steadily despite setbacks or difficulties	The ability to continue or repeat a task in order to achieve a goal
Virtues of regard	Fair-mindedness - concern that all get their due (including oneself) in cooperative arrangements of mutual benefit	Tolerance - willingness to allow others to lead a life based on a certain set of beliefs differing from one's own	Truthfulness/Honesty - telling someone what you know to be true in the context of a direct inquiry
Virtues of respect	Respect - regard for the worth of others	Self-respect-regard for self-worth	Humility - respect for one's limitations
Social virtues	Politeness	Charisma	Unpretentiousness
Friendliness		Sportsmanship	Cleanliness
Virtues of kindness	Kindness - regard for those who are within an individual's ability to help	Generosity - giving to those in need	Forgiveness - willingness to overlook transgressions made against you	Compassion - empathy and understanding for the suffering of others
Vices	Anger – emotional response related to one's psychological interpretation of having been threatened. Often it indicates when one's basic boundaries are violated. Some have a learned tendency to react to anger through retaliation. Anger may be utilized effectively when utilized to set boundaries or escape	Jealousy – emotion, and the word typically refers to the negative thoughts and feelings of insecurity, fear, and anxiety over an anticipated loss of something of great personal value, particularly in reference to a human connection. Jealousy often consists of a combination of emotions such as anger, resentment,	Sloth –Laziness, disinclination to activity or exertion despite having the ability to do so. It is often used as a pejorative; related terms for a person seen to be lazy include couch potato, slacker, and bludger.

	from dangerous situations	inadequacy, helplessness and disgust.	
Seven Deadly Sin	1. Lust –emotion or feeling of intense desire in the body. The lust can take any form such as the lust for knowledge, the lust for sex or the lust for power. It can take such mundane forms as the lust for food as distinct from the need for food.	2. Gluttony – over-indulgence and over-consumption of food, drink, or wealth items to the point of extravagance or waste. In some Christian denominations, it is considered one of the seven deadly sins—a misplaced desire of food or its withholding from the needy	3. Greed – also known as avarice, cupidity, or covetousness, is the inordinate desire to possess wealth, goods, or objects of abstract value with the intention to keep it for one's self, far beyond the dictates of basic survival and comfort. It is applied to a markedly high desire for and pursuit of wealth, status, and power.
4. Sloth – spiritual or emotional apathy, neglecting what God has spoken, and being physically and emotionally inactive. It can also be either an outright refusal or merely a carelessness in the performance of one's obligations, especially spiritual, moral or legal obligations. Sloth can also indicate a wasting due to lack of use, concerning a person, place, thing, skill, or intangible ideal that would require maintenance, refinement, or support to continue to exist.	5. Wrath – also known as "rage", may be described as inordinate and uncontrolled feelings of hatred and anger. Wrath, in its purest form, presents with self-destructiveness, violence, and hate that may provoke feuds that can go on for centuries. Wrath may persist long after the person who did another a grievous wrong is dead. Feelings of anger can manifest in different ways, including impatience, revenge, and self-destructive behavior, such as drug abuse or suicide.	6. Envy – emotion which "occurs when a person lacks another's superior quality, achievement, or possession and either desires it or wishes that the other lacked it"	7. Pride – inflated sense of one's personal status or accomplishments, often used synonymously with hubris.

DESign Science in The New Paradigm Age

Body with its hint being *space*, answers the questions; who, where, what? Mind with its hint being *time* answers the questions when, how, how long? Spirit with its hint being energy answers the questions about states of being, consciousness and creativity. Action and energy are in these states or forms of expression.

Select the states that you relate to or that you think describe who you truly are now (Highlight in RED), those you are missing (Highlight in Blue) and those you think you need (in our favorite color). What else can we use these words for?

Vanity – excessive belief in one's own abilities or attractiveness to others.
Let's see if we can arrange these other constructs into their Body, Mind, Spirit, categories in the Matrix and this
Other; List of emotions, Acceptance, Altruism, Assertiveness, Attention, Autonomy,
Awareness,
Balance, Benevolence,
Candor, Cautiousness, Charity, Chivalry, Citizenship, Commitment, **Confidence**,
Conscientiousness, **Consciousness** Consideration, Cooperativeness, Courteousness, Creativity,
Dependability, Detachment, **Determination**, Diligence, Discernment,
Endurance, Equanimity,
Fairness, Faithfulness, Fidelity, Freedom, Flexibility, Foresight (psychology),
Gentleness, Goodness, **Gratitude**,
Helpfulness, Honor, Happiness, Hospitality, Humor,
Impartiality, Independence, Individualism, Integrity, Intuition, Inventiveness,
Justice,

Knowledge,
Logic, Loyalty,
Meekness, Mercy, Mindfulness, Moderation, Modesty, Morality, **Muchness**
Nonviolence,
Obedience, Openness, Order,
Peacefulness, Perseverance, Philomathy, Piety, Potential, Prudence, Purity
Reason, Readiness, Remembrance, Resilience, Respectfulness, Responsibility, Restraint,
Self-reliance, Sensitivity, Service, Sharing, Sincerity, Silence, Solidarity, Spirituality,
Stability, **Suchness,** Subsidiarity, Tactfulness, Tenacity, Thoughtfulness, Trustworthiness,
Understanding, Unity,
Vigilance, Virtue (ethics)
Wealth, Wisdom

Harmful traits (Emotional Suicide)
Abjection, Crime, Self-abasement, self-absorbed, Self-abuse, Self-blame, Self-criticism, Self-deception, Self-deprecation, Self-envy, Self-handicapping, Self-harm, Self-hatred, Self-immolation, Self-loathing, Self-pity, Self-propaganda, Self-punishment, Self-righteousness, Self-sacrifice, Self-serving, Self-victimization, Sexual self-objectification, **Stress***, Suicide.*

How do you Get What You Want; is what you want, what you need and does it align with your body, mind and spirit (creativity) or Consciousness in harmony?

Quantum theory tells us that there are unlimited possibilities for those who are, not just passionate and motivated, but have clear intensions, elevated emotions and focused visions. In the world of conjuring this might be somewhat useful but in the world of D^{E}Sign there are other skills needed for manifestation. Your life is created by more than just your mind. Your emotions and creative spirit are essential to the creative process on all levels of understanding we are experiencing. What you have attained in life, whether you intended it to be for you, your family, your community or the world, was conceived by thought as a response to stimuli and desire (D^{E}SIRE™). The principle at work here is *MENTALISM*. We attract what we think about. There are creative forces and agents involved in all the forms of manifestation of ideas. Some are overt, direct and physical. They are behavioral and psychological and other are 'spiritual' or creative involving covert processes we are not aware of. Humans operate on three; overt, inert and covert creative dimensions.

The word *'desire'* here is interesting. If we study the root of de-'sire', we discover a deeper relationship with this construct that relates to the definition of *"SIRE"*. A 'sire' is the male parent of an animal, especially a stallion or bull kept for breeding. Can we extract FROM this then that the reproduction process might be implicated in this context? Can creativity be reproductive and does it employ the metaphor of "sex" as a natural principle?

In the 'passive state of becoming' or Imagineering anything in our' hands off wishful thinking' habit or *paradigm*, the outer world reflects the inner world. To me this idea of reflection implies little direct or an inadequate amount of active participation. Satisfied and fulfilled people know what they want and *design ways* to get (obtain or achieve) it. Dissatisfied and unfulfilled people need to first acquire knowledge and implement it to design manifest their destiny and visions. Believing that knowledge (alone) is power without implementing it with clear intention and elevation does not 'animate' matter into desired form.

So what is it that you want?
Do you know what you want? Have you been exposed to it? Have you written your goals down? Do you focus on them and work towards them each and every day? Most people don't, and they can't find their joy. They haven't spent the time deciding what it is they want and secondly they don't write it out to affirm and define it. There are a reasons why people don't create goals; here are some of them.

1. Lack of exposure or familiarity with similar desires other successful people have executed. Have they shared their story with you? Do your family and friends share their values and their goals with one another and with you?
2. Goal creation and setting are not taught in most institutions.
3. The thought of having setting and acting on achievable goals is unfamiliar to most people. Goals have been forced upon them, their goals have been set by other people and traditions they belong to. They've been made to do things that other people want them to do.
4. They just wish and daydream. They don't make clear, specific, well thought out goals. They do not understand the spiritual dynamics of life and optimizing their fullest potential.
5. They are conditioned by fear. They are unaware of fear, stress and failure being fight or flight responses to the unknown. Traits that are deeply rooted in our limbic brain that were once useful. They are scared to set goals they cannot define and articulate. Fear is false

evidence appearing to be real. Risk can be calculated and taken with confidence. Failure is our best teacher. Every successful individual has failed several times.
6. Rejection stimulates anxiety, stress and fear generally from toxic people, places, situations and experiences. Please avoid them all. They zap your life juices.

Now is your time to sit down and get *very* clear about your wants, needs and desires; affirm and write them down in the format laid out on these pages. What do I truly want and the complement of that thought is what will it cost me if I do not accomplish or acquire it? Use your journal for anything you want or desire you experience. They don't have to be in any particular order. Don't censor them. They can be big or small. The point here is just broadcast *'BLAST'* it into the universe. Do not think of the logistics or 'the how' at this time. Just let the stream of conscious desires flow! What the ones that reflect who you truly are appear you will surely know! Peace!

The tongue, even though among the smallest of all the members of the body can do the most damage (James 3:5)

WORD-WOUNDS: words like "You can't do that. You'll never amount to nothing. You are so worthless" and others stick long and deep in our minds. The old saying "Sticks and stones may break my bones but words can never hurt me" is not true. Word-Wounds from sticks and stones can heal they wound us deeper than we think and any superficial wounds and with effort they can be healed.

A parent can do great harm to a child and that harm can easily follow them throughout their life. They can either live up (and above) to or down to (and beneath) the words that a parent speaks. Words that are not spoken can be very harmful like never hearing "I love you, you are special," and "I am proud of you."

By our own words we are justified, validated" or are condemned. Words can help or they can harm. We should be measuring our words carefully because we will have to give an account for them someday.
Words reveal what is in the heart, the mind and the spirit. When we express a thought we are projecting our view of ourselves and the world around us. This becomes our personal paradigm and the essence of our communication and our basic nature. Words cut deeply always undermining our integrity, virtues and values leaving long lasting scars.
Words are packets (forms) of information (in mind and in time) empowered by energy. They are encoded by human experiences and interpretations of the world and culture to affect creative thought and action. Vibrational waves of energy transmit meaning. The tongue of the wise brings healing.

Guard our Words
We have two ears and only one mouth for a very good reason; we ought to be listening twice as much as we speak. No one learns anything from speaking but wisdom and knowledge come by hearing. A person who guards their mouth will preserve their life but whomever "opens wide his lips comes to ruin."

Conclusion **Words are represent and define form, body, mind, spirit and consciousness; simply LIFE.** Chief Seattle (more correctly known as Seathl) was a Susquamish chief who lived on the islands of the Puget Sound. As a young warrier, His paradigm for words states that "Words have shadows". We can now expand the code to include many other properties characteristics, behaviors and powers words have that we need to respect.

Are the words we use being healthy or dis-eased? Are we polluting the entire quantum field of human consciousness and all our environments too?

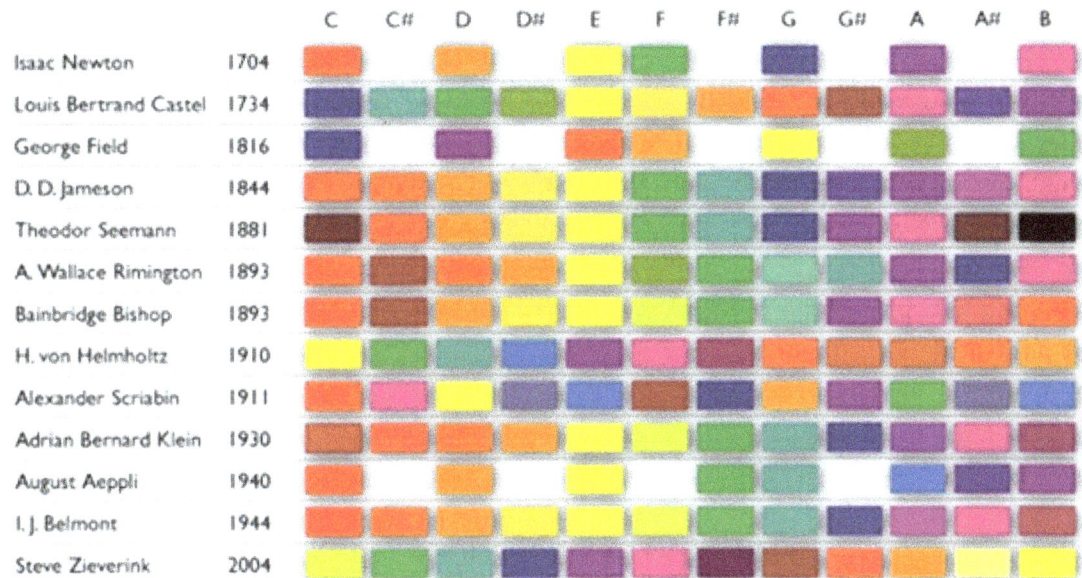

Fig 36

MANIFESTATION: 1 Corinthians 12:4-31 **ESV** "Now there are varieties of gifts, but the same Spirit; and there are varieties of service, but the same Lord; and there are varieties of activities, but it is the same God who empowers them all in everyone. To each is given the manifestation of the Spirit for the common good. For to one is given through the Spirit the utterance of wisdom, and to another the utterance of knowledge according to the same Spirit..."

COLLABORATION: 1 Corinthians 14:26 ESV What then, brothers? When you come together, each one has a hymn, a lesson, a revelation, a tongue, or an interpretation. Let all things be done for building up, through collaboration and co-creation.

Arrange the verbal constructs into their body, mind, spirit FORM with the categories in the matrix or List of emotions.

Acceptance, Altruism, Assertiveness, Attention, Autonomy, Awareness,
Balance, Benevolence,
Candor, Cautiousness, Charity, Chivalry, Citizenship, Commitment, Confidence, Conscientiousness, Consciousness Consideration, Cooperativeness, Courteousness, Creativity,
Dependability, Detachment, Determination, Diligence, Discernment,
Endurance, Equanimity,
Fairness, Faithfulness, Fidelity, Freedom, Flexibility, Foresight (psychology),
Gentleness, Goodness, Gratitude,
Helpfulness, Honor, Happiness, Hospitality, Humor,
Impartiality, Independence, Individualism, Integrity, Intuition, Inventiveness,
Justice,
Knowledge,
Logic, Loyalty,
Meekness, Mercy, Mindfulness, Moderation, Modesty, Morality, Muchness
Nonviolence,
Obedience, Openness, Order,
Peacefulness, Perseverance, Philomathy, Piety, Potential, Prudence, Purity
Reason, Readiness, Remembrance, Resilience, Respectfulness, Responsibility, Restraint,
Self-reliance, Sensitivity, Service, Sharing, Sincerity, Silence, Solidarity, Spirituality, Stability, Suchness, Subsidiarity,
Tactfulness, Tenacity, Thoughtfulness, Trustworthiness,
Understanding, Unity,
Vigilance, Virtue (ethics)
Wealth, Wisdom

EMOTIONAL EXPERIENCES

PURE EMOTIONS	SIMPLE EMOTIONS	COMPLEX EMOTIONS
Fear - nervousness - security	Discovery - anger	Pride - modesty - shame
Togetherness - privacy	Envy - no surprise - expectation	Closeness - detachment - distance
Respect - disrespect	Wonder - surprise - commonplace	Complaint/pain - doing average - pleasure
Appreciation - envy	Happiness - sadness	Caution - boldness - rashness
Love - no love lost - hatred	Amusement - weariness	Patience - mere tolerance - anger
Hope - despair	Courage - timidity - cowardice	Relation - composure - anger
Confusion - being confused	Pity - cruelty	Envy - goodwill

THE BENNETT TENETS:

THE PREMISE IS: DESIGN is a science related to the "methods" we now can deploy by knowing more about the world and how it works. THE PHYSICAL, MENTAL OR PSYCHOLOGICAL SPIRITUAL OR CREATIVE DIMENSIONS OF LIFE MUST BE HARMONIZED. We can no longer focus on materialism and ignore mind and spirit.

We evolve. Needs and methods keep evolving. Humans, needs and methods of satisfying them are evolving constantly. Growth, knowledge, information and innovation are the ingredients of harmony blending known as "Synthesis".

The opportunity cost of not participating responsibly in the transition from information to transformation is too great for us to pay now. There is nothing to pay it with. Participating or collaborating is a much better investment in our survival and destiny. We know some of the possibilities for success in doing the right thing for our civilization that is somehow deep in our sub-consciousness. One of the solutions to our dilemma might just be having the right language to describe, and communicate what we feel, think, see and know. The following concepts could be useful.

These I call 'tenets'; to rhyme with 'Bennett' to be synthesized as "Bennett's Tenets".

The CONSTITUTION for the New Paradigm movement in Synthesis.

BODY	SPIRIT	MIND
The First "Bennett Tenet" of the 'SYNTHESIS'- (PARADIGM SHIFT- RENAISSANCE) FORMULA	(in)Thought, work energy or effort	Is of Body, Mind and Spirit (as consciousness) or (Creative Spirit.
The Second "Bennett Tenet"	is the energy systems of the wisdom traditions 'modern physics, quantum theory', subatomic physics	This is the 'East-West Dichotomy' [with all its dynamic potential to unify the global consciousness].
The Third "Bennett Tenet"	Is deriving profound benefit	in wisdom traditions, esoteric knowledge and metaphysics, and interpretations [of meaning]
The Fourth "Bennett Tenet" into a praxis for 'making the abstract real'	synthesizes tenets 1, 2 &3	(where)Time changes as the patterns of life repeatedly transforms old methods, processes and energies into new technologies.
The Fifth "Bennett Tenet"	Exploiting a new FORM	Vocabulary of Non-Euclidean curved 3D geometric forms.
The Sixth "Bennett Tenet" an idea bank	could keep us busy and prosperous	for a long time to come that

The Seventh "Bennett Tenet Research and Architectural practice	focuses on transforming shipping containers into structures	other modular systems and AUTOMATED construction technologies for holistic communities as a wealth building strategy

This matrix is a linguistic tool that breaks down the "FLOW" of structural elements. Spirit as the verb is now at the center of the 'action' changing the tone of the expressions. Other arrangements are odd.

THE BENNETT TENETS for the New Paradigm Age

The First "Bennett Tenet" in the Synthesis FORMULA recognizes Consciousness as Body, Mind and Spirit (Creative Spirit, thought, work energy or effort.
Questions come from a much deeper place than answers which are buoyed in a sea of materialistic and mythic conditioning and thoughts. Of all questions we continue to ask, on our quest, without satisfactory or clear answers, the most salient ones to me are "Do we have the appropriate language to describe who we are, our place in the universe, our true purpose and how do we relate to these fundamental aspects of consciousness.

Especially when we rely on confusing concatenation posing as truthful communication. We do not upgrade many aspects of knowledge and information to keep up with paradigm changes, thus rendering our language somewhat obsolete. In looking at the definitions of words in our lexicon I find some not relevant to our time.

Who are the wordsmiths responsible for the accuracy of physical, psychological and spiritual experiences and communication? We would be quite surprised to know what forces invent our languages.

If consciousness is defined and accepted as Body, Mind and Spirit; the symmetry principles of living, this would set the understanding of this notion of 'movement' (of the human spirit, mind and body) into rapid and inspired motion. This motion moves in different directions synthesizing information that in turn is correlated with Body, Mind and Spirit in an ever expanding and continuous pure extension, pure duration and pure energy or spirit.

This 'Formula' permeates everything we think, see, say and create. This is a 'new' way or (Dao D^{e}sign) of looking at and applying an evolving design and aesthetic intelligence to a practical manifestation strategy to align with natural symmetry elements of culture and life itself as the core of 'consciousness', leading to higher states of being and becoming.

I attribute these qualities as evidence to a paradigm shift and this movement we need to activate NOW. We all need to be on board with this new agenda.

The Second "Bennett Tenet" is the 'East-West Dichotomy'. This is where the energy systems of the ancient wisdom traditions among various cultures that, if we are open, reveals links to 'modern physics, quantum theory', subatomic physics and many other disciplines offered to us by Paradigm that shift our basic nature. I would imagine this impacts our DNA as well as our neuro genetic makeup. This is all wonderful but forming the relationships to share this knowledge and information to make it more available must be part of the conversation about the new paradigm. Is there a grand paradigm with multiple sub paradigms within it? I would propose the Consciousness be Grand Expression of Infinite, normal awareness and sub-consciousness which then permeates through the filters and densities of manifestation into our material world. "Yesterday's mysteries become today's wisdom, knowledge, information and technologies. It interesting that in the real of real-ity four becomes apparent as the symmetry of nature and manifestation. This aligns with the four forces, four directions and many other examples of the Law of Four (4). This number idea will be explored later as part of the Law of Numbers.

The Third "Bennett Tenet" of the New Paradigm Movement through synthesis is the foundation. It relates (relies) on ancient wisdom traditions, metaphysics, other esoteric knowledge and new interpretations that change constantly. When the correlations with cultural information align awareness is confirmed and there are connection that bind human understanding.

The Fourth "Bennett Tenet" synthesizes tenets 1, 2 &3 into a praxis for 'making the abstract real'. Time changes as the patterns of life are repeatedly transforming old methods, processes and energies into new technology.

The Fifth "Bennett Tenet" Exploiting a new FORM vocabulary of Non-Euclidean curved 3D geometric forms is next. This is the intellectual property asset in the idea bank, the Intellectual Property agency, all services and products to be developed that are available to capable creative professionals through some novel means of technology transfer under development.

The Sixth "Bennett Tenet" is the idea bank that will keep us busy and prosperous for a long time to come. To create an avatar for this movement we can imagine Steve Jobs as Body R. B. Fuller as Mind and Mohandas Karamchand Gandhi as Spirit, fused into one BEING, how could their synthesized consciousness, intelligences and qualities inspire, motivate and help us create this movement.

The Seventh "Bennett Tenet is represented by my Research and Architectural practice. It now focuses on transforming shipping containers into habitable structures, among other modular systems and construction technologies for holistic communities as a wealth building strategy. This is an earth shattering paradigm shift of its own undoing in some circles. The reason for this message is to share this idea with you to explore how to create and implement a strategy for the promulgation of, not just my work, but the wider ken of knowledge I know other designers are currently exploring. We need to redefine the role design plays in our lives and how we communicate.

These Tenets represent a definite economic, material value potential that needs to be assessed, accessed and realized. Realization is the goal of innovation. Innovation is the synthesis engine. Doing this evaluation with the old paradigm methodologies is not the highest and best approach to arriving at the quantitative measure and the qualitative gauges the new paradigm thinking and methodologies can offer or need.

This is an earth shattering paradigm shift of its own undoing in some circles. The reason for this message is to share this idea with you to explore how to create and implement a strategy for the promulgation of, not just my work, but the wider ken of knowledge I know other designers are currently exploring. We need to redefine the role design plays in our lives and how we communicate.

These Tenets represent a definite economic, material value potential that needs to be assessed, accessed and realized. Realization is the goal of innovation. Innovation is the synthesis engine. Doing this evaluation with the old paradigm methodologies is not the highest and best approach to arriving at the quantitative measure and the qualitative gauges the new paradigm thinking and methodologies can offer or need.

D_ESIGN SCIENCE

IN THE NEW PARADIGM AGE

BACK MATTER

The Afterword	Page 187
Notes and Canons	Page 189
References	Page 190
Keywords: index and glossary synthesized	Page 194
Career Resource Guide	Page 195
Creating Community	Page 202
Collaboration Opportunities	Page 206
Epilogue	Page 212
Proverbs & Profherbs	Page 217
The Grand Periodic Qualitative Matrix	Page 224
Key Phrases	Page 238
Illustration Gallery	Page 312
The Design Library for Further readings:	Page 314

THE AFTERWORD

I must share my ideas and joy for what I believe in and do creatively that is critical to our survival as human beings in a world that needs solutions? Here is a mechanism for creating solutions that can transform our lives and our economies. This is what SOURCE compelled me to do but forgot to tell those who might need my help that I even existed. With the many times, avenues and opportunities I presented myself and my goods I was dismissed or not heard.

How can newly discovered information become accessible and useful? The issues we encounter on our journey do not belong to us per say. If we buy into them we have surrendered. Integrity is not to be relinquished or be taken for granted. It resides in the collective soul. The fire can always be re-kindled and acted upon to be the bridge to a shared, loved and passionate vision for a sustainable future.

The goal is to reveal the knowledge, the hidden principles and sources of information to inspire other creative designers, material scientists, producers and developers to recognize their own evolution and heightened consciousness giving them permission to adopt the new Paradigm ideas as viable solutions. Everyone who recognizes this quality in them has a voice, an expression and a flavor to offer the world and world to come. We are developing certain segments of this strategy too. My voice is the tone for three dimensional curved geometric forms. Knowing the logic structures that are the principles of how nature works cannot be monopolized by the Ivory Tower High Priests of an entropic western thought generating system that isolates itself from the rest of the planet's cultures, wisdom traditions and peoples.

Value 'engineering' visioneering and imagneering are the quantitative, qualitative and creative spiritual essences of form generating through "synthesis". It's how we experience Harmony. This supposedly abstract process helps create uses for ALL forms created or not. From this we can derive the value of the ideas for those who need them. Once this exchange is completed and all parties are satisfied economies are created. DO this enough. Do it repeatedly at levels of excellence and your authority will be attained; *don't be late; nor hesitate; Relate, Create a greater you!*

D^{e}sign Science: give us keys to The Unfolding, The Awakening and The Trans-forming of the New Paradigm Movement affecting our lives NOW. D^{e}sign Science is the grand synthesizer of all arts, sciences, knowledge and information evolving from the infinite consciousness or source now being art-iculated as an epochal shift of human consciousness in and through our space, time and energy continuum. It gives us a much deeper understanding of "CONSCIOUSNESS", ourselves and all of our expressions of that life giving force that connects us all to SOURCE. We attain higher states of being, becoming in harmony with all the principles and laws we seek to create our lives, systems of support for life and sustainable processes needed to pursue of desires, live and be happy.

D^{e}sign Science involves applying multidimensional thinking to use the existing resources we have to 'create' original Ideas, Knowledge and Information that aligns with universal laws, strategies and principles of Body, Mind and Spirit to synthesize holistic solutions, expressions and flavors of space, time and energy to get more for our investments and efforts. D^{e}sign Science allows us to access and harmonize the three states of normal, sub and infinite consciousness which we extrapolate into the physiology of matter; things, the psychology of all dynamic behaviors, feelings and the spirituality of thought creativity, direct knowledge, energy and (work) effort, balanced with all natural forces. D$_E$Sign Science involves a totally harmonized and balanced praxis™ that manifests, articulates and expresses solutions, ideas and thoughts through the use of the right materials with the proper applications of creative disciplines and the correspondences of spiritual Laws to create, support and attain higher states of creative excellence, optimized living and self-realization for the freedoms we enjoy.

D$_E$Sign Science uses all media, technologies and creative processes in the manifestation of all of the above that we bring below as true D$_E$Sign Scientists. The oneness of our vision and purpose from now on makes this our sole profession. Ownership of old paradigm definitions now disappear. Who am I

I am a D$_E$Sign Scientist inviting you to celebrate our evolution and freedom revolution in a safe space, with abundant resources, free creative expressions, and all right actions, thought, energy and lasting unconditional love.

There is a schism that can only be bridged by teams of like-minded, conscious, professionals who are experiencing and have survived their own transformations who can align with this vision and mission.

This team is the personification of transition in a "Matrix of Use" that can move towards implementation of the tenets and other policies, principles, value engineering, currency and financial capital formation needs with operations, management functions, marketing intelligence. On the social engagement side, networking with influencers, connected high network folks who are committed to this movement is a (preferred) asset unto itself. Such a collaboration would be incentivized by an autonomous type and level of partnership that goes beyond equity, debt or any other transactional and compensation or remuneration found in the old paradigm.

DₑSign Science in The New Paradigm Age

NOTES & CANONS

One of the many goals of this work is the establishment of a *'body, mind and spirit map'* for the journey of self-discovery, self-effort and self-knowing. We can transcend all the artificial divisions we encounter to enhance our lives. In this case a *'body, mind and spirit gps'* is needed. Fundamental principles and natural laws of the cosmic being in the DNA in us for example, that's ready to unfold and no longer be dormant. The awakening is long overdue.

<u>The Dao De Jing was referred to as a canon that inspired the 'Dao DeSign'</u> (system). It is used as an oriental harmonizer for the *'body, mind and spirit'* triad that in principle is part of a wider swath of oriental wisdom traditions. 'Vasistha's yoga', like the Dao De Jing is the second canon used to inspire the meditation and thought processes from the Hindu tradition. This storehouse of wisdom by Swami Venkatesananda "provides the means to eliminate psychological conditioning".

The next canon is the 'Temple in Man' and the 'Temples of Karnak' by Schwaller De lubicz. This is a three volume series focusing on the study of the geometric, spiritual Design science of temple and celestial architecture. This sets the tone for my Non-Euclidean Geometric inventions and patents. Gauge theories involving the proportions of man as the meter, with space and measure in the western form tradition is art-iculated with the canon of proportions for the human body with the Vitruvian Man by Leonardo da Vinci. The transition from the pharaonic era in Egypt to the Christian or Piscean age employs one of the key canons of mathematics and geometry for temple building; placing man in a hexagon described by the radius of a circle. In so doing "man is placed under the sky of the temple". In the second canon, "man is placed on the sky of the temple as the radius of the circumscribing circle".

I arrived at a similar two dimensional geometric expression by exploring and modeling, with cardboard curving 'interstitial spaces' centered in the voids of spheres in various stacked configurations, starting with the four spheres of tetrahedra. I later tested this 'paradigm shifting thought experiment' by transforming right angle cones (dixie cups) into extensions of curved form vocab-ularies conforming to the Non-Euclidean Geometry of Gauss, Bolyai and Lobachevsky

I received utility and design patents for these inventions. I was more familiar with Schwaller De Lubicz's 'number theory' before discovering these two canons corresponding with the two dimensional, curved triangular resolutions of space that were similar to my work. This research continues to generate insights and innovations in fields beyond my expectation and knowledge. As an artist *first* and an architect I have become an open and curious *'at it dude'* (I misspelt attitude and recreated myself) is who I have become. (My subconscious is smarter than I am.) As a going 'at it dude' there is no monopoly with knowledge I am attracted to. All fields are open and fair game; especially when I recognize and can interpret visual information I find using my trained 'eye' or visual intelligence like I did with the Egyptian canons and my other researches.

DeSign Science in The New Paradigm Age

REFERENCES:

-A-
Edwin a. Abbott: flatland

w. marsham Adams: the Egyptian doctrine of the light born of the virgin
d. g. Adler: the finding of the third eye
nur ankh Amen: the ankh: african origin of electromagnetism
rocky richard Arnold: the smart entrepreneur
khaled Azzam, keith Critchlow, prince of wales's institute of architecture. Study in the geometry of the arch in Islamic architecture (visual, islamic & traditional arts department,benjamin bold): famous problems of geometry and how to solve them

-B-
Richard Bach: jonathan livingston seagull.
michael F. Barnsley: fractals everywhere
roland Barthes: image, music, text
herb g. Bennett: co-author with brian Tracy: transform.
Itzhak, Bentov: stalking the wild pendulum, a brief tour of higher consciousness
william, Blake: poetical works edited by william rossetti, george bell & sons, london, 1891.
karl Blossfeldt: art forms in nature
gregg Braden: the god code
e. a. wallis Budge: the Egyptian book of the dead
brendon Burchard: the motivation manifesto
herbert Busemann: the geometry of geodesics-: convex surfaces

-C-
Fritjof Capra: tao of physics.
Joseph Chilton Pearce: from magical child to magical teen
robert Cialdini: pre-suasion
depak Chopra: the seven spiritual laws of success
george s. Clason: the richest man in babylon-amazon reprint ishi press
paulo Coelho: the alchemist
andrew Collins;'gobekli tepe; genesis of the gods'
theodore a. Cook: the curves of life
richard Courant: dirichlet's principle, conformal mapping and minimal surfaces
stephen r. Covey:The 7 Habits of Highly Effective People
h.s.m. Coxeter: Projective Geometry, Regular Complex Polytopes,
keith Critchlow: islamic patterns, into the hidden environment, time stands still:
the hidden geometry: k. Critchlow, Jon Allen: the whole question of health

REFERENCES (Cont'd):

-D-

The 'Dao DeSign'– inspired by the Dao De Jing the classic Buddhist canon,
Leonardo da Vinci: Vitruvian Man.
edward De Bono; lateral thinking: creativity step by step
schwaller De lubicz:'temple in man', the 'temples of karnak' and a study of numbers.
c. a. Diop: Pigmentation of the ancient Egyptians

-D-

Dr. Joe Dispensa: you are the placebo
clayton w. Dodge: euclidean geometry and transformation
peter Drucker: the social age of transformation

-E-

Tarek El-Bouri, Keith Critchlow, Salmá Samar Damlūji: islamic Art and Architecture:
hans, fischer, Ernst: geometry of classical fields

-F-

Joseph pierce Farrell: manifesting michelangelo
timothy Ferriss: the 4-hour workweek: escape 9-5, live anywhere, and join the new rich
richard buckminster Fuller: nine chains to the moon 1938, synergetics 1975, and it came to pass 1976, critical path 1981.

-G-

Jeremy Gray: janos Bolyai: Non-Euclidean geometry-geometry and the nature of space.
Non-Euclidean geometries of Carl Frederick Gauss (1777-1855), Nikolai Lobachevsky (1792-1856), Janos Bolyai (1802-1860), and Bernhard Riemann (1826-1866).
matila Ghyka: the geometry of art and life
samuel i. Goldberg: curvature and homology

-H-

Ernst Haeckel: kunstformen der natur: art forms in nature (lithographic halftone prints)
graham Hancock: magicians of the gods.
kenya Hara; designing Design
walker evan Harris: the physics of consciousness.
herman Hesse: Siddhartha.
Asa G. Hilliard III, Larry Williams, Nia Damali (Editors): The Teachings of Ptahhotep (The Oldest Book in the World)
alan Holden: shapes, space, and symmetry. Photographs by Doug Kendall
ernest Holmes & willis Kinnear: thoughts are things- Ernest Holmes: creative mind and success-the ernest Holmes papers.

I

REFERENCES (Cont'd)   -J-

James p. Jans: rings and homology
roger A. Johnson: advanced euclidean geometry

:-K-

jerry King: the art of mathematics "Touch [es] the mathematical grandeur that the first geometers contemplated." — *Publishers Weekly*

-L-

Haresh Lalvani; : patterns in hyper-spaces
solomon Lefschetz: algebraic geometry
marc Levinson: The Box

-M-

john Maeda; the Laws of Simplicity (Simplicity: Design, Technology, Business, Life)
edward Malkowski; the spiritual technology of ancient Egypt.
john Martineau: a little book of coincidence.
e.a. Maxwell: deductive geometry
bruce e. Meserve: fundamental concepts of geometry
kōji Miyazaki: An adventure in multidimensional space: the art and geometry of polygons, polyhedra, and polytopes
t. owens More: the science of melanin

-N-

O. Neugebauer: the exact sciences in antiquity
don Norman; the design of everyday things

O--

-P-

Joseph Chilton Pearce: from magical child to magical teen
m. scott Peck: the road less traveled psychiatrist
dan Pedoe: geometry a comprehensive course, geometry and
penney Peirce: leap of perception
luther Pfahler: coordinate geometry
peter Pesic: beyond geometry the visual arts
manfredo Perdigao: riemannian geometry
daniel Pink: Drive: the surprising truth about what motivates us
alfred s. Posamentier: geometry; its elements and structure

Q--

-R-

James Redfield: the celestine prophecy
tony Robbins: awaken the giant within: how to take immediate control of your mental, emotional, physical and financial destiny!

REFERENCES (Cont'd)   -R-

Anne Rooney: the story of mathematics
joe Rosen: symmetry discovered
don miguel Ruiz: the four agreements: a practical guide to personal freedom.
gretchen Rubin: the happiness project: or, why i spent a year trying to sing in the morning, clean my closets, fight right, read aristotle, and generally have more fun

-S-

Harold Scott, MacDonald Coxeter: the beauty of geometry: twelve essays
marc Seifert: transforming the speed of light.
lee Senella: the kundalini experience, kundalini.
rebecca Skloot: the immortal life of henrietta lacks
blake Snyder: save the cat.
barry Spain: analytical conics
lewis Spence: ancient egyptian myths and legends
saul Stahl: geometry from euclid to knots
shlomo Sternberg: curvature in mathematics and physics
robert R. Stoll: set theory and logic
d. m.y. Sommerville: the elements of non-euclidean geometry

-T-

John Thakara: how to thrive in the next economy.
eckhart Tolle; 'the power of now' published by new world library,
a new earth and stillness speaks.
brian Tracy: transform
amos, Tversky, David, Krantz, Suppes, Luce: foundations of measurement

U---

-V-

Swami Venkatesananda: Vasistha's yoga.

-W-

Warren k. Wake: design paradigms: a sourcebook for creative visualization
edward t Walsh: a first course in geometry
d'Arcy Wentworth Thompson: on growth and form
herman Weyl: symmetry
ken Wilber: the eye of spirit
harold e. Wolfe: Introduction to Non-Euclidean geometry
c. r. Wylie, Jr: foundations of geometry
marianne Williamson: return to love "a course in miracles".
h.g. Wells: world brain.

DeSign Science in The New Paradigm Age

X--

-Y-
Paul b. Yale: geometry and symmetry
paramahansa Yogananda: autobiography of a yogi.

-Z-
Gary Zukav: the seat of the soul.

PARADIGM HOOKS:

Paradigm 1 All Wisdom, Intelligence, Knowledge and Information (the WIKI) comes from source.
Paradigm 2. The 'WIKI' with the BODY, MIND, SPIRIT, is the prime life support system formula.
Paradigm 3 Truth and integrity are critical to accessing and sharing the formula.
Paradigm 4 Transforming knowledge through collaboration into the WIKI for communities.
Principle 5. Needs, passion, imagination and Love drive our current external knowledge. Solutions come from within. With our focus, attention and will we manifest reality.
Paradigm 6 Engagement and Empowerment are keys for growth.
Paradigm 7 Respect for life is respect for self, others and source.

KEYWORDS: index and glossary synthesized

A Form Paradigm: The Spiritual Creative thought /Thinking
The Psychological: Time Emotions/ Expressions
The Physiological: Matter Space/ (Gauge) Measure
NEW MAN: The Bio, Electrical, Plasmic, Magnetic Complex
With the KUNDALINI and connected to the 'universal KUNDALINI or CONSCIOUSNESS' as our vital creative source and living processes ART-iculated as (human) expressions of the self-image of all creative forces -THOUGHT-
Observations, participation, and Experimentation: Discovery and other creative strategies.
CONSIDER ART as Creative Process (Mind and Thought Processes) ART–iculated in Space, time and energy. Research, Observation, participation based on a Foundation of mathematics (once a spiritual endeavor now a tool for keeping score), Core Flavors: Muchness, Suchness and Consciousness.
STATE of the ART: STATEMENT- STATE-MEANT-ALIGNMENTS AND COINCIDENTS in harmony with space, time energy and fused into the holistic dynamic of consciousness. Mind, Body and (Creative) Spirit are the three dimensions of consciousness. Is the word 'di-mension' inherently dualistic?
FORM: 'Thought Form'- True Form
Systems of Geometry: Euclidean, Non-Euclidean, Organic, Amorphous (irregular or intuitive)
SYMMETRY PRINCIPLES: Operations and Behaviors, Symbols, Structures, things and objects all
EXPRESSIONS Words, Gestures et al
M O V E M E N T = Vibrations
CREATIVE PROCESS
LOGIC STRUCTURE/S: The Flavor Machine

Periodicity-Number properties:
Physiology-Things that are counted to keep score
The sacred values, qualities and properties of number
VALUE: Psychology- (behavior) of things and materials that are created
Spirituality- Thought Form-Symbolic meaning; Natural Law, SYMMETRY-Logic, Intelligence - Intuition
Flavors Form 3fold realities and expressions of nature like all natural phenomena in the world.
A Periodic matrix follows with the qualitative (suchness), quantitative (muchness) and (Universal Consciousness (Mind) as the foundation for human development and sustainability.
All natural expressions, flavors and forces (realities) have a fundamental 3FOLD Nature (Neter) that corresponds with the three dimensions ('Trimensions') with all symmetry principles harmonized. The creative 'thought-form' processes are governed by (these) laws. They operate in the formulation, manifestation and correlative dynamics of identities, realities, uses and functions within the aesthetic flavors of all natural and artificial expressions. It is for this reason that periodicity exists, meaning that flavors are common to various phyla of realities and expressions that can be blended into a qualitative synthesis. The 'minimum Inventory-maximum Diversity' concept applies. There are few principles and elements that govern all reality.

PROJECTS:
Polyhedra- Generate Euclidean, Non Euclidean, Organic and Irregular (Artificial Forms) following traditions conventions and inventions that continue to unfold as our consciousness continues to expand.
Ancestral and Cultural Adaptations, Applications, Creative life support Systems Industries etc. Design Principles are at the core of all these Logic Structures from the origins of celestial alignments in monolithic ancient technologies to all useful space systems.

CAREER RESOURCE GUIDE
NEW PARADIGM OCCUPATIONS:

Augmented Reality Designer Nominated by Gavin Kelly, co-founder and principal, Artefact

As technologies for augmented reality evolve, they will allow for new information to be layered over the physical world in seamless ways. This will open up an increasing demand for designers who can deliver intuitive and immersive experiences that are tailored to a wide spectrum of industries, from entertainment to education and health care.
Avatar programmer *Nominated by Glen Murphy, director of UX, Android and Chrome*
Our celebrity clients will need help in representing themselves best in virtual scenarios such as VR, mobile games, and movies. This job will entail creating a celebrity's best representation in low-poly, high-polygon variants, and will depend upon rigging a client up for motion capture and text-to-speech emotive output. Some AI-response programming knowledge would be helpful. A version of this job actually exists today (see the digitized actors in L.A Noire), but will become increasingly important and complicated as actors' likenesses become more prevalent in games and VR. As these representations become more mainstream and more powerful, actors will want increasing control of their image, just as they have in every other form of media.
Chief Design Officer or Chief Creative Officer *Nominated by Yves Béhar, founder, fuseproject*
The CDO or CCO will be a job in every company, overseeing the design of a business's every touchpoint and solidifying a fluid visual narrative that can maximize efficiency and purpose. Design is more and more central to the success of the modern business; designers are no longer being brought in at the end of the process to make things look pretty, but rather are providing essential

insights from the ground up. In the future, I see a role on every executive team for a designer—someone whose role it is to ensure that every element of the business is designed well, and designed holistically.

Chief Drone Experience Designer Nominated by Gavin Kelly, co-founder and principal, Artefact
As companies such as Amazon deploy unmanned drones in their business, there will be an increased demand for the design of the entire service experience. For example, what are the end customer interactions? How are fleets managed and maintained? How are risks to the population mitigated? How are privacy concerns addressed? How do we build trust in these semi-autonomous machines? The next big thing is not a thing.

Conductor: Creative Technology Researcher Nominated by Bill Buxton, principal researcher, Microsoft Research Carrying on with the musical analogy, design has typically been preoccupied with creating new instruments. However wonderful any one of those instruments might be, the true potential is only realized when they play well together—essentially as one. It is the creativity and skill of the conductor that is essential to that happening.

The next "big thing" is not a thing. It is a change in the relationship amongst the things. Without the Conductor's input, we are on a fast path to hitting the complexity barrier, since the cumulative complexity of a bunch of simple things—regardless of how delightful, simple and desirable they may be—will soon exceed the ability of humans to cope. It is the Conductor who carries the responsibility for the design of those relationships and ensuring that their collective value significantly exceeds the sum of their individual values, and their cumulative complexity is significantly less than the sum of their individual complexities.

Cybernetic Director Nominated by Matías Duarte, VP, Material Design at Google
Cybernetic directors will be responsible for the creative vision and autonomous execution of highly personalized media services. They will train cybernetic art directors and visual-design bots in the distinct visual language of a brand. They will provide conceptual leadership on creative projects from starting point through execution, and will actively participate in the growth and development of machine-learning infrastructure to keep current with innovations.

Cybernetic directors will need to be well versed in the visual language and traditions of North American audiences and their subcultures.

The job requires at least four years of formal training in visual communication, graphic arts, modern American studies, or equivalent, and at least 10 years of relevant experience working in media, communications or entertainment. Exposure and familiarity with modern popular Western media is a bonus, but not a substitute. Also requires experience with learning systems training and reasonable fluency in HALtalk 9000, Lovelace++, and human-cyborg relations.

In five years machine learning will enable computers to make the kinds of aesthetic choices that humans make today—the more on the production end of the spectrum, the more quickly it will happen. This will enable massively more personalized experiences. Imagine reading a magazine article where the photo editor wasn't just aware of you as part of a broad demographic, but knew your visual fluency and consumption more intimately than your spouse. Yet who teaches the computers to make those creative choices? How do we balance the possibilities of personalization when each article wants to have its own editorial flavor, each publication its own style? Training and directing creative machines will be one of the most exciting and important creative jobs of the future. It's starting today.

Fusionist 'SYNTHESIZER' Nominated by Asta Roseway, principal research designer, Microsoft Research Early technology was, in its most basic form, like a huge block of ice: not very accessible, clunky, and necessitating specialists to handle. Now as technology melts, it will transform from solid to liquid to gas, permeating almost every aspect of our lives and creating a cross-disciplinary opportunities. Such diffusion will become the foundation for future design jobs. The designer's role therefore will be to act as the "fusion" between art, engineering, research, and science. Her ability to think critically while working seamlessly across disciplines, blending together their best aspect, is what will make her a "Fusionist."

While still expertly versed in classical design skills, the fusionist will mix those skills with a "generalist" approach to technology, working across disciplines and interest groups. In many cases, the fusionist may feel like an outlier. The technologies she bridges will require her to expand her own capacities. She'll need to be an expert collaborator and communicator, extending her vocabulary so that she can reverse engineer her vision into discrete items that specialists can act upon. The Fusionist will remain driven by her passion for the Future and her ability to use Design as the unifying vehicle to drive the best experience.

The prospect of artificially made human organs is just around the corner. Who's going to fit these organs to their end user? Designers.

The global challenges that lie ahead can only be solved by a collaboration of minds and vocations, and a diversity of views. The challenge and reward for the Fusionist will be in her ability to communicate, comprehend, and connect all parties through design.

This is already beginning to happen in the emerging fields of biofabrication and wearable technology. Stemming from biotech, biofabrication is a new cross disciplinary movement between the design and science that is generating the next wave of sustainable materials and solutions for our survival. It is not uncommon to see artists and biologists sitting together tackling the same problem. Additionally, wearable technology will see an influx of fashion designers and artists partnered with engineers, in order to create technologies that will go into our fibers and onto our skin. Fusionists will act as the bridges between emerging fields, and their ability to bring all parties together through communication and design will help bring about the greatest experiences.

Human Organ Designer Nominated by Gadi Amit, founder, New Deal Design

Human organ designers will be experts in bio-engineering and design, fitting newly created organs and artificial limbs to humans. They will be fully capable of executing end-to-end design and implementation process for ready-to-use or custom-fit organs; have deep knowledge of the software and hardware involved in bio-electronics; and work within a team tackling multiple biological sub-systems.

We are very close to being able to reproduce artificial biologically fitted tissues. Some of these tissues will come from genetic-engineering, some will be manufactured in bio-reactors, and some will be merged with micro-electronics. The prospect of artificially made human organs is just around the corner. Who's going to design and fit these organs to their end user? Designers will be there, sooner or later.

Intelligent System Designer Nominated by John Rousseau, executive director, Artefact

The intelligent system designer doesn't design discrete objects or experiences, so much as the software systems that make possible the design solutions of others. This designer works as part of a large and diverse network of specialists to create a continually evolving lingua franca of aesthetic production. The systems this person designs will integrate multiple domains, and those domains will themselves be the product of designers, artists, and technologists.

Cybernetic Director Nominated by Matías Duarte, VP, Material Design at Google

Cybernetic directors will be responsible for the creative vision and autonomous execution of highly personalized media services. They will train cybernetic art directors and visual-design bots in the distinct visual language of a brand.

They will provide conceptual leadership on creative projects from starting point through execution, and will actively participate in the growth and development of machine-learning infrastructure to keep current with innovations.

Cybernetic directors will need to be well versed in the visual language and traditions of North American audiences and their subcultures. The job requires at least four years of formal training in visual communication, graphic arts, modern American studies, or equivalent, and at least 10 years of relevant experience working in media, communications or entertainment. Exposure and familiarity with modern popular Western media is a bonus, but not a substitute. Also requires experience with *learning systems training and reasonable fluency in HALtalk 9000, Lovelace++, and human-cyborg relations.*

In five years machine learning will enable computers to make the kinds of aesthetic choices that humans make today.

In five years machine learning will enable computers to make the kinds of aesthetic choices that humans make today—the more on the production end of the spectrum, the more quickly it will happen. This will enable massively more personalized experiences. Imagine reading a magazine article where the photo editor wasn't just aware of you as part of a broad demographic, but knew your visual fluency and consumption more intimately than your spouse. Yet who teaches the computers to make those creative choices? How do we balance the possibilities of personalization when each article wants to have its own editorial flavor, each publication its own style? Training and directing creative machines will be one of the most exciting and important creative jobs of the future. It's starting today.

Director of Concierge Services *Nominated by John Edson, president, Lunar*
Retailers will harness the power of big data to give their most valuable customers a higher level of service than the general public. Smart merchants will start acting more like airlines or credit card issuers and really focus on the small percentage of VIP clients who drive a disproportionate percentage of profit. Concierges will provide the kinds of bespoke services normally associated with high net worth brands like American Express Centurion ("The Black Card"): exclusive perks but also customized products and services designed with an extra level of care to match the individual's tastes.

Embodied Interactions Designer *Nominated by Matt Schoenholz, head of design, Teague*
Screens have demanded a lot of attention from designers over the past 30 years. After all, they have been the source of so much content and so many interactions. They still require our thoughtful attention, but we will also see the rise of software that only rarely manifests on a screen. Or, perhaps it very much manifests on a screen, but the screen is an overlay on reality or it is outright virtual reality. These new modes of interaction require a new type of designer: one that is focused on embodied interactions.

Whether this embodiment is physical or virtual, this new designer is concerned with virtual and augmented reality, as well as the computers embedded into things and spaces. Therefore, this role is expert in interface pattern languages and touch-points that have largely been considered as alternative or merely subservient to screen-based GUIs.

This designer will borrow practices from industrial design and architecture, so that she can model interactions that are oriented in space.

While these new materials force the designer to be intensely concerned with formal and spatial qualities, they belie a behind-the-scenes complexity that is also paramount. The Embodied Interactions Designer must be comfortable wading neck-deep through datasets to mine value while protecting privacy. She must be adept at persuading disparate business stakeholders of a product's value and able to fight for the resources required to design it well. She must have foresight to uncover the biases in algorithms and large-scale systems that can negatively impact people.

Intelligent System Designer *Nominated by John Rousseau, executive director, Artefact*
The intelligent system designer doesn't design discrete objects or experiences, so much as the software systems that make possible the design solutions of others. This designer works as part of a large and diverse network of specialists to create a continually evolving lingua franca of aesthetic production. The systems this person designs will integrate multiple domains, and those domains will themselves be the product of designers, artists, and technologists.

Interventionist *Nominated by Ashlea Powell, location director, Ideo New York*
Interventionists are already in our midsts, we just haven't named the role or cultivated it. As

organizations and their challenges become more networked and complex, it will be harder work to help them digest new ideas and build towards a better future. This is the work of an Interventionist, and it's time that the craft of intervention design takes shape, whether it's designing an experience that creates transformational empathy or hosting a conversation that puts an end to polemics. These designers will come from backgrounds in organizational psychology or behavior change and be experts in facilitating creative conversations, framing unexpected questions, and navigating the uncomfortable.

Machine-Learning Designer, *Nominated by Aaron Shapiro, CEO, Huge*
A machine-learning designer's job will be to construct data models and algorithms that allow companies to create artificially intelligent products. Those products will anticipate the needs of users, and fulfill them before the user ever has to ask. Machine-learning designers must not only be designing the experience, but also ensuring that it uses the best algorithms. Data, design, and artificial intelligence will be the next frontier in digital experience. Companies will compete and win based on the personalization and intelligence in their marketing.

The companies that have the smartest, most individually resonant products and experiences are going to do the best job of attracting and retaining their users. In this world, good AI will become essential to the user experience and the companies with smart experiences will have an exponential advantage over the ones that don't.

Program Director *Nominated by Dave Miller, recruiter, Artefact*
This is the design agency's version of a product manager, and the evolution of its entire project management, engagement, and client services departments. This person is a business strategist; they understand the "who," "what," and "why" behind a project/product; have a deep understanding of what it means to be a designer and a developer; and also have a track record of effecting change and influencing the end product. They are peer-level to a design director and usually come from creative backgrounds. They have practiced design, research, or engineering. They share ownership of a projects success. They handle timelines and client interactions, while creating long-term relationships founded upon their deep industry experience.

Real–time 3-D Designer *Nominated by Dave Miller, recruiter, Artefact*
Virtual and augmented realities are on the forefront of design and technology explorations. Interaction design and game design will collide and integrate. Any design team tasked with creating a full experience in this realm will be in need of a 3-D designer.

Game design as an industry is such a focused discipline and craft: It takes years of practice to operate at a high level. With that in mind, senior level 3-D designers will be pioneers, leaving behind game design and joining product teams to create entertainment and productivity tools with complex interaction problems. We will start to see shifts in school curriculum, where both 3-D and UX disciplines share the same halls and work together to invent the same future.

Sim Designer *Nominated by Rob Girling, co-founder and principal, Artefact*
The sim designer pulls together customer data, behavioral models, and statistical models to design simulated people that can be used to help predict future customer behavior. In this way future products, ad campaigns, software, environments, and services are extensively "experienced" by artificial sim users who give sim reviews, tweets, recommendations, and predicted user data. These simulations help drive improvements into the design of all things before the product is ever realized. But would these sim insights replace talking to real people? I doubt it.

Synthetic biologist/nanotech designer *Nominated by Carl Bass, CEO, Autodesk*
In five to 10 years, we'll see current cancer treatment as totally barbaric. Chemotherapy kills all kinds of cells in the body, not just the cancerous ones. We're already on the path to creating customized medicine, and within five years synthetic biologists will be designing treatment that ties to the DNA of the patient. These medicines will be designed in software and printed on 3-D biological printers.

THE NEW PARADIGM OCCUPATIONS MATRIX

Designers	Nominated by	Need
Augmented Reality Designer	Gavin Kelly, co-founder and principal, Artefact	As technologies for augmented reality evolve, they will allow for new information to be layered over the physical world in seamless ways.
Avatar programmer	Glen Murphy, director of UX, Android and Chrome	Our celebrity clients will need help in representing themselves best in virtual scenarios such as VR, mobile games, and movies.
Chief Design Officer or Chief Creative Officer	Yves Béhar, founder, fuseproject	The CDO or CCO will be a job in every company, overseeing the design of a business's every touchpoint and solidifying a fluid visual narrative that can maximize efficiency and purpose.
Chief Drone Experience Designer	Gavin Kelly, co-founder and principal, Artefact	As companies such as Amazon deploy unmanned drones in their business, there will be an increased demand for the design of the entire service experience
Conductor: Creative Technology Researcher	Bill Buxton, principal researcher, Microsoft Research	Carrying on with the musical analogy, design has typically been preoccupied with creating new instruments.
Cybernetic Director	Matías Duarte, VP, Material Design at Google	Cybernetic directors will be responsible for the creative vision and autonomous execution of highly personalized media services.
Fusionist 'SYNTHESIZER'	Asta Roseway, principal research designer, Microsoft Research	While still expertly versed in classical design skills, the fusionist will mix those skills with a "generalist" approach to technology, working across disciplines and interest groups. In many cases, the fusionist may feel like an outlier.
Human Organ Designer	Gadi Amit, founder, New Deal Design	Human organ designers will be experts in bio-engineering and design, fitting newly created organs and artificial limbs to humans.

Intelligent System Designer	John Rousseau, executive director, Artefact	*The intelligent system designer doesn't design discrete objects or experiences, so much as the software systems that make possible the design solutions of others.*
Cybernetic Director	*Matías Duarte, VP, Material Design at Google*	*Cybernetic directors will be responsible for the creative vision and autonomous execution of highly personalized media services.*
Director of Concierge Services	*John Edson, president, Lunar*	*Retailers will harness the power of big data to give their most valuable customers a higher level of service than the general public.*
Embodied Interactions Designer	*Matt Schoenholz, head of design, Teague*	*Therefore, this role is expert in interface pattern languages and touch-points that have largely been considered as alternative or merely subservient to screen-based GUIs.*
Intelligent System Designer	*John Rousseau, executive director, Artefact*	*The systems this person designs will integrate multiple domains, and those domains will themselves be the product of designers, artists, and technologists.*
Interventionist	*Ashlea Powell, location director, Ideo New York*	*As organizations and their challenges become more networked and complex, it will be harder work to help them digest new ideas and build towards a better future.*
Machine-Learning Designer	*Aaron Shapiro, CEO, Huge*	*A machine-learning designer's job will be to construct data models and algorithms that allow companies to create artificially intelligent products.*
Program Director	*Dave Miller, recruiter, Artefact*	*This person is a business strategist; they understand the "who," "what," and "why" behind a project/product; have a deep understanding of what it means to be a designer and a developer; and also have a track record of effecting change and influencing the end product.*

Real–time 3-D Designer	*Dave Miller, recruiter, Artefact*	*Virtual and augmented realities are on the forefront of design and technology explorations. Interaction design and game design will collide and integrate.*
Sim Designer	*Rob Girling, co-founder and principal, Artefact*	*The sim designer pulls together customer data, behavioral models, and statistical models to design simulated people that can be used to help predict future customer behavior.*
Synthetic biologist/nanotech designer	*Carl Bass, CEO, Autodesk*	*We're already on the path to creating customized medicine, and within five years synthetic biologists will be designing treatment that ties to the DNA of the patient. These medicines will be designed in software and printed on 3-D biological printers.*

POLYGLYPHS™: Polyhedra forms, their colors and elements, used as symbols

HAH   ICH   TRH   HAOTH   OTH   ICDDH   DDH

CREATING COMMUNITY

Our goal is to mine the WIKI and grow a prosperous community worldwide with like minds shared values and vision. D$_E$Sign Science" pays homage to Richard Buckminster Fuller who gave us these ideas in 1957. With the evolution of art, science and technology impacting all we can now relate to and the many Creative Design disciplines of Art, Architecture, engineering etc. They are becoming more complex and it seems that synthesizing the 'MATRIX' of elements into a 'D$_E$Sign Science' is a logical step. This allows us to distill new wisdom, intelligence, knowledge and in-form-ation (a WIKI) to transform the world with more organic solutions and experiences in harmony with the pace at which human consciousness, needs to and in some cases, has evolved. This bridges the gaps between those who are aware and need to be, to level the 'living fields or work, play, self-realization, and community.

This 'Design Science' synthesis distills and simplifies (Simple-Fi) the new knowledge and information to create a holistic; easy to grasp and apply set of ideas, tools and technologies for the new paradigm™'. Simplicity is still a viable goal that nature loves, it seems. This discipline gives us a robust 'Design Information Processing, Management, and Application System' that's a profound transformational technology that will enhance our vision, creativity and life. These creative principles have inspired innovations through decades of research studies, testing applications and planning in applying new ideas and forms vocabularies to architecture art, sculpture and industrial design etc.

I have invented a set of new three dimensional forms, which to me is a clear indication of this New Paradigm Expression of an emerging future, full of new possibilities, we need to embrace. Sharing this knowledge and information will inspire the global community as we seek the visual, verbal and cultural languages for the necessary communication and unification needed. Knowing and trusting the source of this (D$_E$Sign Science) WIKI is critical to all the arts, sciences related fields of endeavor, the people who see these original and important life changing innovations as gifts to empower themselves to tap into our deeper states of becoming as we define, refine and enhance our (constant) creative freedom.

D$_E$Sign Science is developing a Design Information System DIS™ platform with access to the expertise building knowledge and information publishing portal, a constant design and creation research base for an innovative conscious community. What we have attained we ill lovingly offer to inspire and motivate you to gain your own expertise with the right perspectives and vision for your very attainable futures. This 'New Paradigm Movement' will be with us for millennia to come, if the past cycles are any measure. It is therefore necessary to also have a legacy component in place with succession in action. We must make our contributions to its expressions by each 'one teaching one' to control their destiny, regardless of disruptions and distractions. DIS™ gives us all advantages and opportunities during the transformations we might normally not recognize or be conscious of. This is powerful. It's worth the price of admission to DIS™ Future we are creating. Our community loves these new habit forming disciplines that enhance creativity while living in a strange world we can help make familiar. In DIS™ community learning is a shared lifestyle.
All quest-ions reflect your quests and are seen as highly respected requests for directions to help others form their cross pollination of ideas, opportunities and relationship building in true community BMS.

Design professionals need to engage in opportunities to enroll and invest in our educational and training programs, buy our books, consumer products and information products branded with this new paradigm aesthetic, identity and JOY.

There is a schism that can only be bridged by a team of like-minded, conscious professionals who are experiencing or have survived their own transformations who can align with this vision and mission.
This is the personification of transition into a "Matrix of Use" that can move towards implementation of the tenets and other policies, principles, value engineering, currency and financial capital formation needs with operations, management functions, marketing intelligence. On the social engagement side, networking with influencers, connected to high

Networth folks who are committed to this movement is a (preferred) asset unto itself. Such a collaboration would be incentivized by an autonomous type and level of partnership that goes beyond equity, debt or any other transactional and compensation or remuneration found in the old paradigm. Supporting the Artrepreneurs who bring their' New Paradigm gifts is critical.

All currencies, including equity of any kind needed and invested, are to be considered on the basis of their worth, equitable value added or appraised potential. "MONEY" does not speak above all other voices in this cacophony of harmonious sound that resonates from the He-art and excellence of all the key components of Consciousness with its Body, Mind and Spirit dimensions with the rhythm of the universe and all that is now known and verified to be true considered as the New Paradigm and its movement.

Real talented 'actors' have to be identified as activists and action-takers to fill these roles and others that by the nature of transition and change will be needed or be replaced to fill all needs. This is a working team in flux, that remains fresh always moving in cohesion and toward effectiveness to manifest this long term development plan to the best of our abilities. FTGOA.

Develop a "matrix of use" for all sectors of the economy where the plan could be and/or will be applied.

The Null Set	Provider From	Each	Sector With	Specific talents	Resources
Need					
Food					

This all gets put into a space, time, energy continuum and developed into a computer system.

The new paradigm movement™
Neuroplasticity evolutionary wisdom paradigm movements
A knowledge economy based not on old paradigm values that are 'obso-less-sent' but the new paradigm opportunities.
New paradigm knowledge is now available and user friendly. The transformation of data (organized/designed) into information is the key "innovation strategy"; "using care instead of fear to do good"

The question for this quest is: What is "CONSCIOUSNESS?"

1. The Triad of Consciousness: The law off Three (3) and the number system
2. periodicity and the transformation matrix
3. Ancient wisdom traditions, Sciences and Traditions with alternative interpretations of life forces and their Operating Systems
4. The Kundalini vs Western Philosophical Frameworks
5. The Flavor Machine, Ancient 'Archeo-Celestial or Architectural Sites as in 'Gobekli Tepe' Southeastern Turkey.

6. The classic Euclidean Geometry and the Polyhedral vocabulary; the Non-Euclidean
7. Form vocabularies and their applications to design science and technologies
8. Cultural aesthetic expressions Ubiquitous Architecture and holistic communities with the emphasis on wealth creation through housing
9. lifestyle transformations, principles, sciences and technologie
10. "creating an educational system based on new paradigm knowledge
11. Andragogy: the method and practice of teaching adult learners; adult education.

"Much has been written about *Andragogy is the method and practice of teaching adult learners; (adult education)* in general education circles over the past fifty years" but there is no organized system to develop and provide it. Pedagogy is obsolete.

The State Transformation Triad with Universal Consciousness and Symmetry Principles

BODY	MIND	SPIRIT
Stimulus	State	Results
Goal	Desire	Outcome
Statement	Intent	Meaning
Thought	Emotion	Passion
Scene	Mood	Act
Space	Time	Energy
Physiology	Psychology	Spirituality
Vision	Visualize	Imagine
Was (Tense)	Being	Becoming
Expression	Feelings	To Experience
Words	Articulate	Communicate (Vocalize)
Form	Tension	Formulate
Dominion	Personal Power	Create (Creative Energy)
Thing	Balance (forces)	Harmony (Flavors and Essences)
Event	Quality	Flavor
Voice	Tone	Essence (Vibration) (Resonance)
Reflection	Observation	Intuit(ion)
Declaration	Confidence	Faith
Message	Exchange	Resonate (Sharing Values)
Knowledge	Comprehension	Will
Motion	E-Motion	Movement
Reasons	WHY	Evaluate
Concept or Idea	Experience Value	Think (Fulfill Need/Desire)
Meditation	Clarity	Evolve
Brain, CNS, Kundalini, Chakras, Endocrine System, Aura	Living	Self-Realization
Paradigm	Insight	To view
Concept	Perception	To Conceive and Realize

What is said	Why it is said	How To Say it
Aura	Clarity	Faith
Intuition	Sense	Wisdom
Awareness (Conscious)	Subconscious	Superconscious
Intellect	Intelligence	To Know
Task, Deed	Accomplishment	Being Satisfied
Life	Integrity	Happiness

Continue developing this matrix to explore other threefold relationships in the triadic harmony.

Hallucination, Delusion, Agonizing, Hesitation, Indecision, Premeditation, Procrastination

COLLABORATION OPPORTUNITIES:

The vehicle for this is a new paradigm Prize; a gift to conscious creative inventors for the world. This is how it works:

Establish a "New Paradigm Pri$e" for "The global recognition of excellence through practical creative efforts, collaborations and contributions in 'Design Sciences and Visual Intelligence' disciplines responsible for innovative breakthroughs, Philosophies, strategies and concepts to create ideas, exponential and/or disruptive technologies and designs that transforms creative processes, design solutions, technologies, materials and computer technologies and systems for the development, creation and management of any aspect of the inventions. They need to add exponential value to the Quality of Life, preserve the environment and be energy efficient. This allies to in any field of human endeavor, regardless of the patentability of any invention under consideration for **"New Paradigm"** recognition. This extends to qualified individuals, stage one startups, entrepreneurs and educational institutions worldwide.

THOUGHTS: If ancient oriental cultures could believe that destruction was a principle in natural law (and WAR: became a phonetic accident); why did western cultures misinterpret destruction as WAR and not see it as being an organic creative aspect of growth? Could history have pivoted into another direction on this notion?

"Exploring Breakthrough Paradigms to understand consciousness and our higher selves to build better relationships in the human family and all manmade environments. To redefine consciousness, as the body, mind and spirit continuum as the inspiration for making abstract thoughts, natural principles and technologies real. To reclaim and share the emerging creative, intellectual and spiritual capital to be free and totally fulfilled personally, professionally, in community and be on higher purpose. To collaborate and accelerate transformations to harmonious futures for a new world and its peoples embracing the most natural universal attribute of all; change do so with courage and vision. The new imperative is to add value to all human endeavor, with honor, respect and gratitude to all forms of 'design sciences' and arts disciplines with all technologies ;present and future, with deeper 'ubiquitous architectures' encoded with richer identities, new form aesthetic flavors and breakthrough technologies now being and continuing to be manifested and articulated for the greatest good".

In "The Sciences of the Artificial by polymath", by Herbert A. Simon, *the author asserts design to be a **meta-discipline (metaphysics)** of all professions. It is a science.*

"Engineers are not the only professional designers. Everyone designs who devises courses of action aimed at changing existing situations into preferred ones. (Original 'non-existing' situations?)The intellectual activity that produces material artifacts is no different fundamentally from the one that prescribes remedies for a sick patient (healing) or the one that devises a new sales plan for a company or a social welfare policy for a state.
Design, so construed, is the core of all professional training; it is the principal mark that distinguishes the professions from the sciences. Schools of engineering, as well as schools of architecture, business, education, law, and medicine, are all centrally concerned with the process of design." Are courses of action affected by designers only? Courses of action that cause themselves are organic or natural or can be caused by man's interaction with the environment and other systems? Healing is a form of design.
The *meta-discipline for design professions* is a spiritual science. *Evolution* is either manmade or paradigm shifting preferred course of action. Where is "need" in this definition?

THE AGENDA:
1. The Hermetic philosophical principles are in fact principles of consciousness that can be articulated into real world applications.
2. The Bio energetic, magnetic electrical dynamic is a significant element for the nano, micro and macro cosmic creative development processes. It can transform life support, maintenance, healing and wellness protocols, systems, technologies and professions.
3. The Euclidean and Non Euclidean systems of geometry and related disciplines set the tone for new identities and aesthetic expressions for the built environment. They hold in suspense nature's unfolding mysteries that are yet to be discovered.
4. Having new forms to define new concepts and vocabularies of 'Form' and identities for branding this epochal shift is of paramount importance and global cultural significance with great economic value.

5. The intellectual property technology will impact new material sciences, production and engineering technologies offering significant added value and exciting opportunities. The knowledge component of this science, apart from being constructively disruptive, is 'epochal' and exciting.

6. The transformation of conceptual, intellectual, and natural logic structures will create new developments in computing science in calculation, design and manufacturing, management systems and qualitative technologies. New systems for modeling the new paradigm and a new world view that will impact the quality of life significantly.

7. Transforming these concepts into economic opportunities and enterprises can be pursued with the applications of diverse business models each chosen to do the best.

8. This rich knowledge base will prepare professionals and entrepreneurs through education and training programs for generations to come. ¬The quality-of-life design professions include and are not limited to the built environment with emphasis on occupational shifts to emerging technologies and opportunities for community development and wellness sectors.

9. The New Paradigm Research and Development model will focus on future applications of all emerging technologies, expanding knowledge bases and cultivating new opportunities and technologies through collaborations and other mutually beneficial relationships.

10. Esoteric technologies and further research will help pursue future development of concepts beyond boundaries of known technologies to transform knowledge that continues to evolve. Doing this in collaboration with other New Paradigm Scientists and all who support the vision of a positively transformed world.

11. Transforming concepts into solutions is the next task. Developing a well harmonized synthesis for principles, ideas, resources, and applications with all the marketing strategies, collateral and sales systems in place 'to make the abstract real'. The idea bank is created with opportunities and properties in house and with other developers and entrepreneurs.

12. Capital formation and financial development of all forms will be developed, for the good of all.

THE BENNETT CULTURAL ECONOMC DEVELOPMENT MATRIX Copyright 2.2016 H.G. Bennett

		APPLICATIONS	
CONCEPT	OLD PARADIGM	NEW PARADIGM	TECHNOLOGY
Logic Structure	Gobekli-Tepe	The x-z & z-x Flavor Machines	Celestial Architecture Computing Systems O.S and Higher Level Qualitative Reckoning
Philosophical and Esoteric Concepts		Metaphysics and Consciousness Theories	Intelligence & Creativity Enhancement
Geometric Form Inventions	Platonic & Euclidean Polyhedra	Bennett Non-Euclidean Polyhedra	
Architecture and Construction	Ubiquitecture™ in other technologies	ISBU Auto homes and Communities	Housing, Energy and Commercial Space

Industrial Design	Household Accessories Food Forms, Tools Kitchenware ™	New Form Technology applications	New Materials and Forms
ART; production, Publishing & Distribution	Sculpture, Paintings	Digital Images and mixed Media works	
Educational	Systems and Devices	Pedagogy	Andragogy
Literary and Technical Publishing			
Fashion and accessories	Trilliant™ Watches Optix™ Eyeware	Chronotecture	Bespoke Chronotex
Healing & Wellness			Vibration (Sound) and other media
Wisdom, Knowledge & Information	Publishing	Visions & Ideas	Self-Publishing
Paper Engineering			
Toys, Games, Puzzles	Communication Devices	GEOMIFORMZ ™ Media Cubes	
Real Estate	Development		Wealth Building
Environment Organizing Systems		The File Carrier Systems	
Studios and Affiliate Networks	Global Design Studios ™	Global Studio Designs™	
Media Production and Distribution			
Product Development			
Marketing Research & Sales	Internet-Analog-Word/Mouth-WOM	Network Affiliates	e-commerce & Digital
R&D-Licensing and JV	Stealth	Stealth	Intellectual Property
Capital Formation	Crowdfunding	Crowdsourcing™	Cultural Economic Development CED™

"The Lotus River flows between the high mountain and the low delta carrying life from birth to death. In our Piscean story, named after our present epoch, various 'vesica piscis' or fish forms live in its stream. Some, like the salmon, swim against the current upstream to reproduce, answering nature's calling. The bottom feeders are floundering in blissful nescience concerned about meals they missed. The predators, the salmon and the flounders are all in the continuum of 'who eats whom' and are all in/on their DAO to their inevitability.
Are there fish in heaven from all the fishing that's been going on since the original creator of fishers of men did his magic? I am sure Noah saved two, from his mysteriously symbolic genetic cargo, one for himself and one for 'HIMHERIT'.

In the Hindu triumvirate (or Triad) Brahma is the creator of the universe. **Vishnu** is the second god responsible for the upkeep of creation and Shiva is the destroyer of the world. Consciousness the all omnipresent, omniscient cosmic being in it omnivibrational essence causes Visnu to arise. From the cosmic being's heart entire worlds arise.

All that arises have the ability to preserve themselves thru Visnu. Out of Visnu arises Brahama who in turn gives rise to the four expressions of animate and inanimate things; sentient and insentient form. This is all prior to the dissolution. At this point Shiva dances to the music of life and death'; ease and dis-ease; pain and suffering that is played out on the lower register's realm. Here men conjure desires in their hearts building air castles in their minds. The three hearts; the physical, the emotional and spiritual hearts with strong desires and illusions are rendered onto Shiva because the rewards of little or no self-effort and no right action are Shiva's tools.
Self-effort is the result of knowledge from the good books, instructions from wise teachers and one's own passion, focus and will. Here is an equation: $F_s \times E_s = I_y \times SE$; Fruits of endeavors equal the intensity of self-effort. Right actions create positive outcomes.

"Looking further we read: Kundalini is the Goddess of speech, (language, symbols, words) and is praised by all. When awakened she offers illumination (light), the source of all Knowledge and Bliss. She is pure consciousness; the Supreme Force, the Mother of Prana, Agni, Bindu, and Nada. It is by this Sakti that the world exists. Creation, preservation and dissolution are in Her. Only by her Sakti the world is kept up. It is throu, tergh Her Sakti on subtle Prana, Nada is produced. While you utter a continuous sound or chant Dirgha Pranava ! (OM), you will distinctly feel the real vibration starting from the Muladhara Chakra. Through the vibration of this Nada, all the parts of the body function. She maintains the individual soul through the subtle Prana. In every kind of Sadhana the Goddess Kundalini is the object of worship in some form or the other".

This system represents the deepest level of our knowledge of the gross nature and fuels of human physiology without the chemical processes without the electro-magnetic forces, the color, vibrations and the host of flavors involved in the human experience we call life. The Chinese were isolating sex and pituitary hormones from human urine and using them for medicinal purposes by 200 BCE. In 1849 Arnold Berthold pioneered this research. In 1902 William Bayless and Earnest Starling continued the development.

In the western world knowledge feeds our voracious materiality. Prayer is strictly for supplication and revelation is a book in the bible few truly understand. Of all the sentences in this work this is the most dispassionate. Entire libraries have been dedicated to how to do the right thing leaving us on a trajectory to oblivion. From a spiritual (NOT RELIGIOUS) view every word in this text is a practical, emotional and super energy generator to create the lives we were meant to live. Here is another dedication to the library of human failure celebrating Shiva's triumph.

Experiments that are mechanical are STEM; (See, Touch, Estimate and Measure) no longer work. The Thought STEM is now (See, Think, Energize and Meditate) with observation and participation. The search for truth and freedom has an upside and a downside to it. If in the study of the nature of life we do not find truth or freedom the process of seeking could have its rewards. What we discover could soften the pain as changes in life are encountered. Discovering truth and freedom could be the bonus. Wisdom could be gained in either scenario. Like paying attention, which does not cost anything, except when it is not paid attentively. The search for wisdom and freedom cost much leas and the gain is exponentially greater. Physiological, psychological and spiritual transformations follow the eternal creative principles outlined in this work. The goal of this work is to establish a *'body, mind and spirit formula'* for the synthesis process as The Dao of (self-discovery, self-effort and self-knowing) elevating 'DESIGN' to a science.
On second thought maps already exist. Others with broader shoulders have left us their wisdom to help us transcend the artificial diversions we encounter as we continue to enhance our lives. For this a *'body, mind and spirit'* paradigm is needed. Fundamental principles and natural laws of the cosmic being in the DNA in us for example, that's ready to unfold and no longer be dormant. This awakening is long overdue.

We now bring out the big 'canons' to help us fight our spiritual battles.
The Dao De Jing was referred to as first canon to inspire the 'Dao DeSign' (system). It is used as an oriental harmonizer for the *'body, mind and spirit'* triad that in principle is part of a wider swath of oriental wisdom traditions. The 'Vasistha's yoga', like the Dao De Jing is the second canon used to inspire the meditation and thought processes from the Hindu tradition. This storehouse of wisdom by Swami Venkatesananda "provides the means to eliminate psychological conditioning".

HAH   ICH   TRH   HAOTH   OTH   ICDDH   DDH

EPILOGUE: D_ESign Science: in the New Paradigm Age

1. The "Book" Title; D_ESign Science-in the New Paradigm Age
   a. This book is the first in a series of publications devoted to the promulgation of a movement that informs and gives access to and informs all 'designers' in this movement to understand what D_ESign Science is and how it inspires our mission, defines our goals and purpose.
   b. Our intellectual assets come directly from *'SOURCE'* to inspire the highest and best creative expressions that serve our environment for mankind to nurture human growth through body, mind, spirit and energy development.
   c. We have synthesized a new aesthetic (identity) for all our creations in keeping with the 'Body Mind Spirit' principles and the Space, Time Energy Paradigm or STEP formulae.
   d. From the synthesis of these disparate elements in traditional business models comes the 'oneness, cohesion and simplicity' we 'enjoy' and are grateful for sharing. We Simple-fi our vision and mission and will continue to do so as we grow.

The Theoretical Premise: We define the 'BMS' formula as the 'synthesis of consciousness' as our "New definition and ART-iculation" of 'culture'. All creative Systems, technologies, intellectual properties, assets and activities become the 'holistic foundation' and 'guiding symmetry principles' to support all our creative, social and entrepreneurial endeavors. This is our New Paradigm model that 'moves' the research, creation and dissemination of the Direct Wisdom, Intelligence, Knowledge and Information (WIKI) 'using these universal principles we access from source. As creative and respectful spirits we honor all life that's in pursuit of self-realization and higher consciousness. We advocate for all creative, social, economic, cultural and development activities, thoughts and expressions through our research, publishing and production, and the dissemination of information, consumer products, industrial designs, architectural and building technologies, media and communication systems to include and not be limited to the following:

2. Information products in print, digital published media will be made available periodically.
   a. 3D form vocabularies with new aesthetic identities for consumer products, objects and electronic devices, architecture and Art and other inventions yet to be developed.
3. All intellectual properties will be made available through 'agents', to developers, joint venture partners and clients to develop opportunities not aligned our mission and agenda.
4. The people aspect of this campaign becomes a global initiative to explore possible exchanges needed to expand our reach and creative contributions worldwide.
5. This is a cultural development and entrepreneurial movement with a development agenda dedicated to creating business opportunities and models of production, distribution and service delivery related to the new paradigm WIKI technologies. We believe that the future depends on D_ESign Science for the infusion of innovation to create the industry growth and expansion for the prosperity and wellness needed now; for The Good of All (FTGOA). "Design Heals ALL" is our motto. "Making the Abstract Real" is our 'secret sauce'.
6. The book with the WIKI, the business and its agents with the collaborators form the model to create and share the knowledge to promote opportunities that further the cause, growth and prosperity effects of this vision and movement for the good of all (FTGOA).

7. The 'ticket' is the acquisition of the book. With proof of purchase, readers become chartered members of this movement as affiliates. They may choose various levels of membership according to their skills, professional and business acumen and needs.
   a. The social currency aligns with the marketing trends we now engage in.
   b. Supporting the vision and mission is not mandatory but highly recommended with all benefits offered to the 'agency and membership'.
8. The book represents the body of work I have distilled that's now being shared as the process, resource, and product of a synthesis of many creative professions I have enjoyed.
9. The future of design will involve a harmonized blend of Science, Art, and Mathematics; SAM, to manage the complexity that continues to unfold from the new systems and disruptive technologies we will now be prepared to handle with this business model.

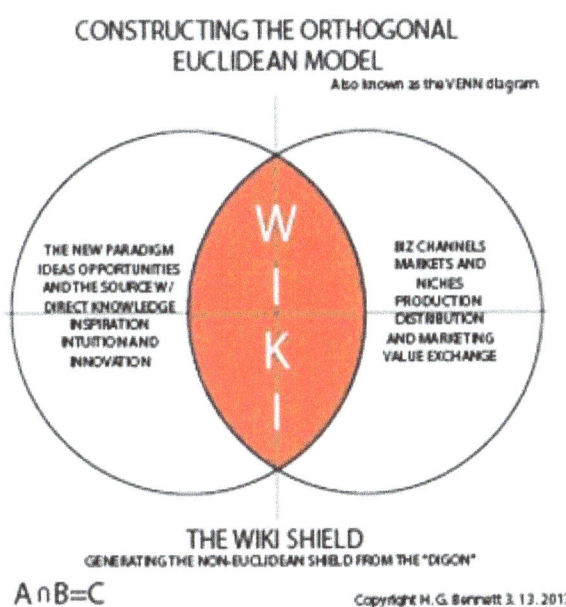

10. Visualization and Meditation are the two recent contributions made to the western world that have had significant impact on self-development starting in the fifties. This is the first step to resolving the East West dichotomy that could bring peace and prosperity to the world.

11. I believe these are the tools for thought processes (not experiments) that can be used to connect to source. The exponential value of this 'creative sector' will be significant, according to all the currency gauges we have.

12. There are 7 tiers of participation, activation and collaboration in this model:

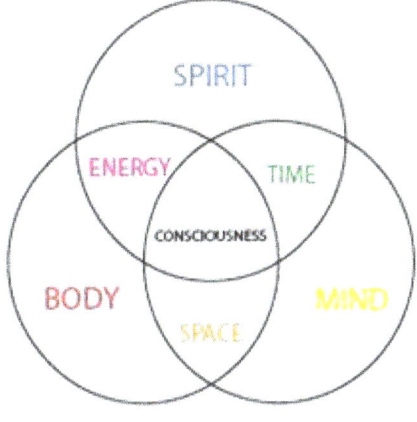

This matrix can be expanded to meet the growth needs and implementation processes.

THE TIERS	CREATIVE TALENT	DISCIPLINES
1 Founders	Finance and Capital Sources	Operations Managers
2 The Agency	Multi-talented talent searches	Data Bases and Associations
3 Consultants,	Vendors and Suppliers	Distributors
4 The 'heart' membership,	Professionals	Genral audience; Students
5 Technology Inventors	Developers	Operators
6 Creatives:	Planners, Artists	Producers ; Fulfillment
7 Joint Venture Partners	Clients	The Market Niches

Non-Euclidean forms become the new aesthetic vocabularies to be exploited in developing all possible applications at different scales and sizes in different materials for uses in different industries and markets.

The various entities are: The Loggia research Group & Babbleon Press Publishing, Bennett Architects RA, Trefoil LLC and Summit Development Industries LLC

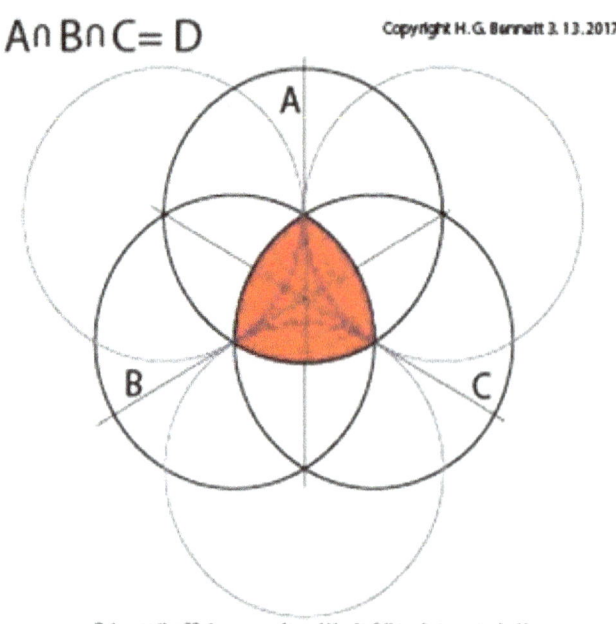

Between the 90 degree angle and the trefoil pentagon construction lies a curved triangle also know as the Reuleaux Triangle or spherical triangle when mapped on the surface of a sphere.

"STRUCTURAL PHILOSOPHY"
Purpose: AUTHOR/S: Channel/s that gain access to 'Direct Knowledge' from Source and is in the flow of Constant Creation and creative manifestation while obeying 'natural laws' using intuitive, creative and spiritual disciplines with 'Design' as the core science or other methodologies we create.

AUTHORITY: Is an organic and natural energy enhanced in the group dynamic rather than with the solo 'expert'. Design as an authority is a ubiquitous discipline when exercised commands attention which confirms the 'authority' inherent in its creative process. I am now synthesizing Art, Architecture and some of the sciences into a 'D$_E$Sign

Science' with its principles to search for the connections and core elements using the new visual intelligence, intuition, metaphysics with knowledge and experience in art, science and inventions related to going beyond the boundaries of '3D form consciousness'. This is then used to create a MATRIX with a quantitative and qualitative periodicity that could organize vast amounts of information and data we are going to need new systems to deal with sooner rather than later.

AUTHENTICATION: This comes from creating viable applications and solutions for various industries, markets and individuals open to new ideas, opportunities and methods of creating prosperity and good works for their clients and themselves in the transformed and emerging world *economy*. Contrary to the non-standard definition of economy as 'the study of the lack of resources', this new economy is one of a-bun-dance; open and available to all who can connect to SOURCE and are in the flow of the WIKI with values to be shared and exchanged. A profound transformation is taking place on every level of life and what supports it now. This turns into a spiritual dimension as our consciousness expands to align with all the changes taking place.

AUTHENTICITY: I am secure in my abilities and confident in my connection to SOURCE and the flow of the 'WIKI' which keeps me humble compassionate and in profound gratitude.

INTEGRITY: is the major system of value exchange. Intellectual property is our currency.

The following tables show the shifts between the old and new paradigms:

OLD PARADIGM	NEW PARADIGM
EFFECT	CAUSE
STIMULI	SENSE DATA (EARLY STAGE INFORMATION
STRESS	TENSION
FOCUS (DISTRACTED)	ATTENTION (ATTRACTING ENERGY TO ACTION)
MEANING	INTENTION
HOARDING	GIVING UNCONDITIONALLY
LOVING SELF	LOVING OTHERS

OLD PARADIGM	NEW PARADIGM
SELF DEVELOPMENT (ENABLING)	SELF-REALIZATION (EMPOWERING & LIBERATING)
PEDAGOGY	ANDRAGOGY
SELF HELP	EMPOWERMENT
ENTREPRENAURESHIP	ARTREPRENAURESHIP
QUANTITY	QUALITY
A JOB FOR LIFE & (A GOLD WATCH)	CREATIVE FREEDOM
DEFERRED LIVING	ACTUALIZING DREAMS
GIVING THANKS	GRAMMERCY (PROFOUND GRATITUDE)
NUCLEAR FAMILY	COMMUNITY
SALT OF THE EARTH	SALTING THE EARTH
EGO & SHGO: ME, MYSELF AND I	CONNECTED TO SOURCE & IN THE FLOW

Here are some issues we need to develop that relate to the Intellectual Properties in this work:

1. The protection and legal issues must be addressed with systems in place to assure the proper and legitimate use of any and all of the ideas disclosed. Publications can be cited in patent application but they are by no means as effective as official patents and copyrights.

2. The relationships will include but not be limited to outright sales, licensing, and other forms of co-creating and partnering on a project basis. They include agents, developers, marketers and manufacturers among analog and digital makers in creative sectors that are form intensive and wanting to be part of this Non-Euclidean aesthetic (movement) and any of the brands we develop.

3. Prepare and make available as a follow up with 'proof of purchase' (POP) of the book for clients to obtain documents and agreements upon request.

4. Trade secret agreements and provisional patents would be the first level forms to consider.

5. Patents will be applied for if and when 'the assignment of rights' are required as part of the memoranda of understanding.

6. This ties in with the TREFOIL LLC website with the url: http://www.brooklyninventor.com

7. The Special Opportunities Division will manage the annual Enterprise Programs for designers to enter an RFP process that will focus on innovative and feasible applications of ideas from the 'Gallery of illustrations' in the back-matter and the "Idea Bank". A supplement related to the RFP will then be released. The prize: the winner becomes the co-creator of the idea as its lead developer (terms to be negotiated).

8. Membership will be granted upon paying a onetime membership fee and the book will be complementary. Other technical information will be available when published.
Putting thoughts into action is the exciting part of this movement creative designers might be looking to collaborate with. Innovative ideas and opportunities are available.

D_ESign Science in The New Paradigm Age

PROVERBS & PROFHERBS:

"The future of design will involve a harmonized blend of Science, Art and Mathematics; SAM, to manage the complexity that continues to unfold from the new systems and disruptive technologies we are not now prepared to handle". It must become a 'D_ESign Science' if we are serious about our evolution.

In the New Paradigm Age our major challenge is keeping pace or ideally overtaking the 'pace of Progress'. Progress is now 'undertaking' us!
"Synthesis" is a way of getting from many to one not with the pain of isolation, but in community.

Through D_ESign Science the fundamental synthesis of innovative 'design' ideas uses the canons and symmetry principles of nature. The challenge we face is reframing the mental scaffolding or old paradigm that does not support the new building blocks we are all co-creating. The New Paradigm Notion (NPN) is a collective energy we are all participating in as we create. This may seem strange to the 'egoic' isolationist and materialist. Here is a candidate for extinction if there ever was one.
Design is a 'spiritual awakening process'; a renaissance it was called, meant to restore and regenerate all that is and will become naturally. This speaks to a kind of 'ubiquity' at the very core of life which is internal with the potential to be expressed in the creative world.

There is a cyclic motion that controls information and innovation that inspires growth. In one phase of the cycle the external environment inspires us. The power of the thoughts that conceived all of it is what we absorb and process. This is what sustains us, doing more good than food, clothing and shelter does for our Body, Mind and Spirit. Is this a paradigm we subscribe to now? What would this idea do for us if we could then process this 'collective internalized thought energy' reframe it with the new (NU) flavor and share it with the world? Our life support systems now have major gaps in almost all aspects of life that seem to be widening. Shifting our mindsets just a few degrees could make a difference. Difference is what we do not handle properly as a mental principle. Similarity is the comfort zone we guard without regard for embracing the value difference makes.

Design is our ability to create whatever we want. Science is the method of doing it for ourselves with materials, tools and gifts we already have but have not 'loved' ourselves enough to try using them. Love is a universal force that holds all the multi-verses together. You may call it light, gravity, the weak or strong force but in the oneness of infinite consciousness it is LOVE. The life force is love. Life must love us to be keeping us in its flow of the three energy fields that 'synthesizes' it for us. They are the physical (Eros) the mental (Philos) and the spiritual (Agape). We seem to be clueless about doing it ourselves. Here again we see the imbalance and the focus on the physical that consumes us. If we use the marketing metaphors in the self-development world this idea of a design science would bring dimensions to new creative levels we can all begin to explore.

Supporting ourselves by loving the life that loves and takes care of us, regardless of the state we have designed for ourselves, is the first step. Our hi-story, my-story or future and the present or the gift we are given needs to be repaid. This is more crucial than paying off the multitrillion dollar US debt. Though the currency is different it will more than compensate for our intransigence. Paying attention is one of the currencies we do not understand. Attention is focused consciousness. It is what directs our energies and our creativity to attract the 'scientiam recta'; direct knowledge we have access to. It is the key to designing our lives. Everyone can afford to pay it (attention). It is one of those powerful gifts that on the surface, seems to cost nothing, so why can't we afford to pay it? It causes all the pain we suffer.

Cycles of life change along with the expressions of the space, time and energy that is encoded in it every 1,000 to 2,000 or 3,000 years. Is there anything about our space, time energy continuum worth keeping? Let's make a time capsule and move on with our new mind agenda and begin designing our new lives. D_ESign Science will help us access the Wisdom Intelligence Knowledge and Information (WIKI) needed. The new forms and skills we 'l-earn' will ART-iculate the new aesthetic reality with the identity that will 'brand' our time and our contributions to humanity in this new era known as the 'new paradigm age.' We have a level playing field opportunity for inclusion into community like we have never seen or experienced before. All of these ideas are the seeds of our time that's been laying fallow too long. It is spring, time for planting new seeds.
Are we going to continue what we are doing now with the same results, or do less with more or do nothing at all and atrophy when we are given (or always had) opportunities to empower ourselves with the ease of an (awkward) awakened consciousness.
All it costs is payed for with attention and love. We need to put design into its most powerful and purposeful context to be the connection to source to which, ironically we are already experiencing, but are not aware of.

There are two spiritual principles the west was recently introduced to in our recent past that have infused our consciousness with methods of developing the human mind and spirit. They are meditation and visualization. These are the disciplines that are the portals to our destiny; the one we create especially for those who are prepared to shift their paradigm and embrace the new WIKI about our reality as spirits on a physical eternally conserved journey. They are the cash registers and the score keeping media for new age wealth.

Meditation is 'free' thoughtless awareness is its reward. All it costs is paying attention. Visualization is mindful and focused visual creative thinking. Meditation opens the portal and visualizations receives the signs that inspire de-signs we process to grow and be healthy. Meditation takes us to the hidden dynamic behind everything, everywhere in spiritual no time, no space but in all energy. There may be realms of 'no energy' if we acquiesce to the quantum theories. Visualization operates in physical space and psychological time and in spiritual space with creative human effort.

We get to an intersection of Mentalism and Visual acuity where thought and action cohere in intelligence activated by meditation (thoughtless awareness) with visualization and all other harmonious 'threefold systems in place' to create 'synthesis'.

Do we create synthesis or anything for the matter? Are we instruments responding to the grand orchestra of the universe and are being betrayed by the 'quality' of our ego, mind and emotions? There is another state with methods, expressions, form, skills and power aligned with will and passion being used to help support the life force which is the ultimate generative, regenerative and eternal dynamic we are privileged to experience. This is the same force that holds multi-verses together. Why are we not in tune with it?

There are some key principles needing enhanced focus, attention, clarity, imagination and thought, to name a few that D$_E$Sign Science offers. Design on all levels is in this unfamiliar intersection of Meditation, Visualization and Attention fueled by thought, imagination and intuition. Let's extrapolate these two fields of information in a matrix to expand our awareness. The questions are about relationships. How does the column elements relate to the row elements?
What is the product of the two terms in 1- intersection and 2 for other ('symmetry') behaviors we define or are observing?

THE MECHANICS OF CONSCIOUSNESS

	MEDITATION,	VISUALIZATION	ATTENTION
Meditation	Thoughtless awareness	Undesirable distractions Guided of unguided thoughts	Return to Thoughtless awareness
Visualization	Thought Processing	Imagination-Innovation	Observation
Attention	Focused Thought Insights	Thought Experimentation	Intense Focus Clarity

	Imagination	Intuition	Thought
Meditation	Creation	Inspiration	Conceptualization
Visualization	Visioneering	Dreams	Cognition
Attention	Creativity	Source	Expression

Every human being is capable of expressing this human dynamic because it is already a part of us and we a part of it. There is no rocket science here. The lack of awareness of this idea comes from our unrealized state of being in existence and stress, not true living. Design is about connecting to source, being in the flow and being the channel for all the goodness that we get out of the way of. This first habit to be absolved is the 'ego'. Knowing how to balance positive and creative aspects of it with all else requires skillful mental work.
Man's ego (and now shego) does not 'pair' with source, nature and self. Many of our current paradigms confront these phenomena under a pretense of dominion. We still do not get the messages that mother earth sends us, for example, with the storms, chaos and 'psychic imbalances'. Native Americans and other cultures have a much better ideas about 'Mother earth'.
Implementing these ideas to support human life with all its desires, needs, wants and wishes requires a methodology one that is designed with physical, mental and spiritual harmony in

heart and mind that is in 'equilibrium' with the symmetry principles, laws of space, time, energy in correspondence with consciousness on all levels. D$_E$Sign Science does this for us. It is meant to give us more than things. It might just return us to ourselves before our 'empty containers' return to the source of their arising.

Body, Mind and Spirit are three distinct vibrations of universal energy.

TRAITS AND HABITS OF A D$_E$SIGN SCIENTIST

BODY	MIND	SPIRIT
DESIGN	VISION	MENTALISM
Formulation/Formation	Polarization	Ideation
Imagination Clarity Incision Form Ex-tension Traction Tension Electricity XLF (7-8hz) The Physical Heart (Eros) Solar Plexus De-Cision Sexual Center and Gender Symmetry: male and female elements in all things The root Matter and will Spirit making the abstract real and with 'thought alone' in ALL three Spaces	Visualization Thought Visual thinking Focus Decision Division Re-tention (memory) Attraction Distraction Spirit making the abstract real and with 'thought alone' Spirit making the abstract real and with 'thought alone' in ALL three Times.	Meditation Thoughtless awareness Attention In-tension Cision Magnetism Spirit making the abstract real and with 'thought alone' in ALL three Energy fields.

DEFINING (DESIGNING) SPACE TIME ENERGY PARADIGMS

Before we create anything we need to understand the simple elements we use and define them and what's behind them, defining them is the first S.T.E.P. Space is pure extension that can be perceived and understood by position and dimension. It can be measured against an established and consistent gauge, e.g. a ruler. It can be viewed in terms of displacement in time. There are three types of time; an objective (or spatial) time, a subjective (or emotional) time and a spiritual time which is timelessness. Time is physical displacement, change and motion of objects (in space or pure extension), and is calibrated by a standard gauge. The experience of internal transformations and changes of state in humans and the events we experience are emotional expressions of 'subjective time'. It is the flow that follows the freedom of thought and needs no gauge because it cannot be measured. The spiritual no-time dimension is the ascended bliss that is neither objective nor subjective and is transformative.

Energy "is" space; space "is" energy. Without energy there is no space and without space there is no energy. Three types of energy correspond with three types of space. They are the physical me-assured Space, the emotional feeling state Space and the Spiritual Space is not three-dimensional; Substance is. Likewise, substance changing and moving "creates" time. Place is well defined and localized space. Displacement is the complement of place.

Sub-stance is *presence* with heavy vibrational characteristics, properties (muchness) and qualities (suchness) like form "mass", resistance, in keeping with the laws of symmetry. Density, integrity are aspects of consciousness. All is vibration which implies movement, in time-space and energy in waves and or particles that switch states to correspond with the environmental conditions they are in. Light, mind and thought (and naturally observers and participators) are important elements of reality. Power is the mechanical advantage and leverage to do work.

The primary element of reality is form; the behavior of the materials that give it its integrity are its psychology. The flavors of its energies are "colors" (visible and not), aesthetics and identity. Matter or energy: particle or wave are expressions of various energies of the "infinite" space, time and prime energy. What 'tricks' us into 'thingifying' space, time and energy and reducing its 'purity' are our materialistic mental proclivities based on the immediate sensorial relationship/s we have with our bodies and the immediate 'false evidence appearing to be real' sensations we experience in space with our bodies. Our friendly overprotective and conditioned responses to the fight or flight limbic warnings overwhelm our immune system leading into stress.
There is a strong positive correlation between materialism and several mental and physical maladies. In other words, people who pursue money and things at the expense of relationships and other meaningful endeavors are more likely to suffer from these stresses. If they are not managed properly they lead to disease and chaos or at least confusion. Living in stress leads to dis-ease. What does this have to do with design you might think?

Just look at the following list of Pain points. Not being recognized, appreciated, respected and rewarded while dedicating one's life to sharing one's creative talents and gifts with others can lead to: 1. Unhappiness 2. Envy and jealousy 3. Depression 4. Social anxiety 5. Passive-aggressiveness 6. Short attention span 7. Poor self- control 8. Feelings of failure 9. Mistrust of others 10. Tendency to mistreat others for personal gain 11. Shorter, more conflicted relationships 12. Feelings of social alienation 13. Exclusion based on Cultural, Social and Political differences 14. Less generosity 15. Narcissism 16. Egoism. The list goes on!
Every one of these stress points can be healed by corresponding design aesthetic or creative spirit with its unique vision and discipline. Some are more general and are in the realms of being physical, psychological or spiritual all requiring forms of knowledge talent and skills energy and thought formulated into methodologies we call 'science and more directly D$_E$Sign Science.

Design here is used as the great interactive opportunity for self-assessment as we all explore the relationships with the stress points as we 'design' the Paradigms that can transform our lives. The design professions have an advantage in being connected to the source, skills and energies needed for adapting to shifting 'PARADIGMS' with many of the correlation to the principles.

Of course, correlation does not suggest causation. The correlation is essential nonetheless, and it's easy to see how synthesis does more to perpetuate than remediate the problems on the list. The goal of this work is to take the positive path by resetting the mental scaffold; to first design and build good habits and healthy paradigms of and for ourselves before we project our creativity onto the world.

Summary

There are three dimensions (states) to everything and in all of consciousness. They come on the 'quixotic 'waves of energies and packets' of interchangeable and transmutable physical things, mental and emotional states experiences and feelings with essences, energies, thoughts, flavors and spiritual expressions defined (and designed) as Body, Mind and Spirit. We interpret the stimuli and inspiration through Physiology, Psychology and Spirituality as the tools of synthesis of culture and the comprehension of consciousness to make connections to higher forms of our selves with the proper application of creative energy, attention and thought.

We use these as the operational dynamics in direct correspondence with space, time, energy and the power of thought, the fin-est fuel we know, with all the symmetry principles and natural laws we use to design our lives, the life support systems, artifacts and environments we need to live not in but with the LOVE Design and the "Scientiam Recta" or direct knowledge bring to our families, our creative communities and to the world.

This is D$_E$Sign Science's purpose and whosoever aligns with it and adheres to its principles is a 'D$_E$Sign Scientist'. The beauty of these principles lies in a compassionate DIY mentality. Welcome to the New Paradigm Age!

 *"Creativity gives rise to the limited out of the limited, to sanity out of madness, to the valuable out of the priceless, to abundance out of nothingness, to the original out of the familiar and to hope out of despair"*-Wallace Huey

THE GRAND PERIODIC QUALITATIVE MATRIX™

THREE GEOMETRIC POLYHEDRA TYPES

EUCLIDEAN POLYHEDRA	NON-EUCLIDEAN BENNETT POLYHEDRA	
Euclidean Geometry	Non-Euclidean	Non-Euclidean

PANEL.1

THE GRAND PERIODIC QUALITATIVE MATRIX™

CONSCIOUSNESS			BODY CENTERS PARTS		LOCATORS	QUALITIES NOTES	FLAVORS OF EXPRESSIONS (TO INFINITY)			
ORDER	TRIAD	Disciplines	NOS	Centers	Periodic symbols	Attributes	Chakras	Colors	Matter	Frequency
2	MIND	Disciplines	7	Crown	333 ⑨	spirituality, holisitic live, light energy, higher consciousness	Sahastrara melatonin	violet (magenta)	thought	963hz
			6	Third eye brow	323 ⑧	intuitive, vision, creativity, for thought & vision, anti-things, charismatic	Ajna somatotropin	indirect indigo	light	852hz
								white		
			5	Throat	313 ⑦	good communication, expression, creativity, inspiration, contentment	Vishuddi	blue	air ether sound	741hz
3	SPIRIT	Normal consciousness spirituality energy wisdom	4	heart	131 ⑤	selflessness, devotion, discerning	Anahata lymphocyts		harmony	
					121 ④	loving, caring, healing	Controlled by Manipura	green smoky	love	639hz
					111 ③	nurturing, balanced			balance	
1	BODY	subconscious physiology space knowledge	3	solar plexus	232 ⑦	Self-esteem, positive self-image, intellect, purpose, alignment with cosmic faith	Manipura Insulin	yellow	fire	528hz
								Deep red		
			2	sexual / gender spleen	212 ⑤	focused emotion, feelings, faith, trusting, natural healing, creativity sexuality, work energy	Swadhisthana oestrogen testosterone	orange	water	417hz
								white		
			1	root	222 ⑥	will, vitality, power, passion,	Muladhara epinephrine	red	earth	396hz
								yellow		

Nadis: Pingala, Sushumna, Ida

PANEL.2

"THE Q-NOME PROJECT" ™

CENTR	HERMETIC PRINCIPLES	MUSICAL NOTES		PLANETS	QUANTUM REALITY	ENDOCRINE SYSTEM	CREATIVE IDEATION DESIGN PROCESS	QUARKS	PLEXI	SINS
				FLAVORS OF EXPRESSIONS (TO INFINITY)						
7	MENTALISM	S I / T I	B	uranus (moon)	ether	pineal gland	thought concept, idea visualization, concentration	ddd	pineal gland	unconscious
6	CORRESPONDENCE: AS ABOVE SO BELOW	L A	A	venus	wave	pituitary gland	imagination originality	dsd	cavenous plexus	envious
5	VIBRATION POLARITY	S O	G	jupiter	subatomic particles	thyroid & parathyroid (gland)	media devmpt verbal and visual communication	dvd	pharyngeal plexus	libelous
4						hypothalamus	aesthetic pleasure reality testing			
	RHYTHM				atoms	Thymus gland	passion attraction "love"		cardiac plexus	duplicitous
		F A	F	saturn			love of design principles compassion	uuu		
3	POLARITY / VIBRATION	M E	E	mercury	molecules	pancreas (gland)	producing technologies for making, work energy, mech advantage	sds	solar plexus	gluttonous
2	GENDER	R E	D	(sun)/venus	matter	testes ovaries sacral	gender and symmetry alignments	sus	prostate plexus	erogenous
1	CAUSE & EFFECT	D O	C	mars	object	adrenals	materials w/ complementary psychology	sss	sacro-coccygeal plexus	avaricious

PANEL.3

HEPTAPARAPARSHINOKH™

FLAVORS OF EXPRESSIONS (TO INFINITY–1) Copyright H. G. Bennett 9. 16. 2016							
EUCLIDEAN POLYHEDRA	NON-EUCLIDEAN BENNETT POLYHEDRA		SYMBOL NAME	AURIC BODIES & THEIR LAYERS	MULTIPLE INTELLIGENCES	QUARKS	LEPTONS
T O I N F I N I T Y			DDH – DODECAHEDRON	the ketheric causal body 7th layer mental	logical mathematical intelligence	up uuu	electrons
			DAH – DELTAHEDRON	the celestial body the 6th layer "being" connect w/ source	visual & spatial intelligence	down ddd / strange	electron neutrino
			OTH – OCTAHEDRON	etheric 5th layer physical forms blue print (strunct energy field)*	verbal intelligence	top	muon
			HAOTH – CUBEOCTAHEDRON	the astral buddhic body 4th layer the litigator love	musical intelligence (rhythmic)	charm	tau neutrino
			TRH – TETRAHEDRON	the mental body 3rd layer thoughts and creativity	interpersonal intelligence	strange / down	tau
			ICH – ICOSAHEDRON	feelings h₂o fluid motion yellow (bright)	intrapersonal intelligence	unity harmony	consciousness
			HAH – HEXAHEDRON (CUBE)	etheric body 1st layer w/ parts & organs geometric force	bodily kinesthetic intelligence	bottom	muon neutrino

*Structural Energy Field

PANEL.4

FLAVORS OF EXPRESSIONS (TO INFINITY-1)						
LEVELS OF CONSCIOUSNESS Copyright H. G. Bennett 9. 16. 2016						
THE COLLECTIVE UNCONSCIOUS	CONCEPTION FECUNDATION INTERNAL & EXTERNAL	CLAIRPRE SCIENCE	WISDOM KNOWING W/OUT REASONING LOCAL OR DEDUCTION AKASHIC ACCESS	CREATE VIA ONENESS VIA THE DISSOLUTION OF THE ILLUSION OF BOUNDARIES/LIMITS	BODY RHYTHM	HALLUCINATION
THE TRANSPERSONAL SELF	GESTATION	clairvoyance	vision seeing insight intuition	see/create big pictures and structures matter	growth	delusion
THE CONSCIOUS SELF EGOTISM "I"	BIRTH INSPIRATION	clairaudience	symbolic intelligence power (breathing) vs. force	use symbol & meaning to create and attract matter, power, force	metabolism	agonizing
THE FRIEND OF CONSCIOUSNESS		compassion love	rapport & attunement — oneness	creative unity empathy & community	immunity	hesitation
THE HIGHER UNCONSCIOUS		clairkinesis	creative intelligence	production w/ energy force & power	digestion	indecision
THE MIDDLE UNCONSCIOUS	*REPRODUCTION	clairempathy	emotional intelligence	manifest * desire emotional connection	development	premeditation
THE LOWER UNCONSCIOUS		clairsentience	body intelligence instinct subconscious	suranimal navigate physical world	actions	procrastination

PANEL .5

INTELLIGENCES	ARCHETYPES	EXISTENCE 7 GREAT POWERS	CONTROLLED SYSTEMS	SUBTLE BODIES	COLORS	REGISTERS	LOCATION
			QUALITATIVE MATRIX				
LOGIC INTELLIGENCE	Jophiel gratitude	Intelligence	body rhythm	the causal body	golden (w/threads)	SPIRITUAL	crown
SPATIAL INTELLIGENCE	Raphael healing the planet & people	Imagination	growth	the celestial body	opalescent pastel colors w/golden silver light		forehead third eye
LINGUISTIC INTELLIGENCE	Michael life purpose	vision	metabolism	the etheric template body	cobalt blue	UPPER	throat
MUSICAL INTELLIGENCE	Chomuel world peace	vibration	immunity	the astral body	rose-tinted colors	MIDDLE RANGE	heart
KINESTHETIC INTELLIGENCE	Uriel divine guidance	doer — doing	digestion	the mental body	bright yellow	LOWER PHYSICAL	solar plexus
INTERPERSONAL INTELLIGENCE	Gabriel creativity	becoming	development	the emotional body	multicolor clouds of energy		lower abdomen
INTRAPERSONAL INTELLIGENCE	Zadkiel spiritual purpose	power/will	actions	the etheric body	light blue/grey		perineum

DeSign Science in The New Paradigm Age

PANEL.6

REGISTERS	LOCATION	PSYCHOLOGICAL FUNCTION	BODY PARTS OTHER	MALFUNCTION	SEED SOUNDS	VOWEL SOUND	
UPPER SPIRITU	Crown	understanding knowing	cns cerebral cortex	depression alienation confusion		aum	systems of the body
	Forehead Third eye	clairvoyance imagination	eyes	blindness headaches nightmares	om	"mmm" as in "man"	organs
	Throat	communication creativity	throat, ears, mouth, arms, hands	thyroids colds flu	ham	"eee" as in "see"	tissues
MIDDLE RANGE	Heart	love balance	lungs, heart, arms, hands	asthma, high blood pressure	sam yam	"ah" as in "father" "ay" as in "play"	cells
LOWER PHYSICAL	Solar plexus	will power	stomach musculature	ulcers, diabetes, hypoglycemia	ram	"ah" as in "father"	molecules
	Lower abdomen perineum	emotion, sexuality survival, grounding	womb, genitals, kidneys, bladder legs, bones, large intestine	impotence, frigidity, uterine or bladder problems obesity, hemorrhoids, constipation	vam lam	"oo" as in "due" "o" as in "rope"	chemicals atoms

PANEL .7
DISSOLUTION

THREEFOLD INTERPOLATION: W/ DISTRIBUTION INTO THE SEVENFOLD MATRIX™

FLAVORS OF EXPRESSIONS (TO INFINITY–1)										Copyright H. G. Bennett 9. 16. 2016
VIBRATION								**MATTER (DISSOLUTION)**		
∞	3						SP		3	9: PURE LIQUID (LLL)	H_2O
2F	2						R				
16F	1						T SSS	8: LGL = (L>G)		
	3						M	7: GLG = (G>L)		
12F	2						N	7'LSL=(L>S) or (s<L)	2	
10F	1						D MMM	6 pure gas G G G		earth elements
	3						B	more gas 5gsg =g>s		
6F	2						O D Y	5'sls= (s>l) or (l<s)	1	
4F		1					BBB	4sgs 1/3 pure solid (SSS)		

more solid less gas	than than	gas solid
less solid	than	gas
more gas	than	solid

Dissolution — distribution
s<g
3 elements →7FLAVORS (SYNTHESIS)
g>s

D_ESign Science in The New Paradigm Age

PANEL .8

By Trisepasis; Tricomisis and (Triadic Separations; Triadic Combinations) Copyright H. G. Bennett 9. 16. 2016
PhenomENA

S P I R I T	①	Superconscious	Thoughts	E	Creation	7 Hermetic Principles		Lydian Instruments	Polyhymnia
						Symmetry laws MENTALISM	Sacred Poetry	Semitone	N/A
									Enterpe
	②		Cognition Intuition		Mentalism	Correspondence	Love	Tone	Erato
	③		Art			Vibration	Tragedy	Tone	MelPomene
M I N D	①	Conscious	Awareness	Ti		Mental (Heart) Rhythm via the brain cells	energy	Love poetry		Terpsichore
	②					Nerve	time	Sacred Dance	Tone	
						Rhythm (Feelings)				
	③					Measured Rhythm	space	Dance Music		
B O D Y	Spiritude ③ (Energy)	Subconscious	Creativity	S	Space	Polarity Opposites		Epic poetry	Semitone	Calliope
	Psychological ②		Feelings Emotions			Gender & Symmetry Alignments	sga	History	Tone	Clio
	Physical ①		Art Work Media		Physical Object	Physical Alignment CAUSE AND EFFECT		Comedy	Tone OCTAVE	Thalia

231

DeSign Science in The New Paradigm Age

PANEL. 9

SYNTHESIS / INTERPOLATIONS / EXTRAPOLATIONAS AS THE SYMMETRY & DISTRIBUTION LAW

⑤	PHENOMENA OR NATURAL EXPRESSION									
HUMAN ANATOMY	**ORGANS**	**HEAD**	**EMBRYO**	**SCIENCES**	**GLANDS**	**BRAIN DIVISION**	**BODILY SYSTEMS FUNCTIONS**	**CHRISTIAN VICES**	**VIRTUES I**	**VIRTUES I**
TOTALITY	brain	ears	brain	logic	pineal	cerebellum	assimilation	pride	faith	pride
5 FOLD SYMMETRY	kidney	eyes	chorionic villi	rhetoric	pituitary	cerebellum	sensation	envy	hope	envy
THORAX	lungs	nostrils	umbilical cord	grammar	thyroid	pons varolii	respiration	Sloth dejection	diligence	sloth
ARMS	heart		heart	music	thymus	medulla oblongata	circulation	anger	charity	anger
ABDOMEN	stomach	mouth	spinal cord	astronomy	pancreas	corpus callosum	evacuation	gluttony	fortitude	gluttony
	genitals		amnion	arithmetic	testes/ ovaries	Spinal cord		lust	justice	lust
LEGS	liver		yolk sac	geometry	adrenals	meninges	reaction	avarice	temperance	liberality

PANEL .10		THOUGHTS SLEEP STILLNESS		Copyright H. G. Bennett 9. 16. 2016
		MINDFUL / GAP		
TAROT SUITS	**QABALA**	**FLOW MEDITATION**		
	KETHER	LIBERATION CONNECTION TO SOURCE		
	BINAH CHOKMAH	CHOICE MAKING JUDGMENT CONSCIOUS CHOICE		
	GEBURA CHESED	PERMISSION TO SPEAK TRUTH TO POWER		
SWORDS	TIPHARETH	GREEN-BLACK PINK SELF-COMPASSION		
WANDS	HOD HITZACH	CITY OF JEWELS- COMPLETION TRANSFORMATION		
CUPS	YESOD	NEW DIRECTION- NEW DAO NEW LIFE		
PENTACLES	MALKUTH	FAMILY-COMMUNITY- STILLNESS		

DeSign Science in The New Paradigm Age

PANEL .11
OPERATIONS

PHENOMENA						
VERB	**ANATOMY**	**ANIMALS**	**EMOTIONS**	**PSYCHOLOGICAL FUNCTION**	**THOUGHTS MIND MAPPING VALUE CREATION**	**FUNCT-IONS**
	BODY PARTS					
I KNOW	CNS CRBL CTX	unicorn	bliss	understanding knowing	I think (need)	rational thinking
I SEE	EYES	owl	dreaming	clairvoyance imagination	I visualize so	imagination
I SPEAK	THROAT EARS MOUTH ARMS HANDS	deer	expansion excitement	communication creativity clairaudience	I speak to myself I speak to others I communicate	problem solving
HEART — I LOVE	LUNGS HEART ARMS HANDS	antelope	love passion compassion	learning by heart love balance intuition	I love my idea or thought I am attracted to it it was me too (desire) it moves my spirit I have, I see and I feel	realism testing compassion
I CAN I WILL	STOMACH MUSCULATURE	ram	anger, joy, laughter	will, power (breathing)	I will do / make it systems & methods	power fuel
I FEEL	WOMB GENITALS KIDNEYS BLADDER	crocodile	desire tears	emotions sexuality	I feel need system technologies	fecundity
I HAVE	LEGS BONES LARGE INTESTINES	elephant	stillness	survival grounding	materials (need) reality	Measuring; the symmetry of "making"

D_ESign Science in The New Paradigm Age

PANEL .12

TRANSLATE THIS QUALITATIVE SYSTEM INTO A QUANTITATIVE TECHNOLOGY™

ESSENSES PSYCHOLOGY	CREATION	GEMSTONES	METALS	HORMONES	KINGDOMS OF CREATION	THE WORLDS
SPIRITUALITY	BRAIN 1. DURA MATER 2. PIA MATER 3. ARACHNOID	sapphire	Mercury	melatonin	light particles	absolute
CREATIVITY	GRANTING ALL DESIRED THINGS	jet	Silver	growth hormone somatotrophin	nuclear particles	all worlds
PHILOSOPHY	COSMIC LOVE	sodalite	Copper	thyroxine	atoms	all suns
HARMONY	EMOTIONAL ROMANTIC LOVE OM: NAHAH SIVAY	emerald	Gold	lymph	molecules	our sun
SELF EFFORT	ENERGY	topaz	Tin	borenaline insulin calvagon	plants	all planets of the solar system
FEELINGS	EMOTIONS	carnelian	Iron	testosterone oestrogen progesterone	animals	moon
FEAR	PRANA, AGUI, BINDU, NADA (SAKTI)	ruby	Lead	epinephrine (thymosin)	man	earth

NOTES: THE 7 METALS ARE: IRON 26, COPPER 29, SILVER 47, TIN 50, GOLD 79, MERCURY 80, LEAD 82
Copyright H. G. Bennett 9. 16. 2016

METALS	SYMBOL	ATOMIC NUMBER	ATOMIC WEIGHT	REMARKS
IRON	Fe	26	55.847	
COPPER	Cu	29	63.54	
SILVER	Ag	47	107.870	
TIN	Sn	50	118.69	
GOLD	Au	79	196.967	
MERCURY	Hg	80	200.59	
LEAD	Pb	82	207.19	

PANEL .13

CHARACTERS AND FLAVORS OF ALL EXPRESSIONS

CHAKRA	ONE	TWO	THREE	FOUR	FIVE	SIX	SEVEN
Sanskrit Name	Muladhara	SvadhIsthana	Manipura	Anahata	Visuddha	Ajna	Sahasrara
Location	Perineum	Lower Abdomen	Solar Plexus	Heart	Throat	Forehead, Third Eye	Top of Head, Crown
Element	Earth	Water	Fire	Air	Ether Sound		
Psychological Function	Survival Grounding	Emotions Sexuality	Will, Power	Love, Balance	Communication, Creativity	Clairvoyance, Imagination	Understanding, Knowing
Emotion	Stillness	Desire, Tears	Anger, Joy, Laughter	Love, Compassion	Expansion, Excitement	Dreaming Astra/level	Bliss
Glands	Adrenals	Ovaries, Prostate, Testicles	Pancreas	Thymus	Hypothalamus, Thyroid	Pineal Gland	Pituitary Gland
Other Assoc'D Body Parts	Legs Bones, Large intestines	Womb, Genitals, Kidneys, Bladder	Stomach, Musculature	Lungs, Heart, Arms, Hands	Throat, Ears, Mouth, Arms, Hands	Eyes	CNS, Cerebral Cortex
Malfunctions	Obesity, Hemorrhoids, Constipation	Impotence, Frigidity, Uterian/Bladder Troubles	Ulcers, Diabetes, Hypoglycemia	Asthma, High Blood Pressure	Thyroid, Colds, Flu	Blindness Headaches Nightmares	Depression, Alienation, Confusion
# Of Petals	4	6	10	12	16	2	1,000+
Colors	Red	Orange	Yellow	Green	Blue	Indigo	Violet
Seed Sound	Lam	Vam	Ram	Sam, Yam	Ham	Om	
Vowel Sound	"Oh" as in "Rope"	"oo" as in "due"	"ah" as in "father"	"ay" as in "play"	"ee" as in "see"	"mmm" – "nnn"	
Tarot Suits	Pentacles	Cups	Wands	Swords			
Verb	I have	I feel	I can	I love	I speak	I see	I know, I am
Qabala	Malkuth	Yesod	Hod, Netzach	Tiphareth	Gebura, Chesed	Binah, Chokmah	Kether
Planets	Earth, Saturn	Moon	Mars, Sun	Venus	Mercury, Neptune	Jupiter	Uranus
Animals	Elephant	Crocodile	Ram	Antelope	Deer	Owl	

PANEL .14

	FIRST Chakra	**SECOND** Chakra	**THIRD** Chakra	**FOURTH** Chakra	**FIFTH** Chakra	**SIXTH** Chakra	**SEVENTH** Chakra
	THE ROOT	**THE SACRAL**	**SOLAR PLEXUS**	**HEART**	**THROAT**	**THIRD EYE**	**THE CROWN**
Sanskrit Name	MULADHARA, meaning root or support	SVADHISTHANA, or sweetness	MANIPURA, or lustrous gem	ANAHATA, or unstruck	VISHUDA, or purification	AJNA, to perceive, or to know	SHASRARA, or thousand-fold
Location	Between the anus and the genitals	Lower abdomen, between navel and genitals	Between navel and base of sternum	At the center of the chest	Centrally at the base of the neck	Above and between the eyebrows	Top of the head or crown of the head
Associated Color	Red	Orange	Yellow	Green	Blue	Indigo	Violet, gold, white
Main Issues	Survival/physical needs like security and shelter	Sexuality, emotional balance	Personal power and self-will; self-esteem	Love and relationships	Communication and self-expression	Intuition, wisdom	Spirituality, depression, confusion, sensitivity to pollutants, epilepsy, Alzheimer's
Endocrine Gland	Adrenals	Ovaries/testes	Pancreas	Thymus	Thyroid and parathyroid	Pineal	Pituitary
Body Parts	Bones, skeletal structure	Sex organs, bladder, prostate, womb	Digestive System, muscles	Heart, lungs, chest and circulation	Mouth, throat, and ears	Eyes, and the base of the skull	Upper skull, cerebral cortex, skin
Astrological Sign	Capricorn	Cancer Scorpio (a passionate tendencies)	Aries, Leo	Libra, Taurus	Gemini, Virgo	Sagittarius, Pisces	Aquarius
Sense	Smell	Taster	Sight	Touch	Sound/Hearing	Sixth Sense	Beyond any known senses

OTHER SEVEFOLD SYSTEMS
The Seven Churches of the Revelation of Jesus Christ to St. John have many meanings.

SEVEN CHURCHES	CHAKRAS: the Hindu system
The Church of EPHESUS	The Church of Ephesus is the Base Chakra or the root muladhara Center.
The Church of SMYRNA	The Church of Smyrna is the Svadhisthana Chakra in the Hindu system, or the Sacral Centre.
The Church of PERGAMOS	The Church of Pergamos is the Solar Plexus Chakra.
The Church of THYATIRA	The Church of Thyatira is the heart or Anahata Chakra.
The Church of SARDIS	The Church of Sardis is the throat or Visuddha Chakra.
The Church of PHILADELPHIA	The church of Philadelphia is the third eye or Ajna (Brow) Chakra.
The Church of LAODICEA	The Church of Laodicea is the Sahasrara or Crown Chakra.

The Book of Genesis (The gene of Isis); The Chakras and the Seven Days of Creation.
Another tradition which expresses the process of re-creation is found hidden in the book of Genesis. Emanuel Swedenborg, the great Swedish mystic, taught that hidden with the seven days of creation was a description of our regeneration. Arcana Celestia, Emanuel Swedenborg. (AC 39) Presented by Julian Websdale, Humans AreFree.com; |From the Institute for Gnostic Studies. (1996-2000). Gnostic Theurgy).

DAYS	CENTER/CHAKRA	ATTRIBUTE
The First Day of Regeneration	The Muladhara Chakra recreation of New Man: the coming of light	The base Chakra and transmutes its desires and drives into something of a higher nature.
Second Day	The Svadhisthana Chakra; division of the things of the Dialectic from those of the Static. Make decisions about what we serve.	Correspondence as above so below. The Gnosis or the World, the instincts or higher things.
The Third Day	Nabhi or Solar Plexus The dry land is concrete know-ledge; Seeds and vegetation are the spiritual wisdom, fruits blossom inner truths.	A 'solid' experience of the Gnosis; various forms of knowledge (seas) are bought together in SYNTHESIS.
The Fourth Day	The Anahata Chakra; The sun; divine love, (Agape), moon; divine faith and stars; divine wisdom appear	The great luminaries are kindled; placed in the internal man; the external receives light from them; man first begins to live.'
The Fifth Day	The Visuddha Chakra. These reflect the facets of the mind for new levels of consciousness to be reached. The Logos regenerates the mind	Rejuvenated and reborn through Transfiguration. The ebb and flow of the unconscious
The Sixth Day	The Chakra is the Ajna center, This marks the appearance of the reborn man.	It is indigo and represents intuition and inner self-mastery
The Seventh Day	The Chakra is Sahasrara, center. It is the birth of the New Man.,	Regeneration/ Integration/Self Realization is complete.

 # KEY PHRASES

ILLUSTRATION GALLERY

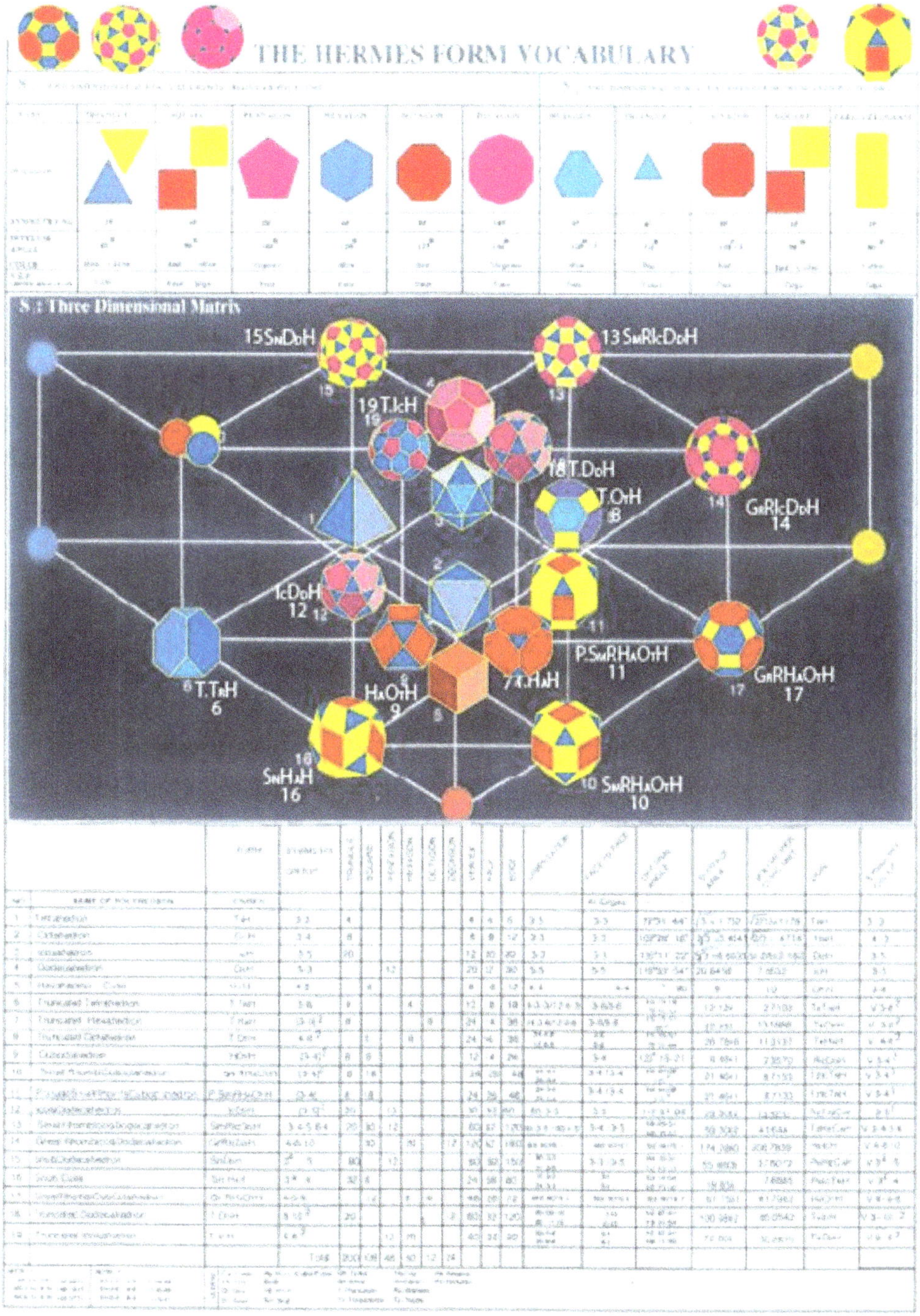

THE 19 STANDARD MODELS OF THE EUCLIDEAN POLYHEDRA

DeSign Science in The New Paradigm Age

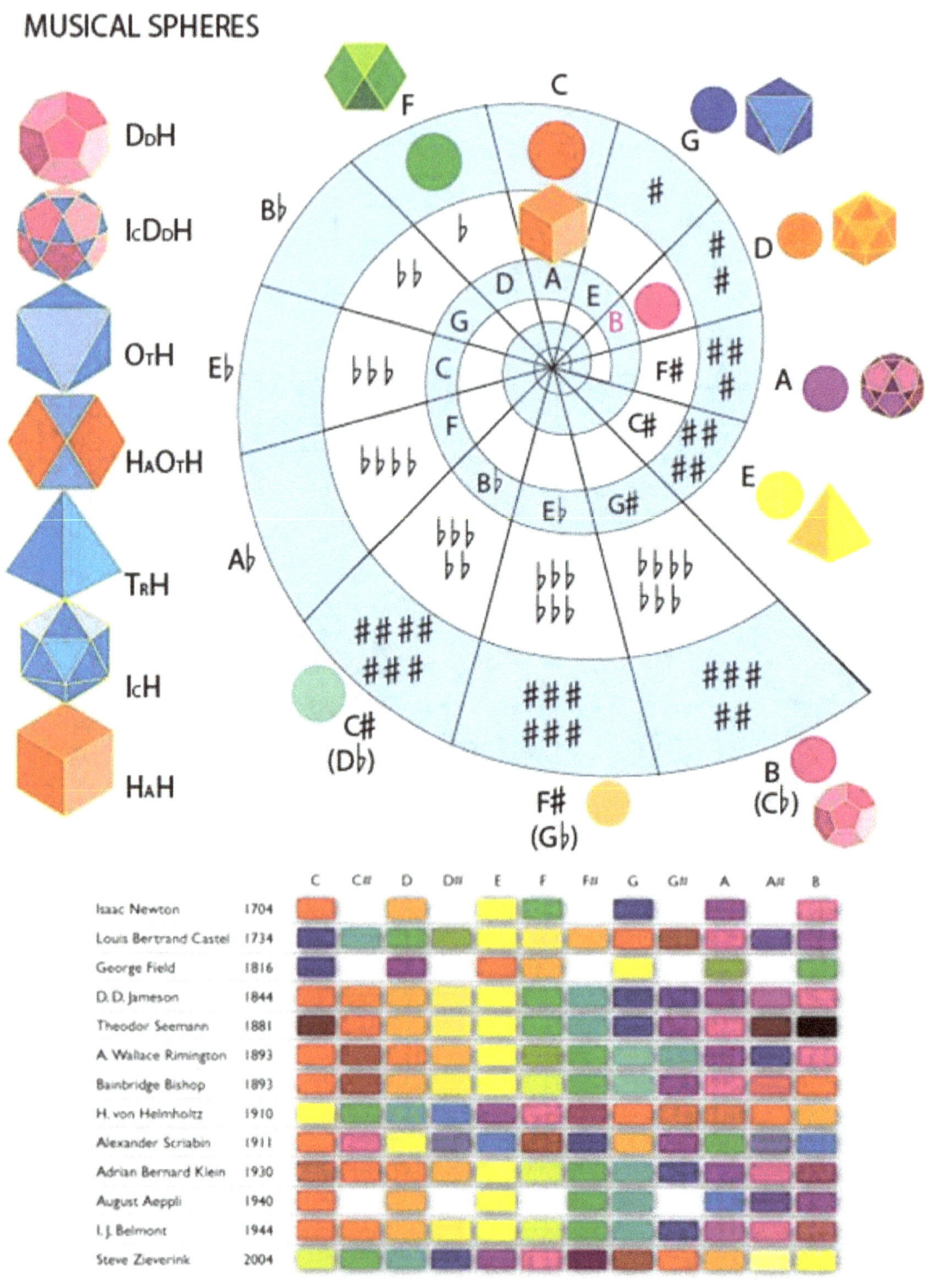

MUSIC-FORM AND COLOR SYNTHESIS

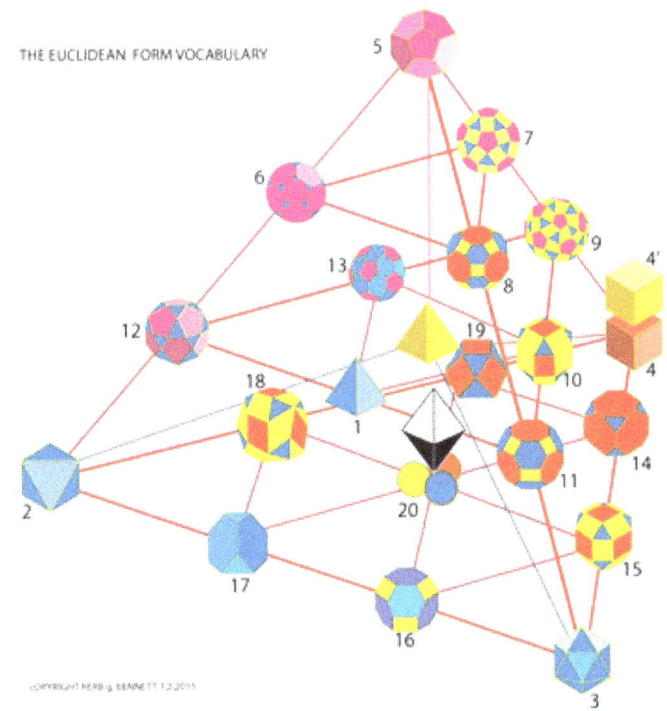

THE TETRAHEDRAL TESSERACT OF THE EUCLIDEAN POLYHEDRA

GOBELKI TEPE MONUMENT

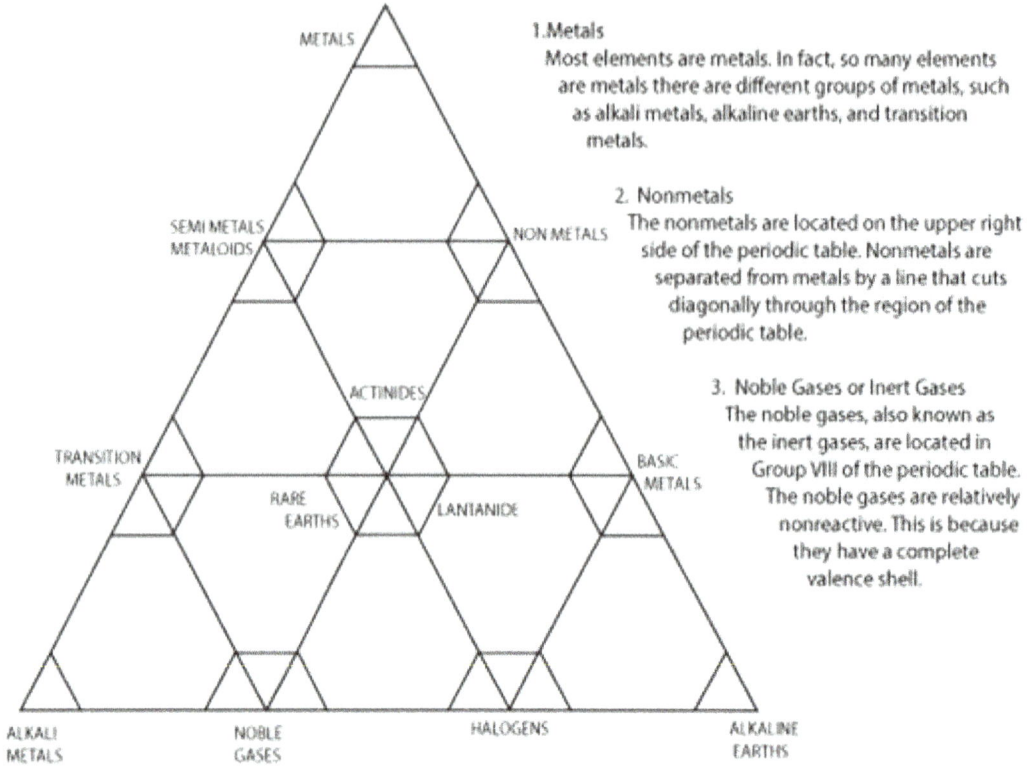

1. Metals
Most elements are metals. In fact, so many elements are metals there are different groups of metals, such as alkali metals, alkaline earths, and transition metals.

2. Nonmetals
The nonmetals are located on the upper right side of the periodic table. Nonmetals are separated from metals by a line that cuts diagonally through the region of the periodic table.

3. Noble Gases or Inert Gases
The noble gases, also known as the inert gases, are located in Group VIII of the periodic table. The noble gases are relatively nonreactive. This is because they have a complete valence shell.

4. Halogens
The halogens are located in Group VIIA of the periodic table. Sometimes the halogens are considered to be a particular set of nonmetals. These reactive elements have seven valence electrons.

5. Semimetals or Metalloids
The metalloids or semimetals are located along the line between the metals and nonmetals in the periodic table. The electronegativities and ionization energies of the metalloids are between those of the metals and nonmetals, so the metalloids exhibit characteristics of both classes.

6. Alkali Metals
The alkali metals are the elements located in Group IA of the periodic table. The alkali metals exhibit many of the physical properties common to metals, although their densities are lower than those of other metals.

7. Alkaline Earths
The alkaline earths are the elements located in Group IIA of the periodic table. The alkaline earths possess many of the characteristic properties of metals. Alkaline earths have low electron affinities and low electronegativities. As with the alkali metals, the properties depend on the ease with which electrons are lost.

8. Basic Metals
Metals are excellent electric and thermal conductors, exhibit high luster and density, and are malleable and ductile.

9. Transition Metals
The transition metals are located in groups IB to VIIIB of the periodic table. These elements are very hard, with high melting points and boiling points. The transition metals have high electrical conductivity and malleability and low ionization energies.

10. Rare Earths
The rare earths are metals found in the two rows of elements located below the main body of the periodic table. There are two blocks of rare earths, the lanthanide series and the actinide series. In a way, the rare earths are special transition metals, possessing many of the properties of these elements.

11. Lanthanides
The lanthanides are metals that are located in block 5d of the periodic table. The first 5d transition element is either lanthanum or lutetium, depending on how you interpret the periodic trends of the elements.

12. Actinides
The electronic configurations of the actinides utilize the f sublevel. Depending on your interpretation of the periodicity of the elements, the series begins with actinium, thorium, or even lawrencium. All of the actinides are dense radioactive metals that are highly electropositive.

THE SUCHNESS OF METALS

SYNTHESIS OF PLANETS, METALS AND THE CHAKRA ENERGY CENTERS IN MAN

Planetary Body:	Metal:	Western Chakra correspondence:
The Sun (Sol):	Gold	Heart Chakra
The Moon (Luna):	Silver	Brow Chakra - Third-Eye
Mercury:	- Mercury (original, I know)	Crown Chakra
Venus	Copper	Throat Chakra
Mars:	Iron	Sacral Chakra
Jupiter	Tin	Solar-Plexus Chakra
Saturn	Lead	Root Chakra

GOBEKLI TEPE IN SOUTHEATERN TURKEY 12,000 YEARS OLD

SYNTHESIS OF THE LOGIC STRUCTURE OVERLAID ON THE CONCEPTUAL MODEL OF GOBEKLI TEPE

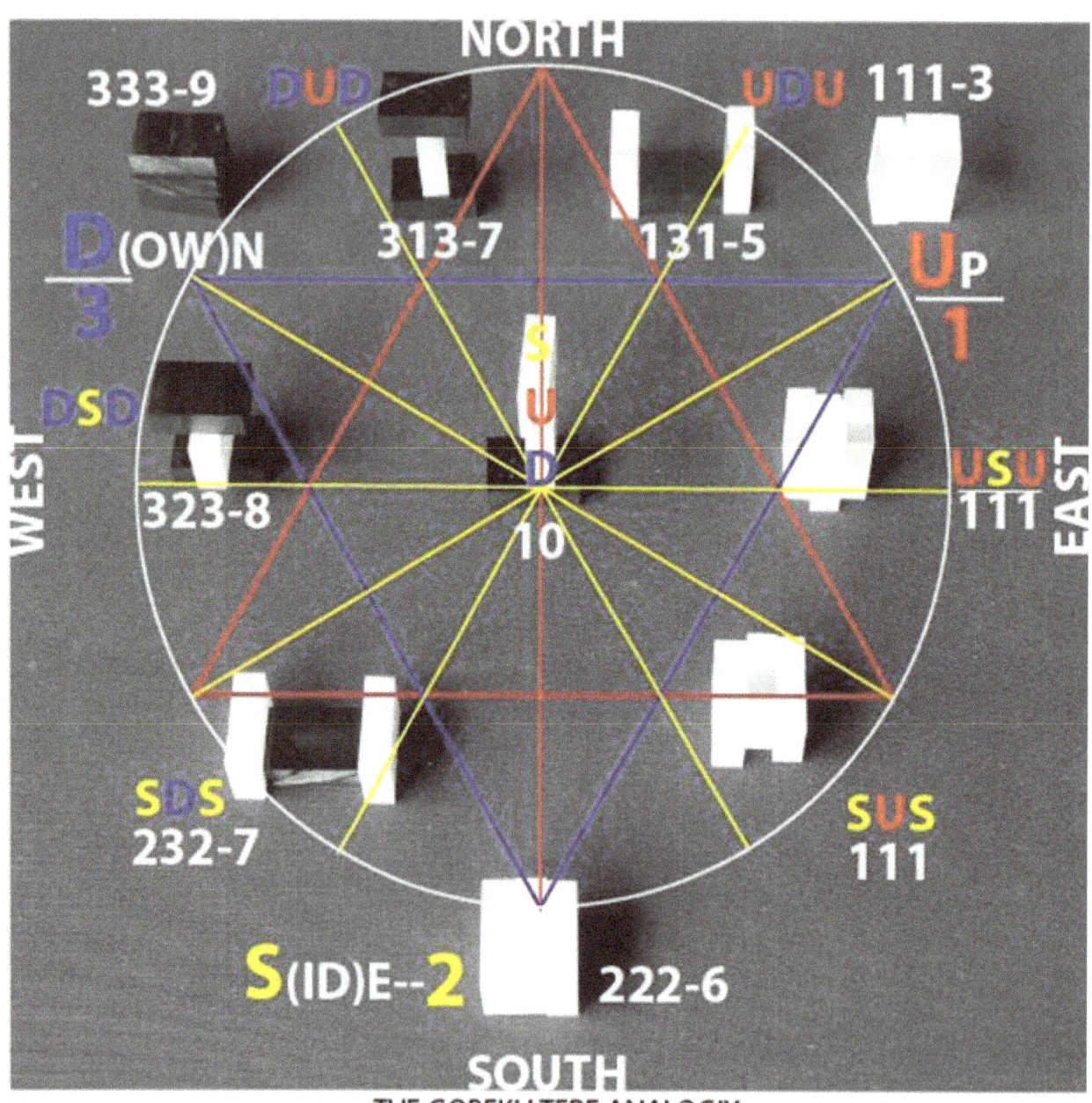

NUMBER-COLOR FORM-ORIENTATION [SPACE] AND LPOSITION IN ONE MATRIX

D_ESign Science in The New Paradigm Age

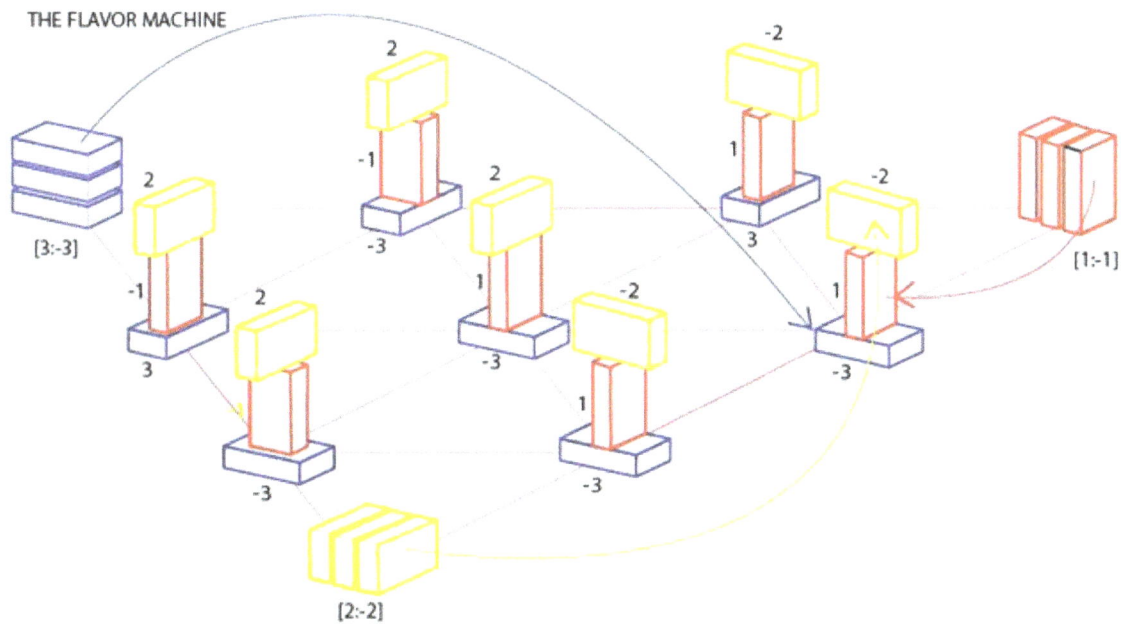

THE FLAVOR MACHINE

COMBINATORIAL LOGIC STRUCTURES

SINGULAR ELEMENTS USED IN A UNIQUE QUALITATIVE DISTRIBUTION MATRIX
TO PRODUCE THE SIX UNIQUE COMBINATIONS OF THREE (3) UNITS

copyright Herbert Glenn Bennett 9/1/2015

THE LOGIC STRUCTURE CONCEPTUAL MODEL & DIAGRAM   Copyright Adger Cowans

DeSign Science in The New Paradigm Age

DeSign Science in The New Paradigm Age

	TIME	ENERGY	SPACE
	PSYCHOLOGY/MIND	SPIRITUALITY/SPIRIT	PHYSIOLOGY/BODY
	SUBCONSCIOUS	**SUPERCONSCIOUS**	**CONSCIOUS**
SUBCONSCIOUS	SUBCONSCIOUS / INSTINCT / SUBCONSCIOUS	SUPERCONSCIOUS / INSPIRATION / SUBCONSCIOUS	CONSCIOUS / INTUITION / SUBCONSCIOUS
SUPERCONSCIOUS	SUBCONSCIOUS / INTELLIGENCE / SUPERCONSCIOUS	SUPERCONSCIOUS / THOUGHT / SUPERCONSCIOUS	CONSCIOUS / MOTIVATION / SUPERCONSCIOUS
CONSCIOUS	SUBCONSCIOUS / EMOTIONS / CONSCIOUS	SUPERCONSCIOUS / CREATIVITY / CONSCIOUS	CONSCIOUS / AWARENESS / CONSCIOUS

THE CONSCIOUSNESS MATRIX

THE SUN BEHIND THE SUN, OSIRIS, AND THE GREAT PYRAMID
The Principle of All Things. Beyond the Sun in the direction of the Dog Star lies that incorruptible flame or Sun, Principle of All Things, willing obedience from our own Sun but which is but a manifestation of its relegated force. The existence of the Sun behind the Sun has been known in all ages, as well as the fact that its influence is most potent upon earth during that period every 2000 years when it is in conjunction with the Sun of our solar system. Then gathering to itself the power of its own Source and transmitting it through our Sun to this planet, it is said to send the Sons of God into the consciousness of the earth sphere, that a new world of thought and emotion may be born in the minds of men for the stimulation of humanity's spiritual evolution. Such a manifestation marks the beginning or end of an epoch upon the earth by the radiation of that divine consciousnss known as the Christ Ray or Paraclete.

DeSign Science in The New Paradigm Age

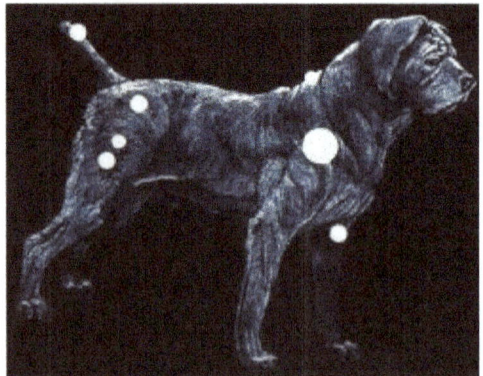

To the Egyptians the Sun behind the Sun was known as Osiris (and also as Amen-Ra, The Hidden Sun), said to be the husband of Isis (Nature) and the parent of Horus (the Sun), symbolically represented as a hawk because that bird flies nearest the Sun. This ancient people knew that once every year the Parent Sun is in line with the Dog Star. Therefore, the Great Pyramid was so constructed that, at this sacred moment, the light of the Dog Star fell upon the square "Stone of God" at the upper end of the Great Gallery, descending upon the head of the high priest, who received the Super Solar Force and sought through his own perfected Solar Body to transmit to other Initiates this added stimulation for the evolution of their Godhood. This then was the purpose of the "`Stone of God,' whereon in the Ritual, Osiris sits to bestow upon him (the illuminate) the Atf crown or celestial light." "North and South of that crown is love," proclaims an Egyptian hymn. "And thus throughout the teaching of Egypt the visible light was but the shadow of the invisible Light; and in the wisdom of the ancient country the measures of Truth were the years of the Most High (Marshall Adams, The Book of the Master," page 141-2)."

Modern science partially confirms these facts as to the significance of the Great Pyramid, but lacks the key to them. Dr. Percival Lowell, in a recent essay entitled "Precession and the Pyramids," says-- "The Great Pyramid was in fact a great observatory, the most superb one every erected," and "The Great Gallery's floor exactly included every possible position of the Sun's shadow at noon from the year's beginning to its end. We thus reach the remarkable result that the gallery was a gigantic gnomon or sundial telling, not like ordinary sundials the hour of the day, but on a more impressive scale, the seasons of the year."

Excerpted from the Comte de Gabalis, originally by the Abbe N. do Montfauconde Villars (1670)

Copyright Adger Cowans

THE CONIC TRANSFORMATION OF THE DELTAHEDR

THE TREFOIL PLATE AND DELTAHEDRA

THE EARTH STAR (FUSED NON-EUCLIDEAN TETRAHEDRA)

CERAMIC EARTHSTAR TEAPOT (NOEL COPELAND COLLABORATION AND PRODUCTION)

THE ENDOCRINE SYSTEM

THE CHAKRAS CORRESPOND WITH THIS SYSTEM AS THE ELECTRICAL AND BIOLOGICAL OF A UNIVERSAL PHENOMENON

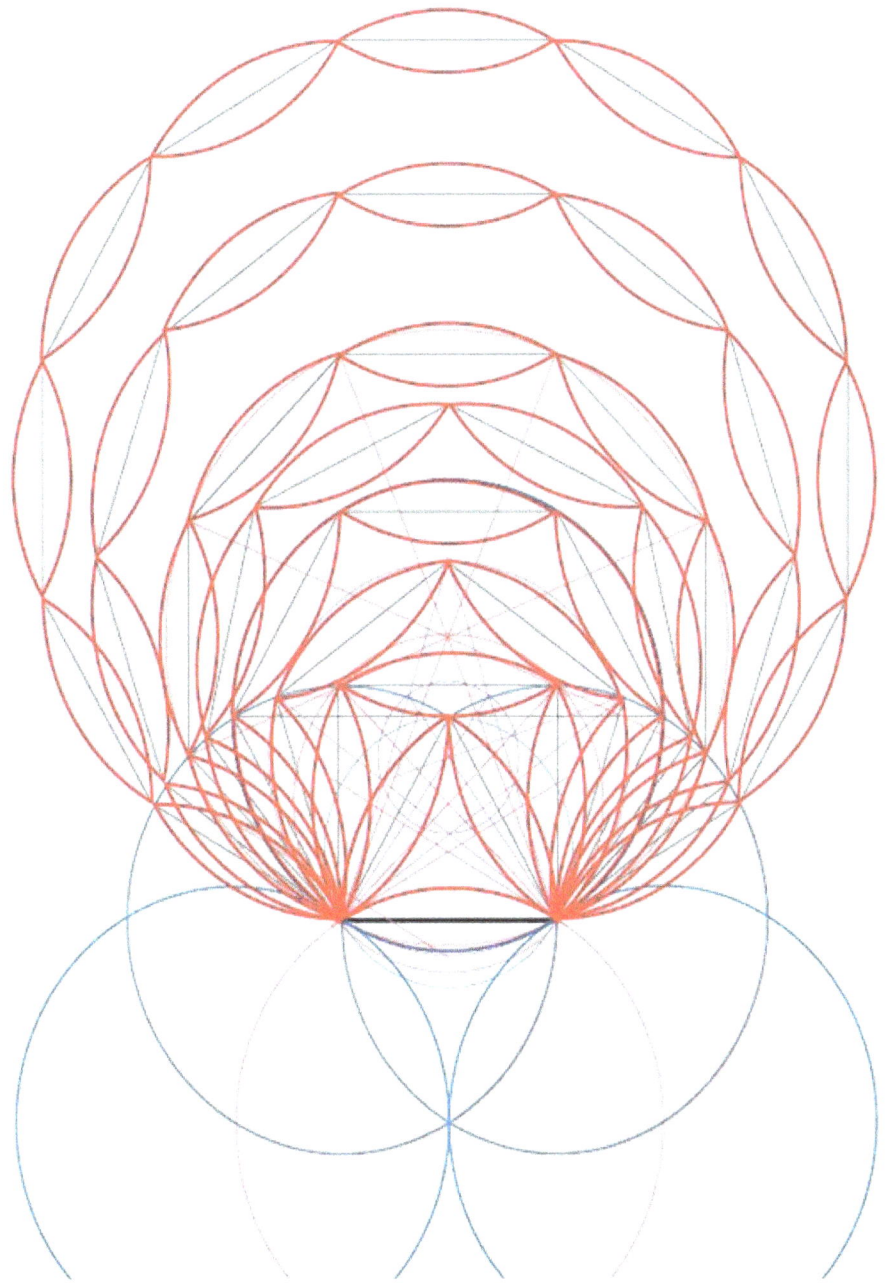

THE NON-EUCLIDEAN POLYGON SET GENERATED FROM THE TREFOIL PATTERN OF INTERSECTING CIRCLES OF THE SAME RADII.
THEIR EDGES ARE ALL RESOLVED WITH DIGONS OF THE THREE SYTEMS OF GEOMETRY

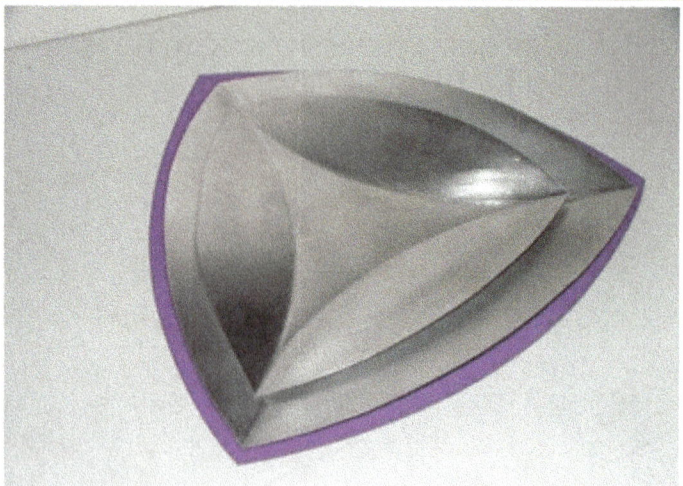

TREFOIL METAL PLATES OR SIMILAR ARTICLES

PLUSH FLIP CUBE TOY

MEDIA CUBE

PLUSH DODECA DOLL SET

DeSign Science in The New Paradigm Age

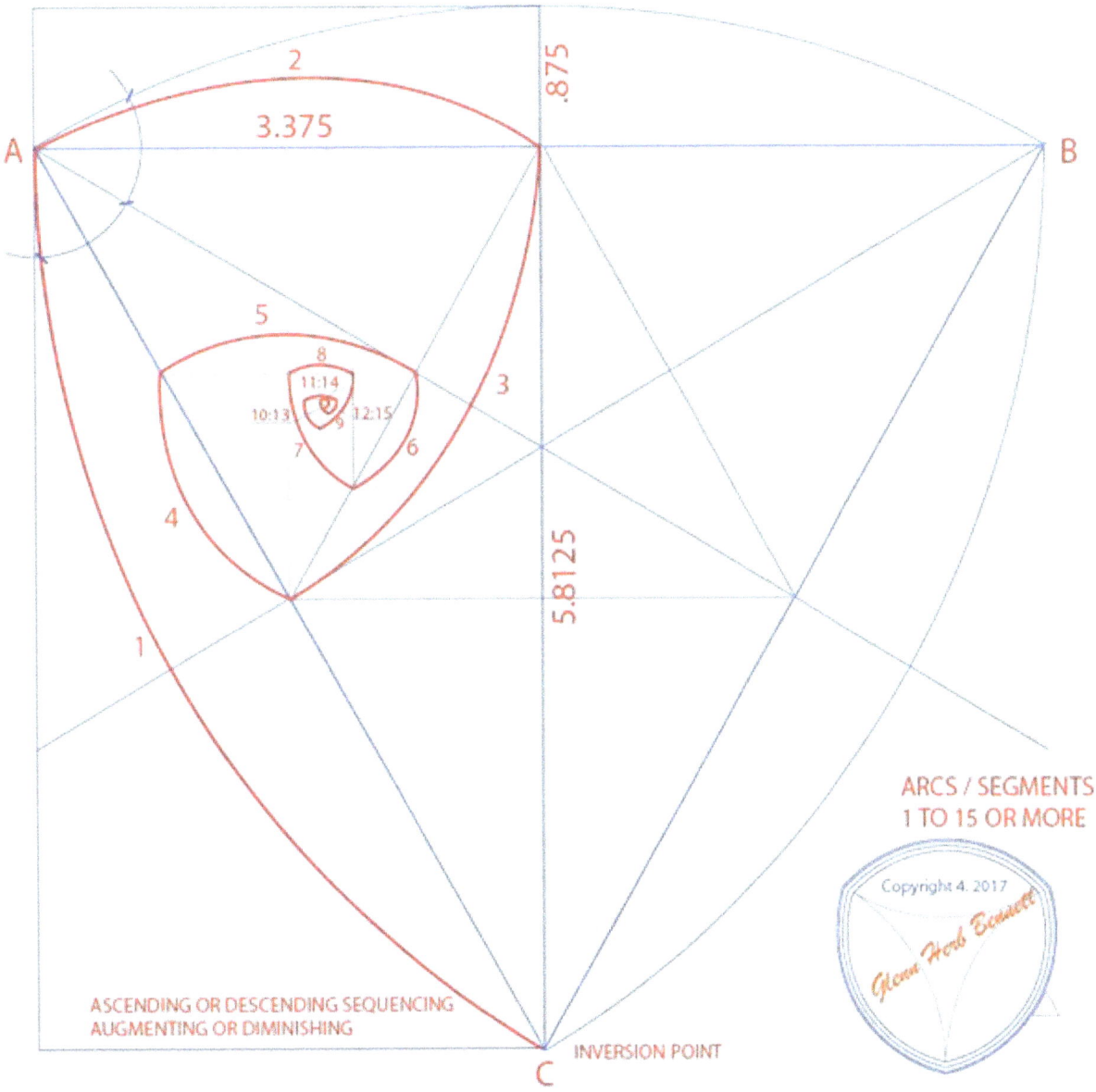

A SPIRAL OF TREFOIL SYMMETRY

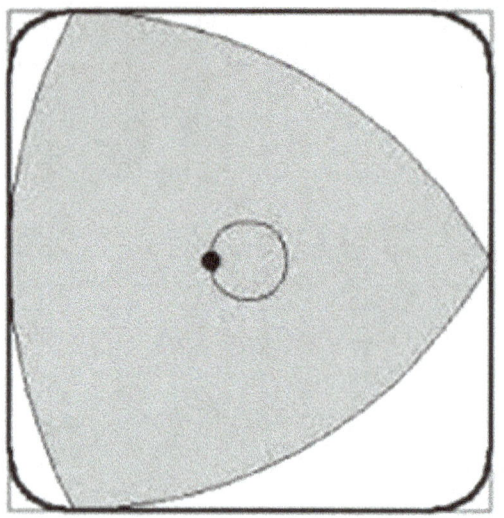

REULEAUX TRIANGLE –TREFOIL LOGO

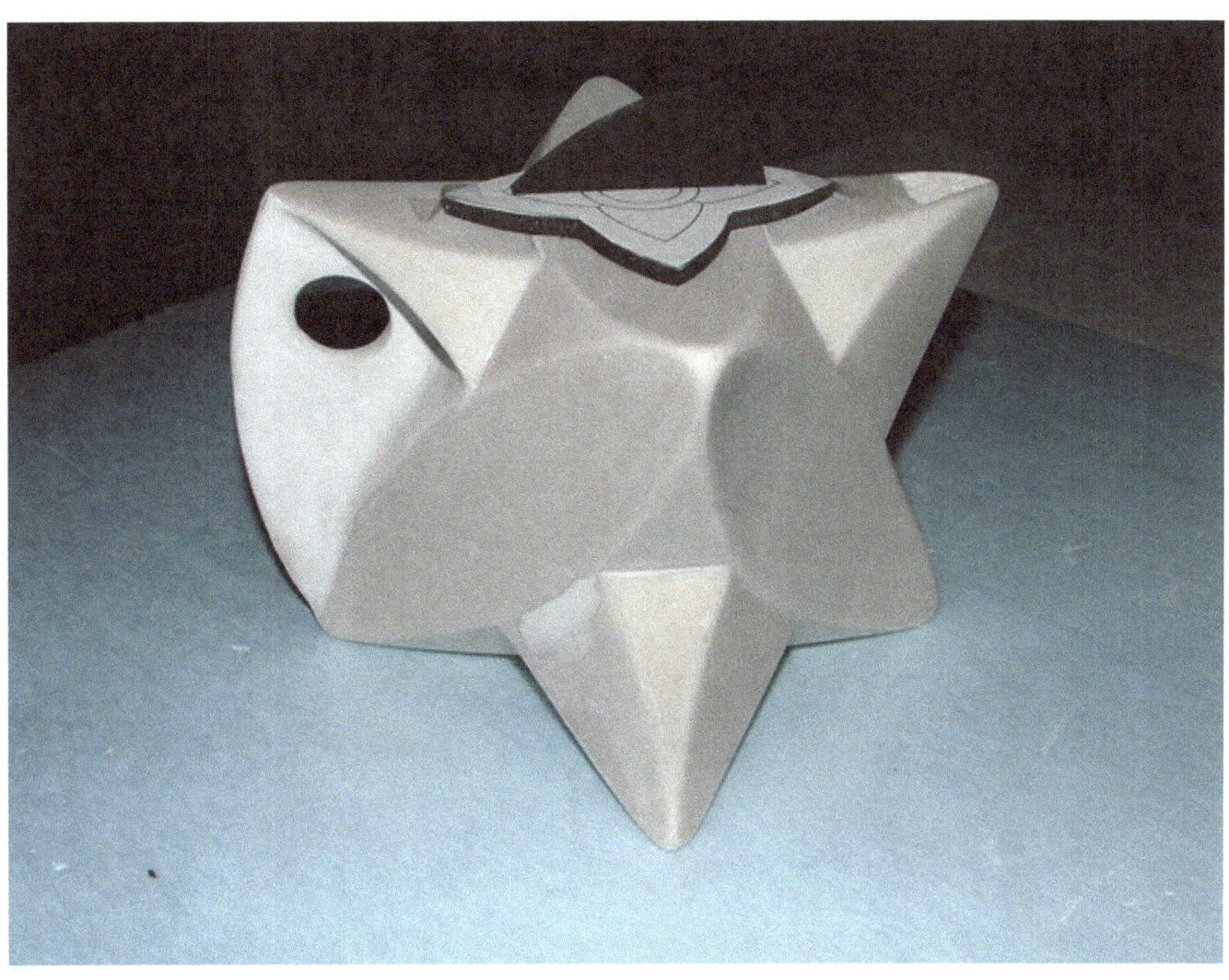

EARTHSTAR CERAMIC TEAPOT (SCULPTURE HOUSE CASTING NYC COLLABORATION AND PRODUCTION)

TETRAHEDRON WITH DELTAHEDRON ELEMENTS

MODEL CONTAINER

DELTAHEDRA [D$_A$H] CONSTELLATIONS

STAR SCRAPER TOWER CREATED WITH STACKED EARTH STARS AND A TETRAHEDRAL SPACE FRAME

PORT NON EUCLIDEON

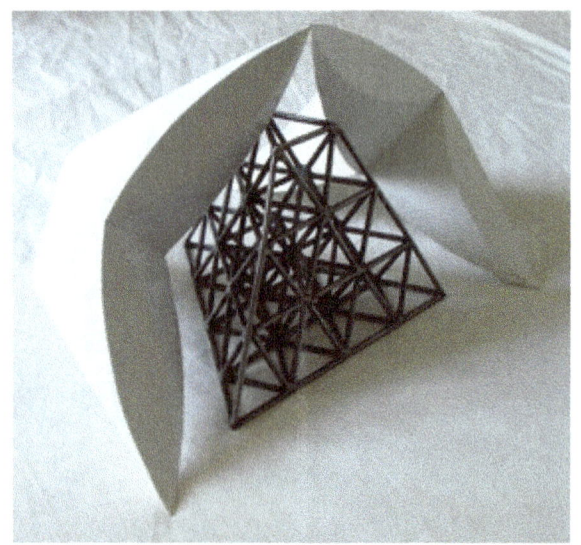
DELTAHEDRON SPACEFRAME

ARCH-CUBOID MODULES

A FOLDED TOWER TORI

[A MEDITATION FOR PEACEFUL CENTERING]

BLISS: AN ACRYLIC PAINTING ON A 4'X8' CANVAS   Copyright H. G. Bennett

DeSign Science in The New Paradigm Age

ELEMENTS FROM MONOPRINTS DONE IN ACRYLIC DIGITIZED AND ENHANCED IN VARIOUS COMPOSITIONS FOLLOWING STRICT SYMMETRY RULES TO GENERATE IMAGES PRINTED OF VARIOUS TWO AND THREE DIMENSIONAL MEDIA

DIGITAL ART FROM THE WALLPAPER EXTRACTION SERIESCopyright H. G. Bennett

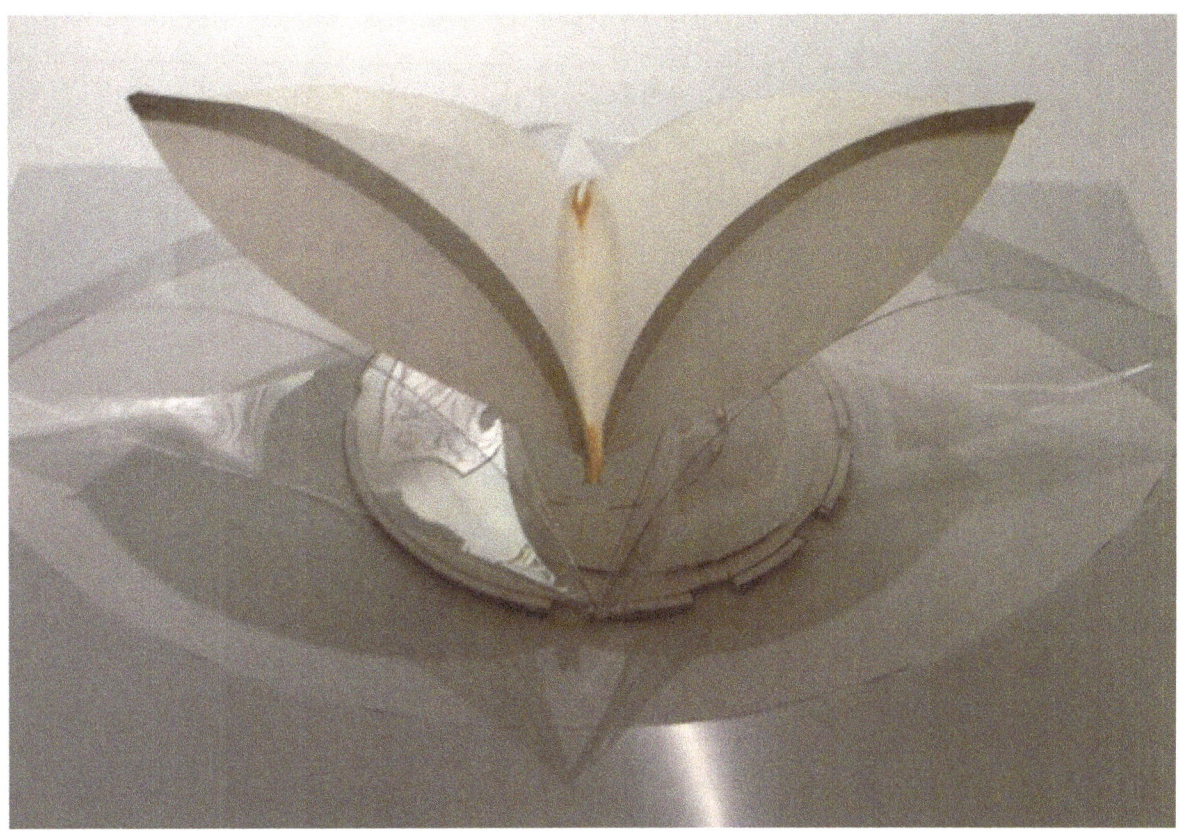

DELTAHEDRON TRANSFORMED INTO STRUCTURAL ARCHITECTONIC ELEMENT

EARTHSTAR MADE OF FUSED TETRAHEDRAL FORMS FORMING CUBIC DUALS

THE VIBRATIONAL 'CHLADNI' PATTERNS OF SOUND IMBUED INTO THE NON-EUCLIDEAN PROPORTIONS OF 3D FORMS

THE TETRAHEDRON AND OCTAHEDRON

MAKING ART-INTO AN-"ISM" (ANARTISM)

BENNETT TORI BENT INTO THREE AND FOUR FOLD "CLOVERS"

THE STRUCTURE OF AN-ISM" (ANARTISM)

THE STRUCTURE OF AN-ISM" (ANARTISM)

STRUCTURAL NON-EUCLIDEAN ARCHITECTONIC FORM ELEMENTS NO 1

STRUCTURAL NON-EUCLIDEAN ARCHITECTONIC FORM ELEMENTS NO2

STACKED STRUCTURAL NON-EUCLIDEAN DELTAHEDRAL FORM ELEMENTS

THE DELTA STAR

THE EGYPTIAN CANON WITH MAN'S TREFOIL PROPORTION FOR SACRED SCIENCE AND TEMPLE ARCHITECTURE

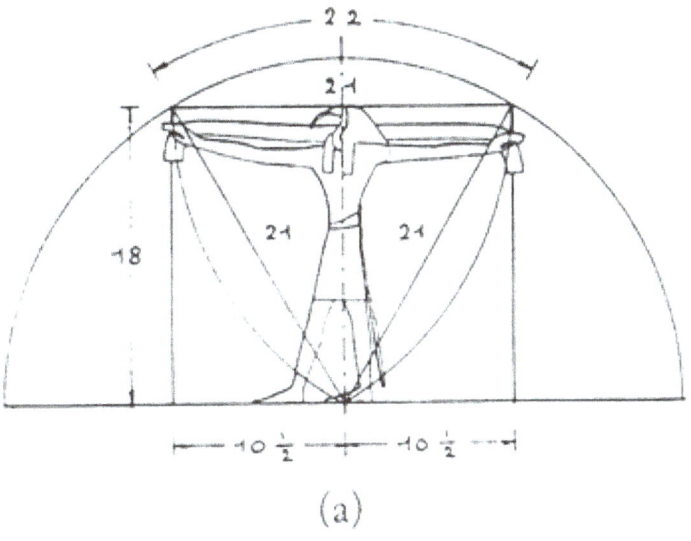
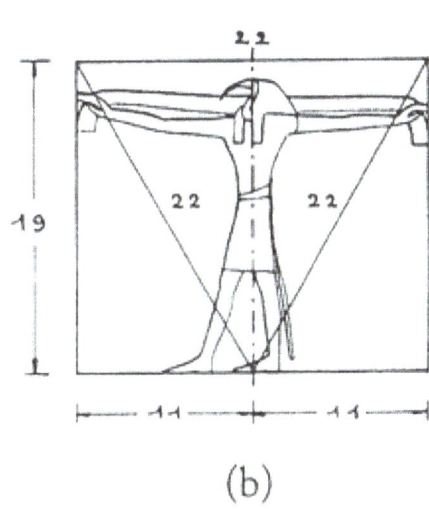

(a)　　　　(b)

EUCLIDEAN EARTHSTAR AND MAPS

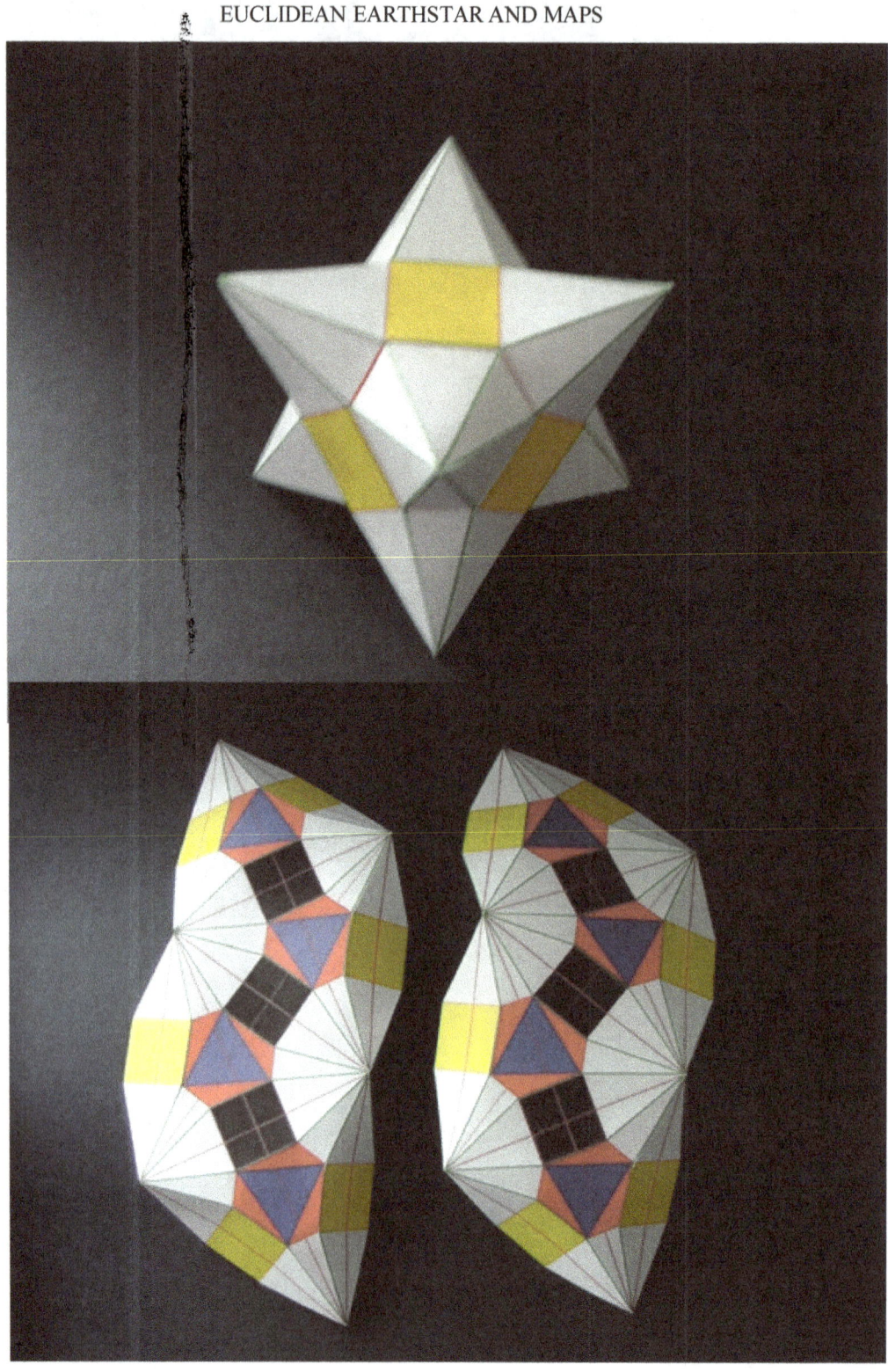

DeSign Science in The New Paradigm Age

DUAL OCTAHEDRON AND CUBE

D_ESign Science in The New Paradigm Age

NESTED TETRAHEDRA

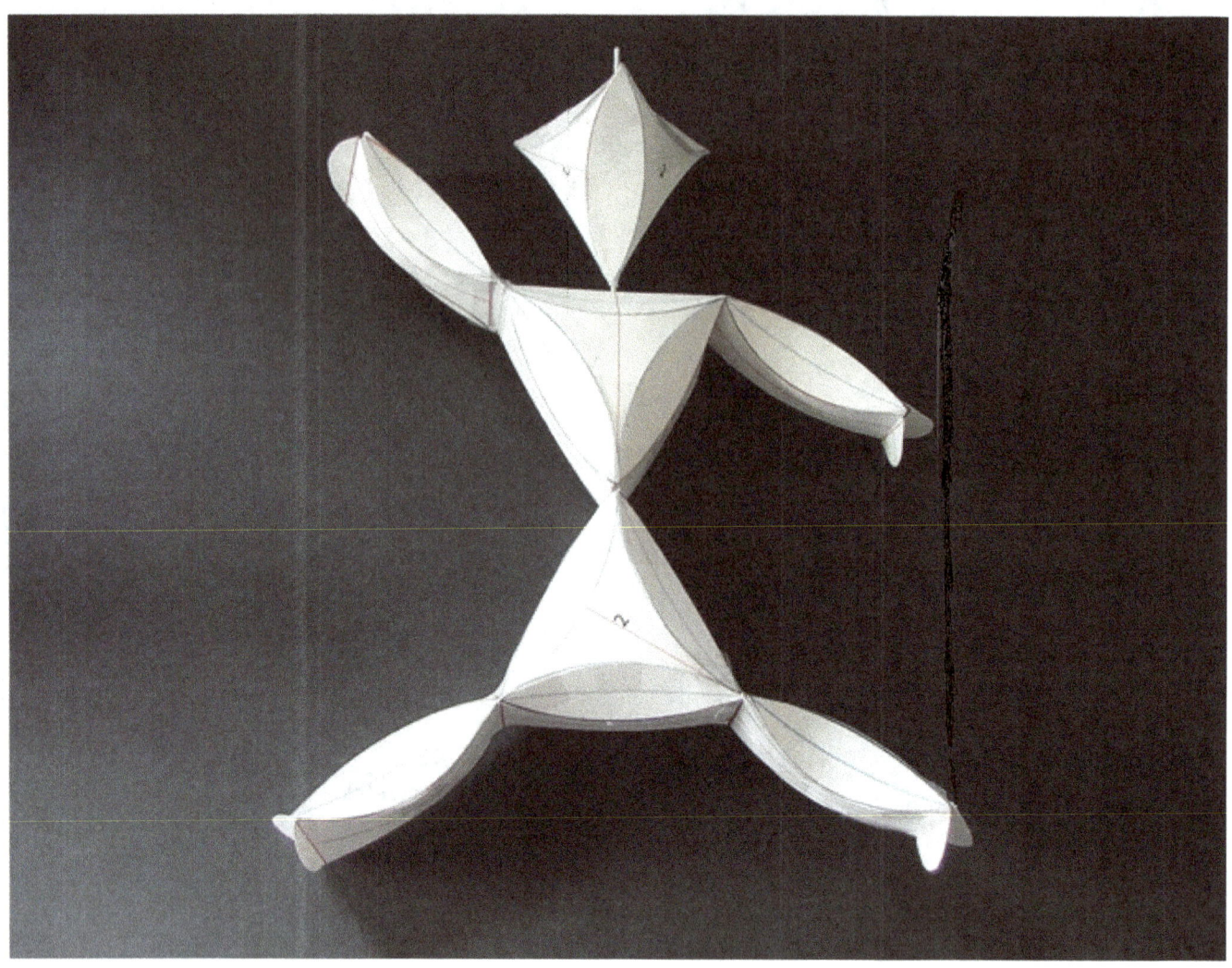

TETRAMAN

DeSign Science in The New Paradigm Age

TWO FIVE PART CUBES WITH INTERIOR TETRAHEDRA SURROUNDED BY TWO SETS OF TRIANGULAR PRISMS
Copyright Adger Cowans

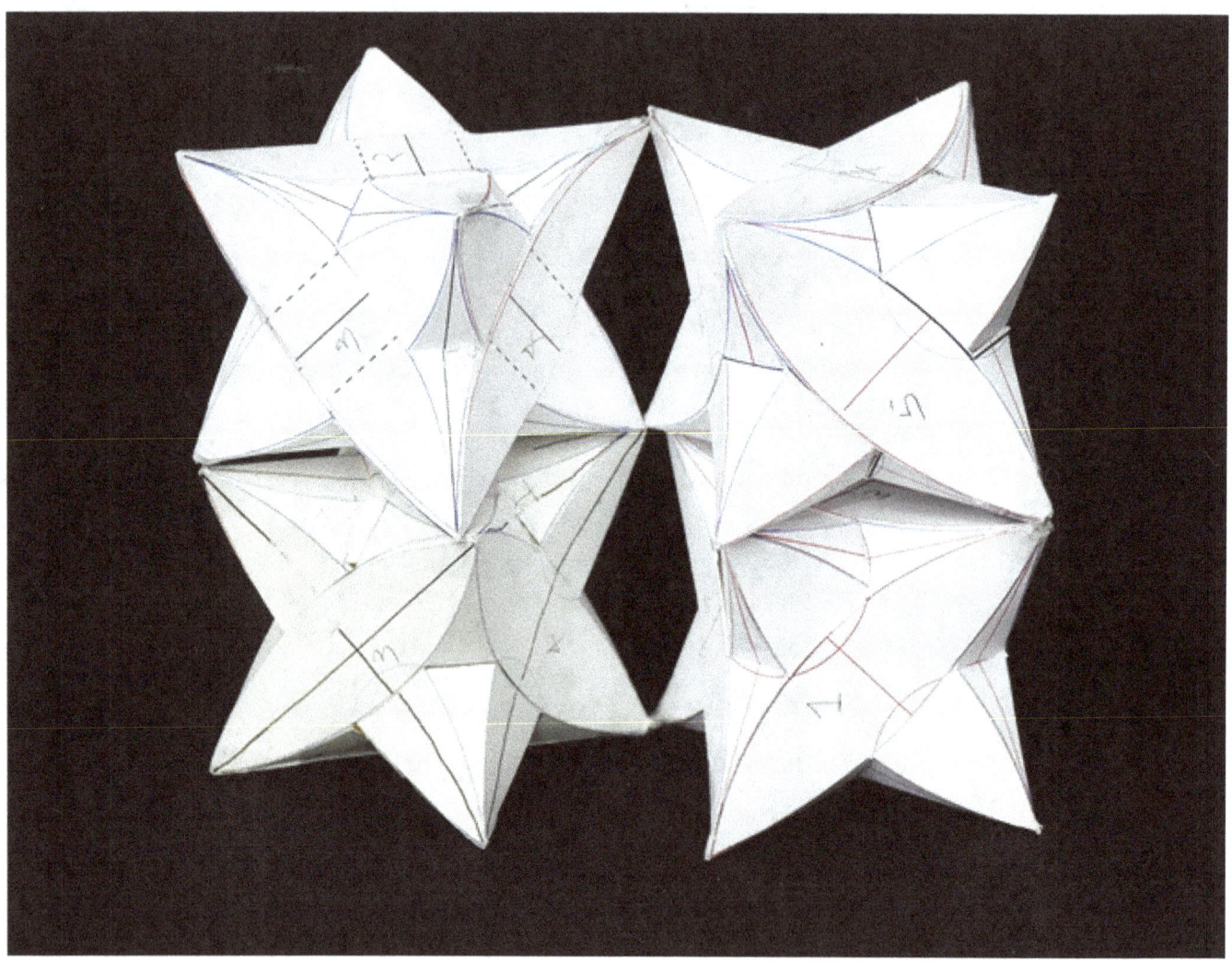

EARTHSTAR FLIP FORM 1   Copyright Adger Cowans

EARTHSTAR FLIP FORM 2　　　　　　　　　　　　　　　　Copyright Adger Cowans

EARTHSTAR FLIP FORM 3   Copyright Adger Cowans

MEDIA CUBE 1   Copyright Adger Cowans

MEDIA CUBE 2 — Copyright Adger Cowans

MEDIA CUBE 3 — Copyright Adger Cowans

DeSign Science in The New Paradigm Age

MEDIA CUBE 4　　　　　　　　　　　　　　　　　　　　　　　　　Copyright Adger Cowans

EARTHSTAR AND SLIDE CUBE   Copyright Adger Cowans

CLOSED SLIDE CUBE WITH EARTHSTAR

Copyright Adger Cowans

SLIDE CUBE AND EARTHSTAR

Copyright Adger Cowans

THE SLIDE CUBE FRAME

Copyright Herb Bennett

THE FLIP CUBE COLOR FORMS   Copyright Adger Cowans

POSITION 1

POSITION 2

POSITION 3

POSITION 4

DeSign Science in The New Paradigm Age

THE FLIP CUBE MAP   Copyright Adger Cowans

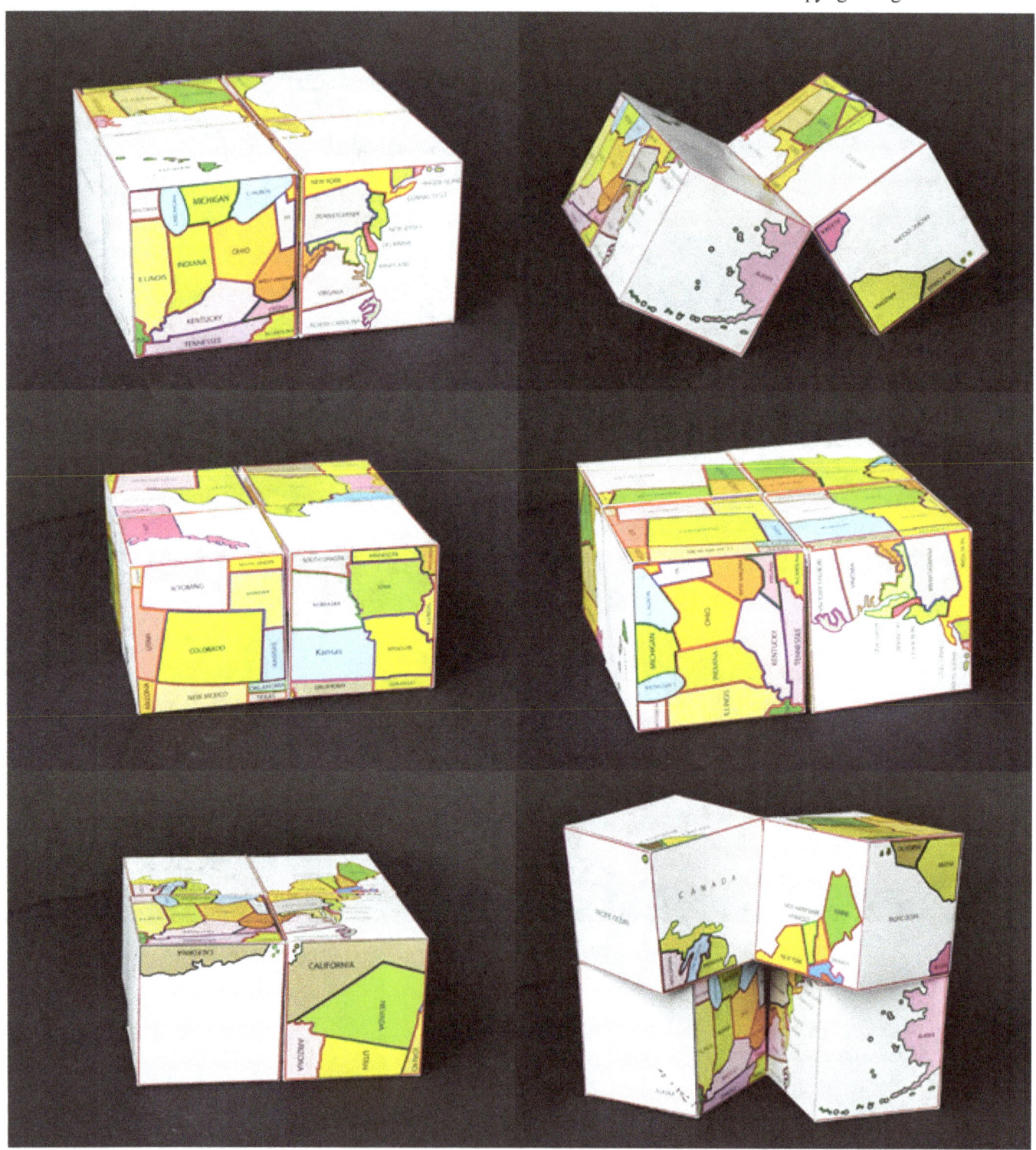

DeSign Science in The New Paradigm Age

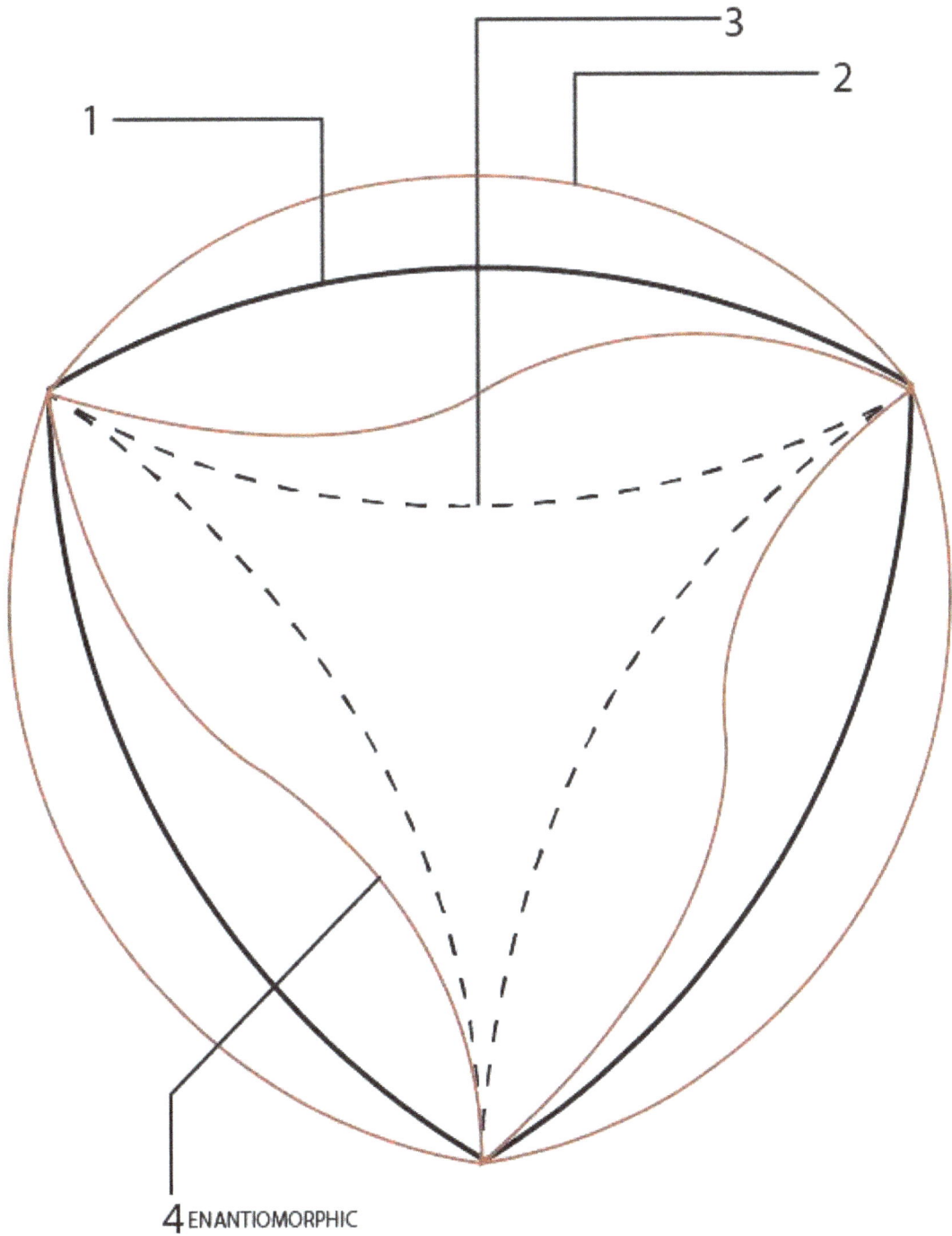

4 ENANTIOMORPHIC

TREFOIL (TRILLIANT) SHAPES

VARIATIONS ON THE TREFOIL [TRIANGLE]

THE BRONZE EARTHSTAR [TERRASTAR]

INVERTED DELTAHEDRA DEFINING A TETRAHEDRAL SPACE

COMBINED FORM ELEMENTS DEFINING SPACE

DeSign Science in The New Paradigm Age

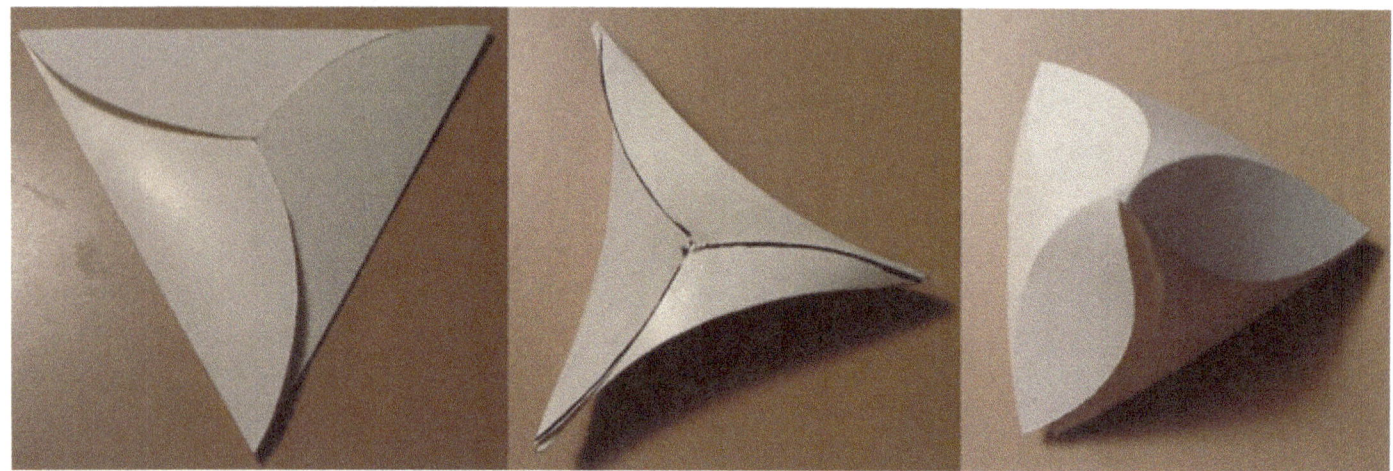

THE TRANSFORMATION OF TWO TO THREE DIMENSIONAL FORM VOCABULARIES

PRACTICAL APPLICATIONS OF A NEW IDENTITY AND AESTHETIC LANGUAGE

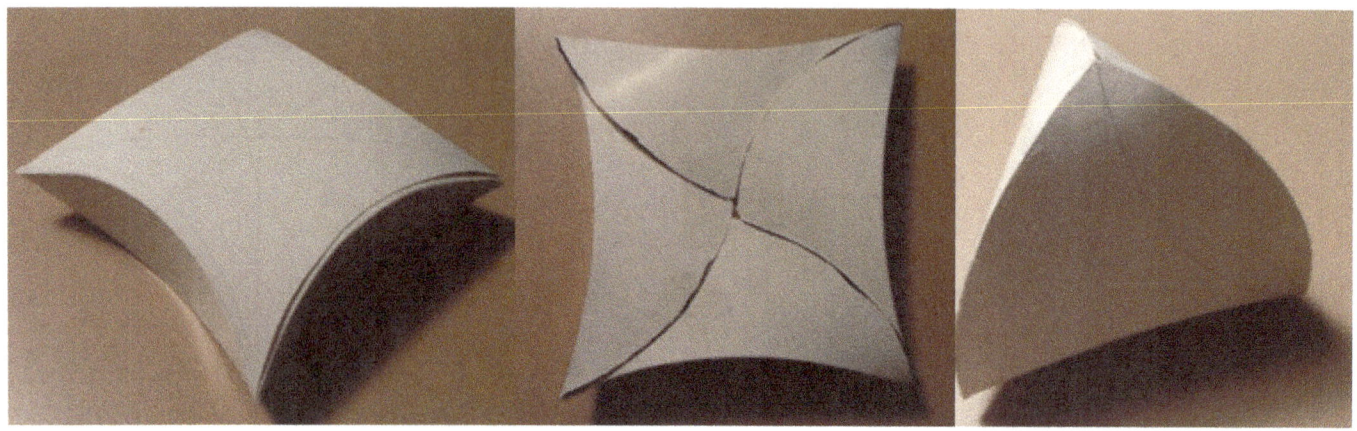

 THE STAR TREFOIL DINNERWARE COLLECTION

Here is an interpretative; non-descriptive geometry with another premise that is not about *'SPACE'*. It is a synthesizer of thought quantities, qualities and flavors that can generate and predict results of formations, combinations and formulations using the principles of the threefold symmetry (of nature) in polarization, ideation and formation to arrive at a transformative phase of creation as thought processes found in D$_E$Sign Science.

In the current paradigm Descriptive geometry's primary function is to translate an object in three dimensions into a two dimensional representation of that object. Each such representation is called a **view**

With 3D modeling however this becomes irrelevant to the visualization and realization processes since the "view" is not needed. New definitions and parameters are needed. The ideal is "the holistic and instant manifestation" of 3D form with ability to generations and variations of the vocabulary following strict symmetry rules. What type of "metry" is this? Al·lom·e·try: is the growth of body parts at different rates, resulting in a change of body proportions. Morphology is related to this function. A holometry with roots in holistic and hologram could be a candidate for this instantaneous manifestation of form.

D_ESign Science in The New Paradigm Age

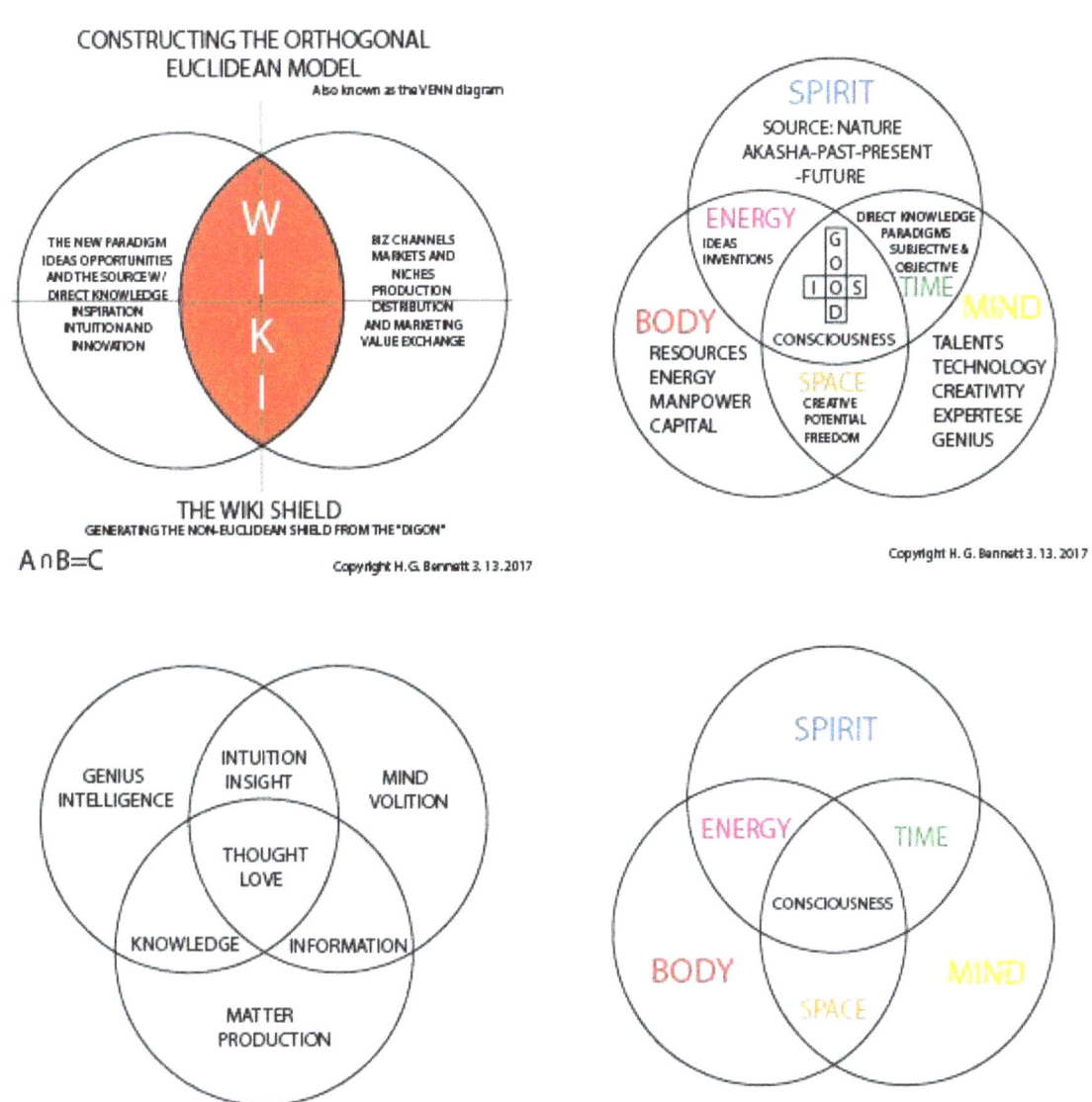

S₃ THREEFOLD THOUGHT FORM LOGIC STRUCTURES

D_ESign Science in The New Paradigm Age

NON-EUCLIDEAN FORMS

FIG 1   ARCUATED NON-EUCLIDEAN PRISM-BOX

FIG 2   DUAL TETRAHEDRAL CUBE

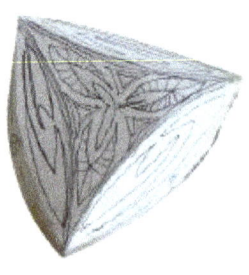

FIG 3   CLOSED ARCUATED TRIPRISM-BOX

FIG 4   DUAL FOLDED CUBE

FIG 5   ARCUATED TRIPRISM

FIG 6   TREFOIL PLATES OR SIMILAR OBJECTS

Copyright H. G. Bennett 3. 12. 2017

NON-EUCLIDEAN FORMS

WINDSWEPT FORMS

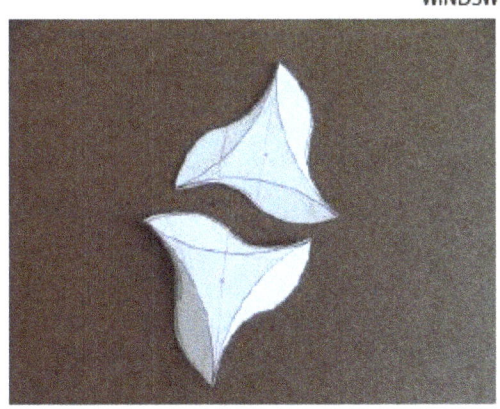

FIG 1

FIG 2

WINDSWEPT FORMS

FIG 3

FIG 4

NON-EUCLIDEAN FORM CLUSTERS

FIG 5

FIG 6

Copyright H. G. Bennett 3. 12. 2017

NON-EUCLIDEAN FORMS

GEOMETRIC THOUGHT FORMS

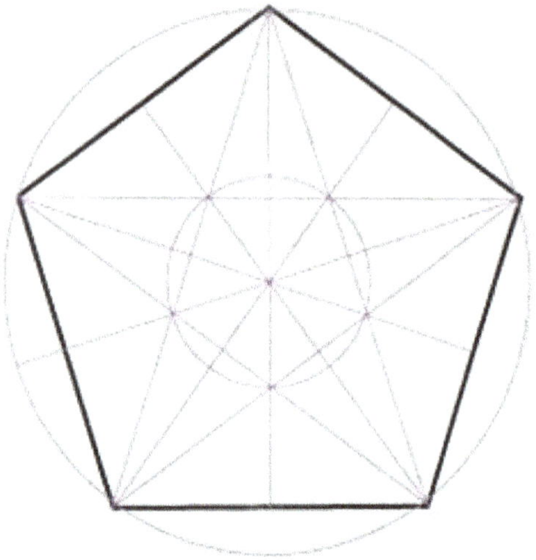

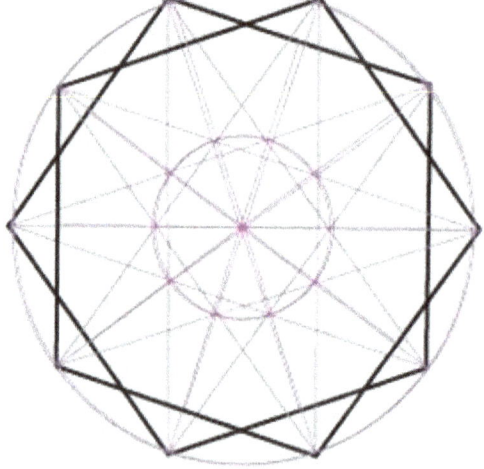

FIVEFOLD FREQUENCY

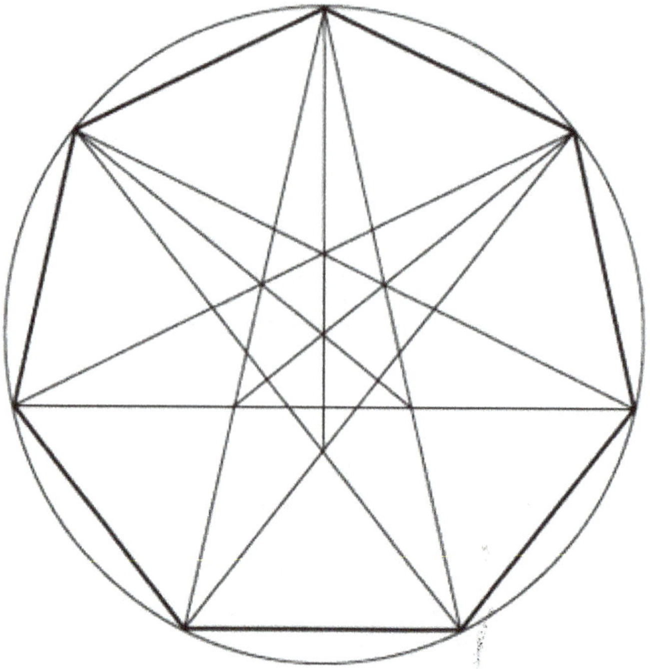

SEVENFOLD FREQUENCY

TWENTYFOLD FREQUENCY

TENFOLD FREQUENCY

Copyright H. G. Bennett 3. 19. 17

DeSign Science in The New Paradigm Age

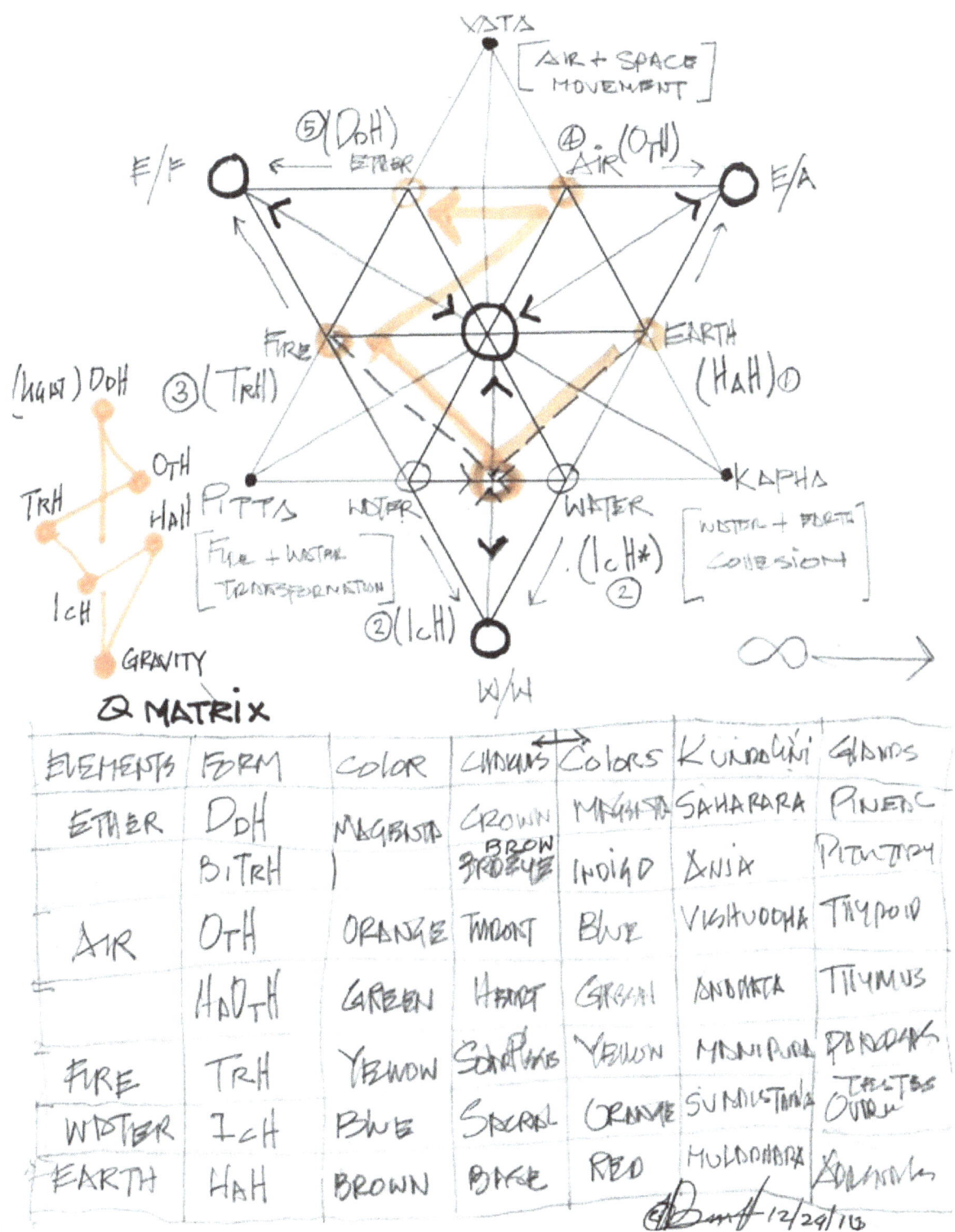

THE 'Q' MA/A/TRIX [OF OSIRIS AND ISIS]

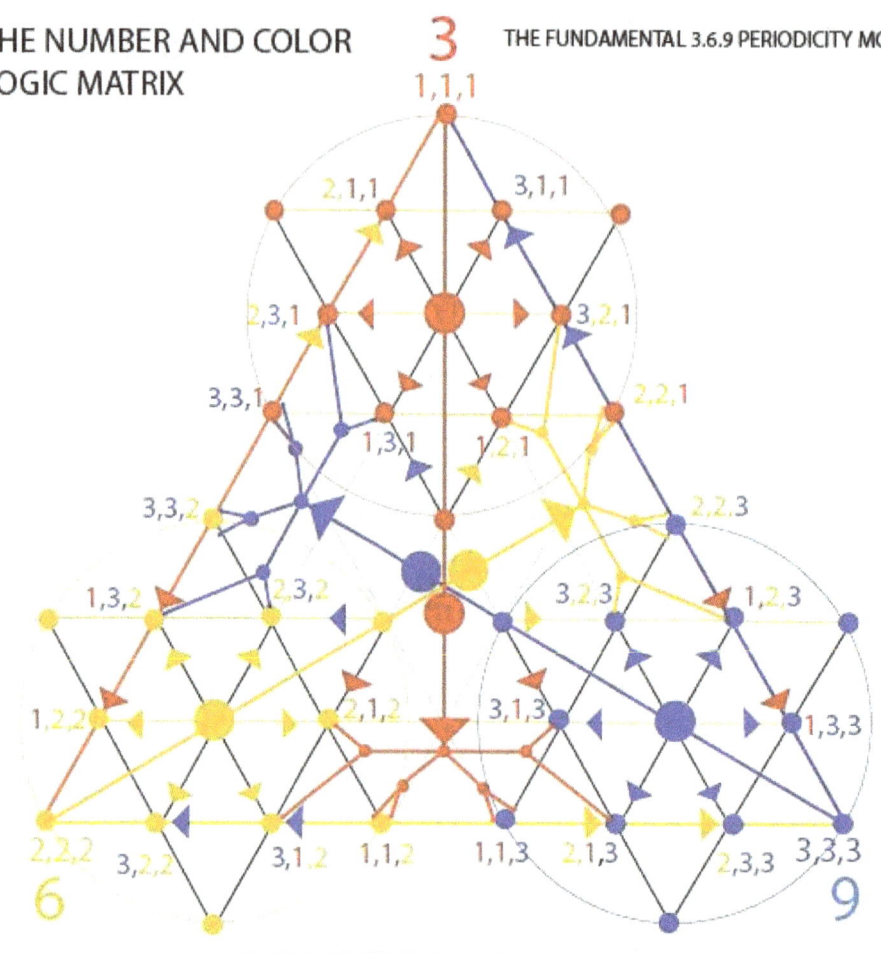

THE NUMBER AND COLOR LOGIC MATRIX

THE FUNDAMENTAL 3.6.9 PERIODICITY MODEL

BILATERAL SYMMETRY FOR SYMBOLS AND FLAVORS

Copyright H. G. Bennett 3. 19. 17

This 'Triadic Maat-rix' is the essence of a *visual mathematics, logic structure and other systems yet to be Maat or Ma'at (Egyptian mꜥ3t) refers to both the ancient Egyptian concepts of truth, balance, order, harmony, law, morality, and justice, and the personification of these concepts as a goddess regulating the stars, seasons, and the actions of both mortals and the deities, who set the order of the universe from chaos to order. This is the canon for a new geometry beyond what we call sacred. It is holistic and a 'HOLOMETRY'.*

Discovered. They go beyond the traditional 'Geo (physical) Metric' principles and gauge theories we now use to synthesize new holistic reckoning strategies and methods. The evolving phenomena being discovered now require a different level of reckoning.

It is the method of SYNTHESIS in the discipline of DESign Science of all creative fields of Wisdom, Intelligence, knowledge and Information, past, present and future.

Body: The most familiar expression is its role in counting and keeping score and for settling accounts bill or account, or its settlement.

Mind: It is a person's view, opinion, judgment, evaluation, estimate which all relate to the idea of "paradigm".

Spirit: Reckoning is the action or process of calculating or estimating something based on observations and thought processes that would harmonize the Body, Mind, Spirit triad and the calculation methods we know.

NON-EUCLIDEAN POLYGONS

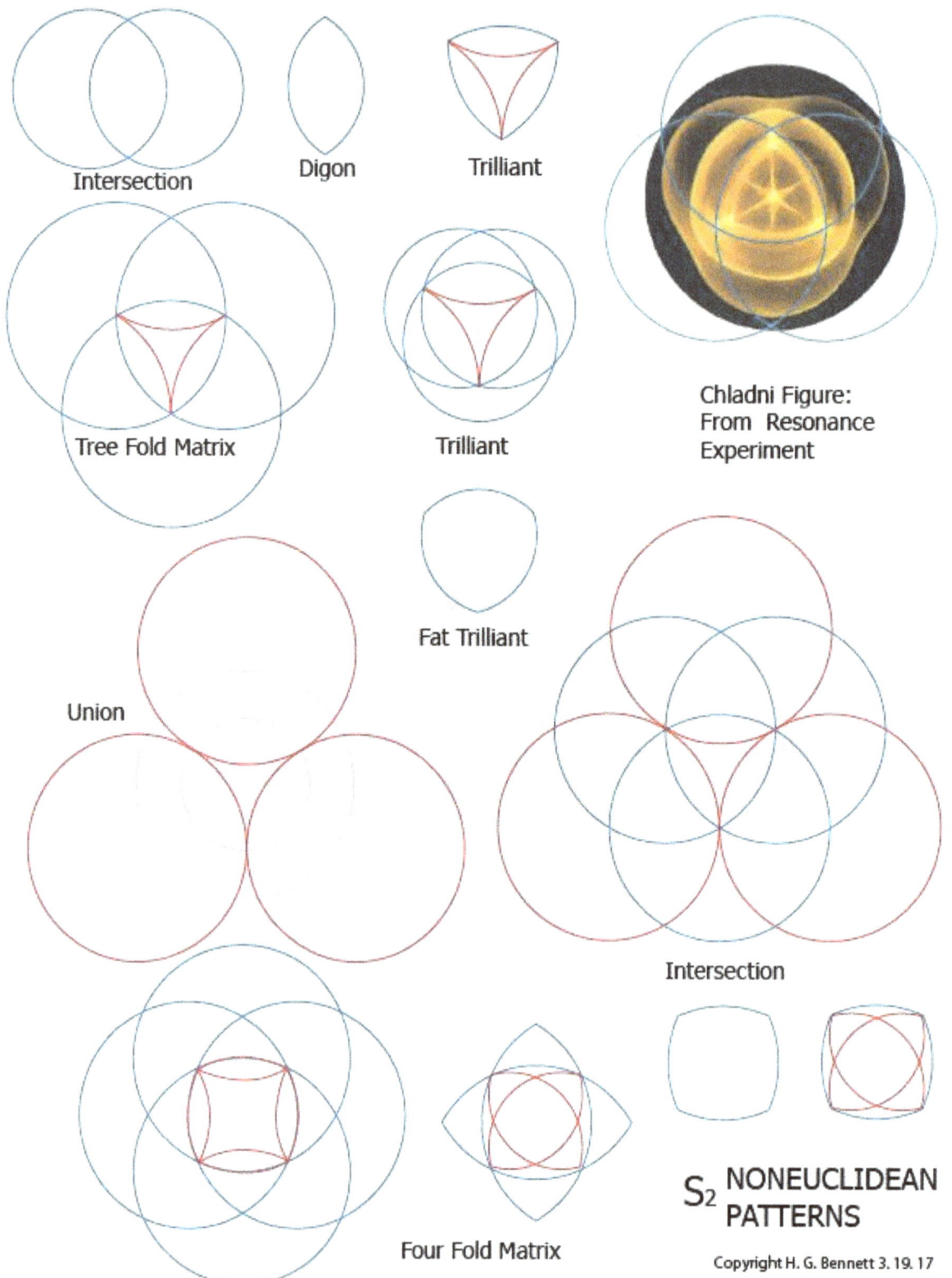

DIGON BLANK PANEL

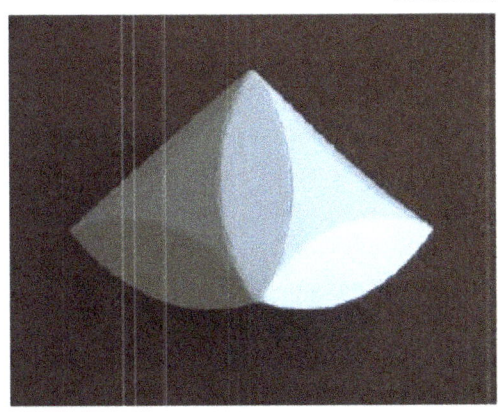

FIG 1

FIG 2

TETRAHEDRAL BLANKS

FIG 3   TETRAHEDRON

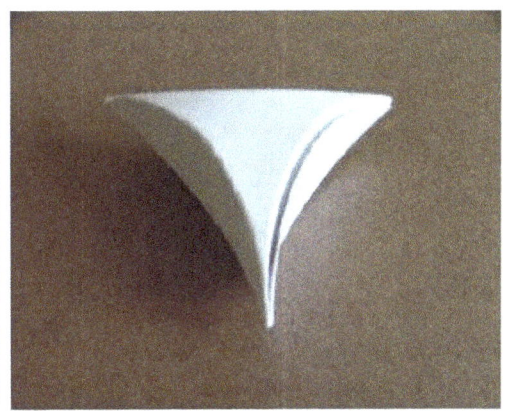

FIG 4   DELTAHEDRON

FIG 5   TETRAHEDRON BLANKS

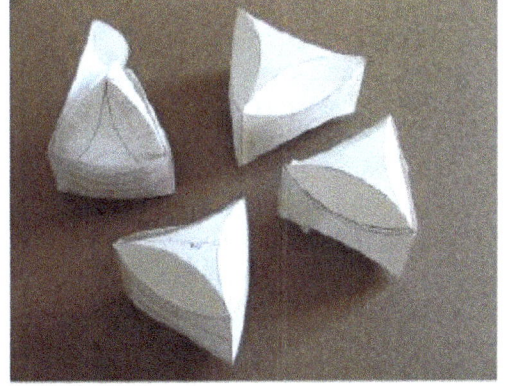

FIG 6   ARCUATED PRISM-BOX

Copyright H. G. Bennett 3. 12. 2017

PRIME NON-EUCLIDEAN FORMS

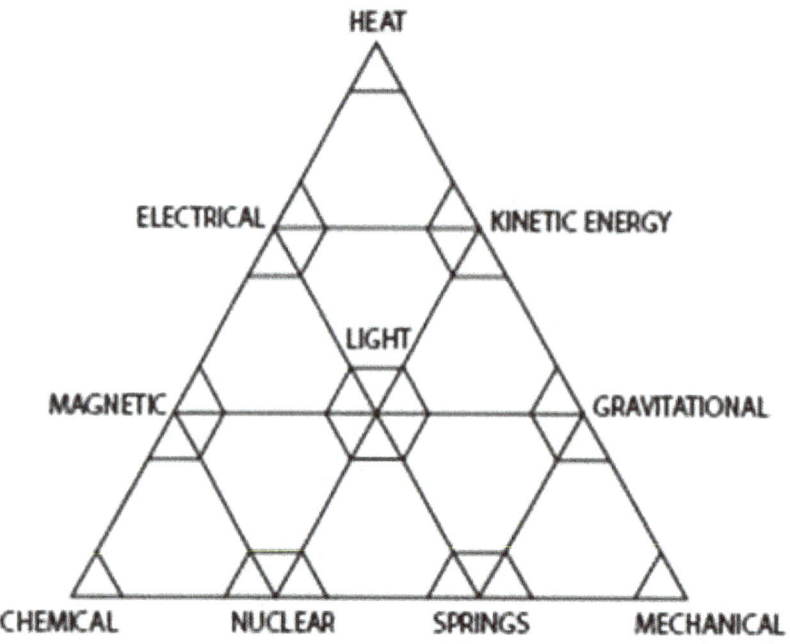

EPI-GEOMETRY: REALMS OF CONSCIOUSNESS BEYOND [NON EUCLIDEAN] GEOMETRY
A dream deferred by the Harlem Renaissance poet **Langston Hughes**. Written in 1951, this poem was the inspiration for Lorraine Hansberry's classic play A Raisin in the Sun.
What happens to a dream deferred?
Does it dry up like a raisin in the sun?
Or fester like a sore--And then run?
Does it stink like rotten meat?
Or crust and sugar over--like a syrupy sweet?
Maybe it just sags like a heavy load.
Or does it explode?
Langston Hughes   D_ESign Science

The tone and the quest implied in this poem resonates with the story of my search for the Non-Euclidean Geometry and the vision I have for sharing it with the world. The deferment of the dream, between the development stage and where we are in the new paradigm age (renaissance), is basically the same range of emotional states ART-iculated here. These questions are derived from those who are in touch with the ideas of their time and choose to express them in their voice as they seek and speak to plumb the depths of our humanity and values. The story of life can be told as a 'design story' with time as the rhythm of events and accomplishments we participate in. We do not ask for risks we take them and are emboldened by the courage they offer us when we connect to source and are in the flow of the energy that coalesces into the synthesis of our life and purpose throught our vision.

DeSign Science in The New Paradigm Age

Geometry is the voice that tells the tale. We are the instruments that must be connect to the source to be in tune with the wisdom to encode the message. The response to the message is the world we create. That requires an expression that is best described by geometry. It echoes much more than materialistic sensibilities and values we get trapped in.

The Language of form started with the recognition of physical space. Once thought to be rendered by straight lines, flat surfaces and fixed, it turned out to be (a space-time-energy dynamic) in a curved space redefined by general theory of relativity, such as dramatically shown by the 1919 solar eclipse measurements that corroborated all assumptions first recognized by the giants of Non-Euclidean geometries; Gauss, Lobachevsky, Bolyai and Riemann. Their rigorous application of mathematical possibilities developed new paradigm ideas that transformed geometry into an elaborate branch of generalized mathematical concepts in the nineteenth century. Disciplines like astronomy, astrophysics pointed to the possible geometrization of physics and the many emerging disciplines we see today. Up to this point there was no tangible constructed proof of any of these theories. Nothing was built nor created with it, still amorphous and philosophical.

The diffusion of this innovation imbued fields of scientific investigation (scientification) that was influenced by the prevailing ideologies and even the religious dogma of the day. Albert, not believing that God did not roll his dice, is one example.

What was the dream deferred that to this day still is not realized? Non-Euclidean geometry has a rich aesthetic potential to transform the world. There could be no bigger dream than this. There has never been such a profound transformation on this planet for very logic reasons. The major one being the 'Tower of Babel' syndrome we are still affected by.

The metaphor used here is meant to include all dimensions of division and confusion derived from basically the collective ego and 'shego' over time, wrapped in one convenient and powerful symbol.

In astronomy Friedrich Wilhelm Bessel **Born**: July 22, 1784, in Minden, Germany believed that the geometry of his day was incomplete needing correction that disappears in the case that the sum of angles in a plane triangle = (equal) 180°." He thought that "Non-Euclidean geometry would be the *true* geometry, while the Euclidean would be the *practical expression*, at least for figures on the earth." But earth was not the only realm that these ideas would relate to.

The 'dreamers' of non-Euclidean geometry were Hungarian mathematicians János Bolyai and the Russian Nikolai Ivanovich Lobachevsky, both of whom (contrary to Gauss) published their own discoveries that *a geometry different from and as valid as Euclid's is possible*. While Bolyai's sole work on what he called "absolute geometry" dates from 1831, Lobachevsky's first study was from 1829, followed in 1840 by a booklet in German, *Geometrischen Untersuchungen zur Theorie der Parallellinien*. Although working independently and in almost complete isolation, the two theories were remarkably similar. Synchronicity is the name of this concept.

DIFFERENT TYPE OF GEOMETRIES

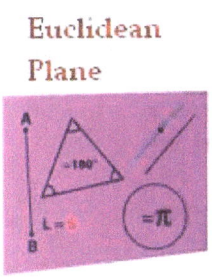

Euclidean Plane

Zero Curvature
Euclidian geometry

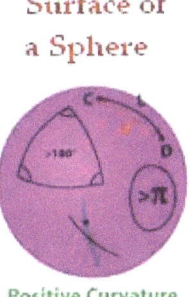

Surface of a Sphere

Positive Curvature
Elliptic geometry

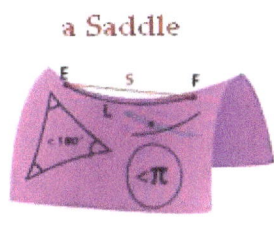

Surface of a Saddle

Negative Curvature
Hyperbolic geometry

The ideas of Gauss, Lobachevsky and Bolyai circulated slowly in the mathematical community. It gained attention when the Italian mathematician Eugenio Beltrami in 1868 presented it that the Non-Euclidean *paradigm* and its geometries truly entered the world of mathematics and then resulted in a revolution in geometry

The British mathematician William Kingdon Clifford translated the lecture into English in 1873. It came to be seen as a visionary address on the possible geometrization of physics. Riemann's thoughts also became known to the English-speaking world through articles by Helmholtz that appeared in *Mind*, a new quarterly journal for philosophy and psychology founded in 1876.12

In 1870 Clifford introduced Riemann's ideas to a British audience and wrote, "there are very *different kinds of space of three dimensions;* and that we can only find out *by experience* to which of these kinds the space in which we live belongs." He extended Riemann's ideas by speculating that physical phenomena could be fully reduced to properties of space curvature varying between one portion of space to another. *Heat, light and magnetism might be mere names for 'tiny variations in the curvature of space'*, he boldly hypothesized.

Clifford pointed out that within the framework of Riemannian geometry *the association (correlation) between limitedness and finite extent* was invalid. He emphasized, such as earlier contributors to Non-Euclidean geometry had done, *that the 'geometrical structure of space' was a question of empirical facts and not of metaphysics.*

The Chladni vibration model with Non-Euclidean Geometry elements.

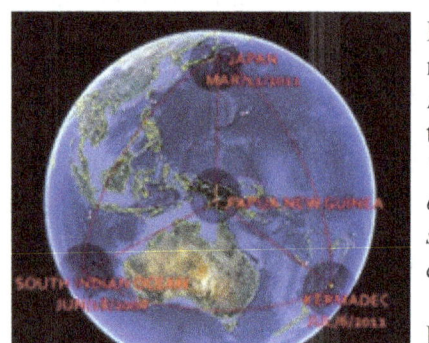

For roughly the next 140 year this paradigm shifting event in geometry remained an academic 'ivory tower' event. There were sporadic attempts at ART-iculating these ideas in real world applications in philosophy, technology, science itself, the design professions of art, architecture.
"Non-Euclidean geometry is a true practical geometry with a 'real' system of form generation, for figures on the earth that obey all natural laws and symmetry principles of space, time and energy in the Body, Mind Spirit continuum of consciousness itself as well."

In Egyptian temple architecture, reducing three dimensional forms to their two dimensional geometric patterns of building components was a major 'design science strategy' for simplified solutions to all complex temple forms. This two dimensional pattern generating process is used to generate a vocabulary of forms from three dimensional Non-Euclidean geometry.

Flat S_2 two dimensional blanks are created from a basic threefold pattern to create blanks that are folded into complex structures that would ordinarily require computer modeling and D_3 three technologies and advanced mathematical knowledge.

Simple methods create complex innovative solutions that can then be used in various fields of design, engineering and production. The vocabulary that's created expresses a unique identity and a resulting 'aesthetic' with curved spaces as the predominant feature of the forms. The classic 'Polyhedra vocabulary' serves to reality test this Non-Euclidean form theory. In the Euclidean Polyhedra, using a cube for example, the Edges are straight, the faces are flat and the vertices are 'mostly' equidistant.

In the Non-Euclidean vocabularies edges are curved and doubled to form 'digons', faces are imploding or exploding and vertices coincide with those of the classic vocabulary. Polyhedra were first invented by the Egyptian temple builders who needed to count the number of blocks needed for their monuments. This knowledge was then taken to Greece by Plato to be the 'Platonic solids'.

The materials used to build structures using Euclidean form designs are generally rigid framing systems and flat sheets or modular units like bricks and stones.

This invention was a paradigm shift that would take time to find its niche in the post, 'post and lintel' technology employed around the world. Richard Buckminster Fuller, the American Architect and visionary, championed the technological advancements in this space. In 1957 he was the original proponent of Design Science that inspired social and cultural movements in the 60's.

The relationship between the observer and the observed goes beyond the physical space or location. It involves the psychological, emotional (time) and spiritual or energy (dimensions of consciousness) operating in and around both the observer and the observed in space, time and energy. If the observer is located at Papua New Guinea central location on the planet the geometry perceived will be relative to the height of the observer and the curvature of the earth that is displacing the prime space of the entire universe. The Egyptians had this 'perspective' factored into their understanding of the world and their 'strategic' location on it.

The premise for Non-Euclidean geometry started in the same place and with the same idea; *constructing a 90 degree angle*. It is my opinion that there was a major east-west paradigm shift in the application of this development here and sides were chosen, as it were. Each side in the East-West dichotomy represented the meaning of what this idea represented according to their philosophical, cultural and religious traditions. This divide paved the way for each part to create traditions that their design, science and mathematics would then lead them to. Euclid stayed in the '90 degree construction' realm and exploited the iterations of 'his' technique and the emerging technology that followed and spread to the world.

The oriental cultures integrated their systems into their lifestyles in keeping with the aesthetic traditions of their ken. They both expressed their rhythm, faith and belief in the Body, Mind Spirit expressions of 'consciousness' either knowingly, viscerally or spiritually that related to
religion or religious belief according to their guiding principles which informed many if not all their 'canons'. There were time differences during which the growth and development took place along with interactions and exchanges and some were not very peaceful.

What was handed down to us has not yet been resolved on many of the levels needed for the bridges to be built permanently and peacefully. A number of new actors have encroached on this cultural axis making matters more complicated. The term East-West Dichotomy is a cultural metaphor with many dimensions to it not needed for this demonstration and argument.

DeSign Science in The New Paradigm Age

Sample 1. Expresses the intersection of circles used in some oriental fabrics representing the "Harmonic Expansion" of **Riemannian geometry**; the differential geometry of Riemannian manifolds, smooth manifolds with a Riemannian metric, i.e. with an inner product on the tangent space at each point that varies smoothly from point to point for Riemannian Space patterns in all dimensions.

Sample 1.

Sample 2. Expresses the intersection of circles used in some oriental fabrics representing the "Harmonic Contraction" of the **Bolyai, Lobachevsky geometry** János Bolyai or Johann Bolyai, was one of the founders of non-Euclidean geometry along with Lobachevski— a geometry that differs from Euclidean geometry in its definition of parallel lines to define complementary spaces found in the Riemannian Space in Sample 1.

Between the 90 degree angle and the trefoil pentagon construction lies a curved triangle also known as the Reuleaux Triangle. A spherical triangle when mapped on the surface of a sphere. The polyhedra approximates creating spheres using available flat materials before modern forming techniques were available to artisans. The world of sacred geometry was considered the highest form of endeavor to engage in man's early development.

Sample 2.

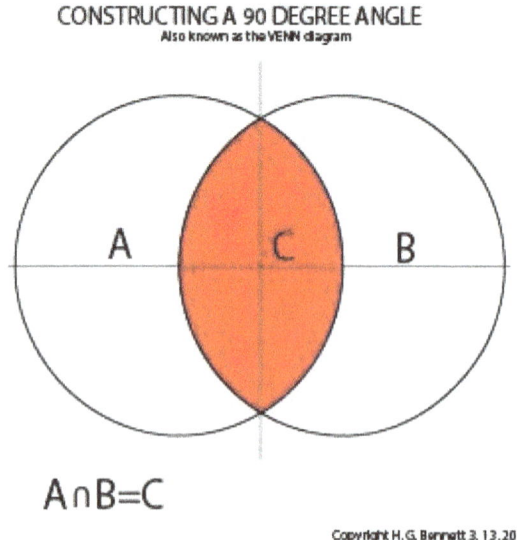

CONSTRUCTING A 90 DEGREE ANGLE
Also known as the VENN diagram

A ∩ B = C

Copyright H. G. Bennett 3. 13.2017

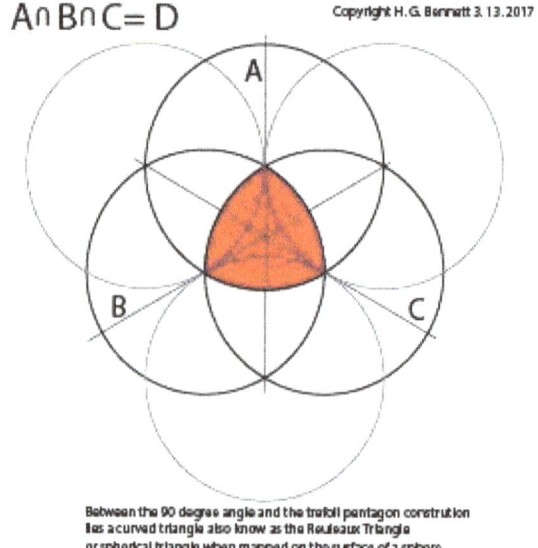

A ∩ B ∩ C = D

Copyright H. G. Bennett 3. 13.2017

Between the 90 degree angle and the trefoil pentagon constrution lies a curved triangle also know as the Reuleaux Triangle or spherical triangle when mapped on the surface of a sphere.

DIAGRAMS THAT REPRESENT LOGIC STRUCTURES, SYNTHESIS OR EXTRAPOLATIONS OF CONCEPTS AND IDEAS.

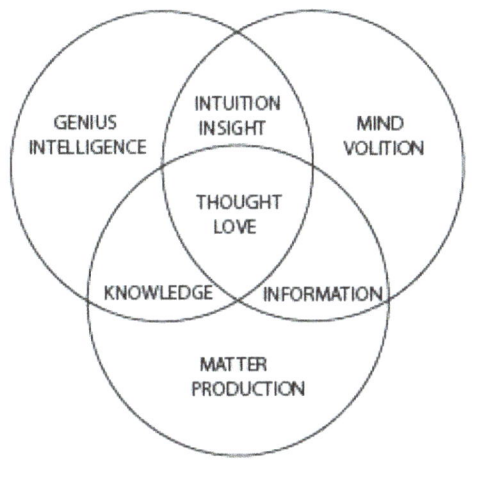

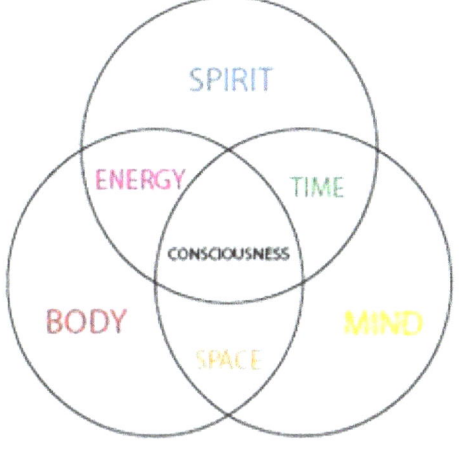

Copyright H. G. Bennett 3. 13. 2017

ILLUSTRATION GALLERY

| IMAGE N0 | TITLE |

I-1 THE 19 STANDARD MODELS OF THE EUCLIDEAN POLYHEDRA
I-2 MUSICAL SPHERES: MUSIC-FORM AND COLOR SYNTHESIS
I-3 THE TETRAHEDRAL TESSERACT OF THE EUCLIDEANPOLYHEDRA
I-4 GOBELKI TEPE MONUMENT
I-5 THE SUCHNESS OF METALS
I-6 SYNTHESIS OF PLANETS, METALS AND THE CHAKRA ENERGY CENTERS IN MAN
I-7 GOBEKLI TEPE IN SOUTHEATERN TURKEY 12,000 YEARS OLD
I-8 SYNTHESIS OF THE LOGIC STRUCTURE OVERLAID ON THE CONCEPTUAL MODEL
OF GOBEKLI TEPE

I-9 THE LOGIC STRUCTURE CONCEPTUAL MODEL & DIAGRAM
I-10 THE CONSCIOUSNESS MATRIX
I-11 THE SUN BEHIND THE SUN, OSIRIS, AND THE GREAT PYRAMID
I-12 THE CONIC TRANSFORMATION OF THE DELTAHEDRA
I-13 THE TREFOIL PLATE AND DELTAHEDRA
I-14 THE EARTH STAR (FUSED NON-EUCLIDEAN TETRAHEDRA)
I-15 CERAMIC EARTHSTAR TEAPOT (NOEL COPELAND COLLABORATION
AND PRODUCTION)

I-16 THE E NDOCRINE SYSTEM
I-17 THE NON-EUCLIDEAN POLYGON SET GENERATED FROM THE TREFOIL PATTERN OF INTERSECTING
CIRCLES OF THE SAME RADII.

I-18 TREFOIL METAL PLATES OR SIMILAR ARTICLES
I-19 A SPIRAL OF TREFOIL SYMMETRY
I-20 REULEAUX TRIANGLE –TREFOIL LOGO
I-21 EARTHSTAR CERAMIC TEAPOT (SCULPTURE HOUSE CASTING NYC COLLABORATION
AND PRODUCTION)

I-22 TETRAHEDRON WITH DELTAHEDRON ELEMENTS
I-23 MODEL CONTAINER AND DELTAHEDRA [D_AH] CONSTELLATIONS
I-24 STAR SCRAPER TOWER CREATED WITH STACKED EARTH STARS AND A TETRAHEDRAL
SPACE FRAME

I-25 PORT NON EUCLIDEON
I-26 DELTAHEDRON SPACEFRAME; ARCH-CUBOID MODULES AND A FOLDED TOWER TORI
I-27 BLISS: [A MEDITATION FOR PEACEFUL CENTERING]
I-28 DIGITAL ART FROM THE WALLPAPER EXTRACTION SERIES
I-29 DELTAHEDRON TRANSFORMED INTO 'A WINGED' STRUCTURAL ARCHITECTONIC ELEMENTAND
I-30 EARTHSTAR MADE OF FUSED TETRAHEDRAL FORMS
I-31 THE VIBRATIONAL 'CHLADNI' PATTERNS OF SOUND IMBUED INTO THE NON-EUCLIDEAN
I-32 PROPORTIONS OF 3D FORMS WITH THE TETRAHEDRON AND OCTAHEDRON
I-33 THE TETRAHEDRON AND OCTAHEDRON
I-34 BENNETT TORI BENT INTO THREE AND FOUR FOLD "CLOVERS"

I-35 THE STRUCTURE OF AN-ISM" (ANARTISM)
THE REPEATABLE STRUCTURAL ELEMENT OF AN-ISM" (ANARTISM)
STRUCTURAL NON-EUCLIDEAN ARCHITECTONIC FORM ELEMENTS NO1
STRUCTURAL NON-EUCLIDEAN ARCHITECTONIC FORM ELEMENTS NO2
STACKED STRUCTURAL NON-EUCLIDEAN DELTAHEDRAL FORM ELEMENTS
THE DELTA STAR- THE EGYPTIAN CANON WITH MAN'S TREFOIL PROPORTION FOR
SACRED SCIENCE AND TEMPLE ARCHITECTURE
EUCLIDEAN EARTHSTAR AND MAPS
DUAL OCTAHEDRON AND CUBE
STACKED FENESTRATED OCTAPRISMS
NESTED TETRAHEDRA
TETRAMAN
TWO FIVE PART CUBES WITH INTERIOR TETRAHEDRA SURROUNDED BY TWO SETS
OF TRIANGULAR PRISMS
EARTHSTAR FLIP FORM 1
EARTHSTAR FLIP FORM 2
EARTHSTAR FLIP FORM 3
MEDIA CUBE 1
MEDIA CUBE 2
MEDIA CUBE 3
MEDIA CUBE 4
EARTHSTAR AND SLIDE CUBE
CLOSED SLIDE CUBE WITH EARTHSTAR
SLIDE CUBE AND EARTHSTAR
THE SLIDE CUBE FRAME
THE FLIP CUBE COLOR FORMS IN 4 POSITIONS
THE FLIP CUBE MAP
VARIATIONS ON THE TREFOIL [TRIANGLE
THE BRONZE EARTHSTAR [TERRASTAR]
INVERTED DELTAHEDRA DEFINING A TETRAHEDRAL SPACE
COMBINED FORM ELEMENTS DEFINING SPACE
THE TRANSFORMATION OF TWO TO THREE DIMENSIONAL FORM VOCABULARIES
PRACTICAL APPLICATIONS OF A NEW IDENTITY AND AESTHETIC LANGUAGE
THE STAR TREFOIL DINNERWARE COLLECTION
A HOLOMETRY WITH ROOTS IN HOLISTIC AND HOLOGRAM
S_3 THREEFOLD THOUGHT FORM LOGIC STRUCTURES
NON-EUCLIDEAN FORMS
GEOMETRIC THOUGHT FORMS
THE 'Q' MA/A/TRIX [OF OSIRIS AND ISIS]
THE NUMBER AND COLOR LOGIC MATRIX
NON-EUCLIDEAN POLYGONS
NONEUCLIDEAN S_2 PATTERNS
PRIME NON-EUCLIDEAN FORMS
EPI-GEOMETRY: REALMS OF CONSCIOUSNESS BEYOND [NON EUCLIDEAN]
GEOMETRY
THE DESIGN LIBRARY FOR FURTHER READINGS:

DeSign Science in The New Paradigm Age

THE DESIGN LIBRARY FOR FURTHER READINGS:

THINKING

1) *The Design of Everyday Things* by Don Norman
2) *The Laws of Simplicity (Simplicity: Design, Technology, Business, Life)* by John Maeda
3) *Fab: The Coming Revolution on Your Desktop–from Personal Computers to Personal Fabrication* by Neil Gershenfeld
4) *Designing Design* by Kenya Hara
5) *Universal Principles of Design* by William Lidwell, Kritina Holden and Jill Butler.
6) *Cradle to Cradle: Remaking the Way We Make Things* by William McDonough and Michael Braungart.
7) *It's Not How Good You Are, Its How Good You Want to Be: The World's Best Selling Book* by Paul Arden
8) *The Lovemarks Effect: Winning in the Consumer Revolution* by Kevin Roberts
9) *Small Is the New Big: and 183 Other Riffs, Rants, and Remarkable Business Ideas* by Seth Godin.
10) *Design (Tom Peters Essentials)* by Tom Peters.
11) *Journals from the Design Management Institute* by DMI members.
12) *The Creative Priority : Putting Innovation to Work in Your Business* by Jerry Hirshberg
13) *Designing Interactions* by Bill Moggridge.
14) *Lateral Thinking: Creativity Step by Step* by Edward De Bono.
15) *What They Don't Teach You At Harvard Business School: Notes From A Street-Smart Executive* by Mark H. McCormack.
16) *The 48 Laws of Power* by Robert Greene.
17) *The Art of Innovation: Lessons in Creativity from IDEO, America's Leading Design Firm* by Tom Kelley.

PROCESS

18) *Design Secrets: Products 1 and 2: 50 Real-Life Product Design Projects Uncovered* by Lynn Haller and Cheryl Dangel Cullen, and edited by Industrial Designers Society of America.
19) *Process: 50 Product Designs from Concept to Manufacture* by Jennifer Hudson.
20) *Manufacturing Processes for Design Professionals* by Rob Thompson.
21) *Biomimicry: Innovation Inspired by Nature* by Janine M. Benyus
22) *Product Design and Development* by Karl T. Ulrich and Steven D. Eppinger.
23) *Managing the Design Factory* by Donald G. Reinertsen.

DESIGNER SKILLS

24) *Presentation Techniques* by Dick Powell.
25) *Creative Marker Techniques: In Combination With Mixed Media* by Yoshiharu Shimizu
26) *Sketching: Drawing Techniques for Product Designers* by Koos Eissen and Roselien Steur.
27) *Architecture: Form, Space, & Order* by Francis D. K. Ching.
28) *Elements of Design: Rowena Reed Kostellow and the Structure of Visual Relationships* by Gail Greet Hannah.
29) *Basic Visual Concepts and Principles For Artists, Architects And Designers* by Charles Wallschlaeger and Cynthia Busic-Snyder.
30) *Digital Lighting and Rendering (2nd Edition)* by Jeremy Birn.
31 Thinkertoys (Michael Michalko)

DeSign Science in The New Paradigm Age

32 The Creative Habit: Learn It and Use It for Life (Twyla Tharp with Mark Reiter)
33 Geometry of Design: Kimberly Elam
34The Industrial Design Reader: Carma Gorman
35Design Basics: S. Pentak and A. Lauer
36Design in the USA (Jeffrey Meikle)
History of Modern Design (David Raizman)
Design Studies: A Reader (Hazel Clark and David Brody)
Design as Future-Making (Susan Yelavich and Barbara Adams)
Product Design (Rodgers and Milton)
The Design Process (Karl Aspelund)
Toward a New Interior: An Anthology of Interior Design Theory (Lois Weinthal)
Graphic Design: Now in Production (Ian Albinson and Rob Giampietro)
The New Basics (Ellen Lupton and Jennifer Cole Phillips)
Digital Design Essentials Rajesh Lal
Don't Make Me Think, Revisited Steve Krug
About Face: The Essentials of Interaction Design Alan Cooper, Robert Reimann, David Cronin, Christopher Noessel
The Best Interface Is No Interface Golden Krishna
Designing Interfaces Jenifer Tidwell
Simple and Usable Web, Mobile, and Interaction Design Giles Colborne
Evil by Design Chris Nodder
Designing with the Mind in Mind Jeff Johnson
UI is Communication Everett N McKay
Serious Creativity, Edward de Bono
Cracking Creativity, Michael Michalko
A Technique For Producing Ideas, James Webb Young
Making Ideas Happen, Scott Belsky
Applied Imagination, Alex Osborn
The Art of Innovation, Tom Kelley
The Power of Positive Deviance, Richard Pascale & Jerry and Monique Sternin
Biomimicry; innovation inspired by nature, Janine Benyus
Patterns in Nature: Philip Ball,- Why the Natural World Looks the Way It Does
Black Hole Blues and Other Songs from Outer Space: Janna Levin
Mapping the Heavens: The Radical Scientific Ideas That Reveal the Cosmos: Priyamvada Natarajan
Welcome to the Universe: An Astrophysical Tour: Neil deGrasse Tyson, Michael A. Strauss, and J. Richard Gott.
Hidden Figures: Margot Le Shetterly
The American Dream and the Untold Story of the Black Women Mathematicians Who Helped Win the Space Race
Rise of the Rocket Girls: Nathalia Holt
The Women Who Propelled Us, from Missiles to the Moon to Mars
The Only Rule Is It Has to Work: Ben Lindbergh and Sam Miller
Our Wild Experiment Building a New Kind of Baseball Team
The Invention of Nature: Andrea Wulf; Alexander von Humboldt's New World
The Gene: Siddhartha Mukherjee. An Intimate History
Are We Smart Enough to Know How Smart Animals Are? : Frans de Waal
The Wasp That Brainwashed the Caterpillar: Matt Simon-
Thoughts on Design (Paul Rand)
Originally published as an essay in 1947, Thoughts on Design is still fiercely relevant to today's designers.
Tibor Kalman: Perverse Optimist (Peter Hall and Michael Bierut)
This joint work designed by Pentagram partner Michael Bierut and edited by writer Peter Hall pays tribute to the graphic design of Tibor Kalman.

DeSign Science in The New Paradigm Age

Design as Art (Bruno Munari)
Thinking With Type (Ellen Lupton)
The Visual Display of Quantitative Information (Edward Tufte)
Drawing is Thinking (Milton Glaser)
Graphic Design: A Concise History (Richard Hollis)
Type and Image: The Language of Graphic Design Paperback(Philip B. Meggs)
Meggs' History of Graphic Design (Philip B. Meggs, Alston W. Purvis)
Anatomy of Design: Uncovering the Influences and Inspiration in Modern Graphic Design (Steven Heller and Mirko Ilic)
The Design of Everyday Things (Don Norman)
Change by Design (Tim Brown)
Designing for Growth (Jeanne Liedtke)
Next Generation Business Strategies for the Base of the Pyramid(Ted London and Stu Hart)
Vision in Motion (Laszlo Moholy-Nagy)
Part of the original Bauhaus school, László Moholy-Nagy illustrates where design, art, and science meet.
World Changing: A User's Guide for the 21st Century (Alex Steffen)
Humble Masterpieces: 100 Everyday Marvels of Design (Paola Antonelli)
In the Bubble: Designing in a Complex World (John Thackara)
Sustainable Design: Explanations in Theory and Practice (Stuart Walker)
Art and Visual Perception: A Psychology of the Creative Eye (Rudolf Arnheim)
Originally published in 1974, this book explores where art and psychology collide.
Weird Ideas that Work, Robert I. Sutton

BOOKS ON CREATIVE THINKING

Serious Creativity, Edward de Bono
Cracking Creativity, Michael Michalko
The Medici Effect, Frans Johansson
A Technique For Producing Ideas, James Webb Young
Making Ideas Happen, Scott Belsky
Applied Imagination, Alex Osborn
Weird Ideas that Work, Robert I. Sutton
The Art of Innovation, Tom Kelley
Biomimicry; innovation inspired by nature, Janine Benyus
The Power of Positive Deviance, Richard Pascale & Jerry and Monique Sternin
Nir Eyal: Hooked: How to Build Habit-Forming Products
100 Things Every Designer Needs to Know About People, Susan Weinschenk
Designing for the Digital Age by Kim Goodwin
Living with Complexity by Donald A. Norman
The Design of Everyday Things by Donald A. Norman.
Lance Wyman: The Monograph edited by Adrian Shaughnessy
Damn Good Advice (For People with Talent!) by George Lois
Herb Lubalin American Graphic Designer by Adrian Shaughnessy
George Nelson: Architect / Writer / Designer / Teacher by Stanley Abercrombie
Saul Bass: A Life in Film and Design Hardcover
Dieter Rams: As Little Design as Possible by Sophie Lovell
Kern and Burn by Tim Hoover

DESign Science in The New Paradigm Age

Steve Jobs by Walter Isaacson Isaacson
Design Forward by Hartmut Esslinger
Design Is the Problem by Nathan Shedroff
Obey the Giant by Rick
World without Words by Michael Evamy Evamy shows how simple design can explain so much without words and how culture is moving toward a world without words.
Predictably Irrational by Dan Ariely
The Craft of Words By The Standardistas
The Creative Habit by Twyla Tharp
Thinking, Fast and Slow by Daniel Kahneman Nobel Prize-winning psychologist.
Geek-Art: An Anthology by Thomas Olivri
The Icon Handbook by Jon Hicks
Logo Creed by Bill Gardner
Logo Design Love by David Airey
Monogram logo by Leterme Dowling and Counter-Print
Symbol by Angus Hyland and Steven Bateman
Articulating Design Decisions by Tom Greever
Insanely Simple: The Obsession That Drives Apple's Success by Ken Segall
Why We Buy: The Science of Shopping by Paco Underhill
How Google Works by Jonathan Rosenberg and Eric Schmidt
Design is a Job by Mike Monteiro This easy read explains how to build your business backbone while feeding your creative talent.
Drawn to Business by Bill Beachy of GoMedia.us
It's Not How Good You Are, It's How Good You Want To Be by Paul Arden
Creativity, Inc. by Ed Catmull and Amy Wallace
Manage Your Day-To-Day by Jocelyn K. Glei and 99U
Remote by Jason Fried and David Heinemeier Hansson
Seductive Interaction Design: Creating Playful, Fun, and Effective User Experiences by Stephen P. Anderson
Don't Make Me Think by Steve KrugIn ,
About Face by Alan Cooper, Robert Reimann, David Cronin and Christopher Noessel
Interaction Design by Yvonne Rogers, Helen Sharp and Jenny Preece
Product Design for the Web by Randy J. HuntEtsy's
Simple and Usable by Giles Colborne
Designing for Emotion by Aarron Walter
Designing with the Mind in Mind by Jeff Johnson
You're Designing It All Wrong! by Tal Florentin
Above the Fold by Brian D. Miller
The Visual Display of Quantitative Information by Edward R. Tufte
The Animator's Survival Kit by Richard Williams
Grid Systems in Graphic Design by Josef Müller-Brockmann
Design, Form, and Chaos by Paul Rand
Data Points by Nathan Yau

DeSign Science in The New Paradigm Age

Form+Code in Design, Art, and Architecture by Casey Reas
Graphic Design Before Graphic Designers by David Jury
How To Be a Graphic Designer Without Losing Your Soul by Adrian Shaughnessy
Meggs' History of Graphic Design by Philip B. Meggs and Alston W. Purvis
Drawing Ideas : Mark Baskinger and William
Sascha Michael Trinkaus, CEO of Trinkaus Creative Consultants
90 Degrees by Andrew Kim
Designing Brand Identity by Alina Wheeler
Designing Design by Kenya Hara
Designing News by Francesco Franchi
Interaction of Color by Josef Albers
Just Enough Research by Erika HallHall presents an easy
The Shape of Design by Frank Chimero
Sass for Web Designers by Dan Cederholm
Universal Principles of Design by William Lidwell, Kritina Holden and Jill
Art as Therapy by Alain de Botton.
The Noble Approach by Tod Polson
Scripts: Elegant Lettering from Design's Golden Age by Steven Heller and Louise Fili
Designing With Web Standards by Jeffrey Zeldman
Responsive Web Design by Ethan Marcotte
What They Didn't Teach You In Design School: The Essential Guide to Growing Your Design Career by Phil Cleaver
Life's a Pitch: How to Sell Yourself and Your Brilliant Ideas by Roger Mavity
Everything I Know by Paul Jarvis
Creativity for Sale by Jason SurfrApp The man behind IWearYourShirt and BuyMyLastName shares a practical guide on how to turn passions into profits.
Execute by Josh Long and Drew Wilson
The Good Creative by Paul Jarvis
Steal Like An Artist by Austin Kleon
Show Your Work! by Austin Kleon
Where Good Ideas Come From by Steven Johnson Johnson identifies the seven key patterns behind genuine innovation, and traces them across time and disciplines.

MAGAZINES

Computer Arts Magazine The industry-leading title for graphic designers, Computer Arts is the magazine for people who believe design matters.
Offscreen Magazine Offscreen is an independent magazine about people who use the internet and technology to be creative, solve problems, and build successful businesses.
Smashing Magazine and its publications; Smashing Magazine is a website and blog that offers resources, books, and advice to web developers and web designers.
DIS Magazine; The self-described "post-internet lifestyle magazine" has an editorial mission to interrogate and collapse hierarchies.
Dwell Magazine; Dwell is an American magazine devoted to modern architecture and design.

AUTHOR
Margaret Kelsey
Content + community at InVision. Newly Bostonian.
Keep up with me on Twitter and Google+.

YOU MIGHT ALSO LIKE...

1 "Isaac Julien: Riot," By Giuliana Bruno, Paul Gilroy, Stuart Hall, and bell hook with Isaac Julien, et al.
2 "Kerry James Marshall: Painting and Other Stuff," Edited by Navi Haq, with Dieter Roelstraete and Okwui Enwezor
3 "Charles Gaines: Gridwork 1974-1989," Edited by Naima J. Keith, with a foreword by Thelma Golden and contributions by Courtney J. Martin, et al.
4 "Nick Cave: Epitome," By Andrew Bolton, Elvira Dyangani Ose and Nato Thompson with Nick Cave
5 "Charles Gaines: Gridwork 1974-1989," Edited by Naima J. Keith, with a foreword by Thelma Golden and contributions by Courtney J. Martin, et al.
6 "Nick Cave: Epitome," By Andrew Bolton, Elvira Dyangani Ose and Nato Thompson with Nick Cave
7 "LaToya Ruby Frazier: The Notion of Family," by LaToya Ruby Frazier, with Dennis C. Dickerson, Laura Wexler and Dawoud
8 "Black Artists in British Art: A History Since the 1950s," By Eddie Chambers
9 "Chris Ofili: Night and Day," Edited by Massimiliano Gioni with Gary Carrion-Murayari and Margot Norton
10 "Prospect.3: Notes for Now," By Franklin Sirmans, with contributions by Rita Gonzalez, Christine Y. Kim, Mary A. McCay, et al.
11 "Lynette Yiadom-Boakye," with contributions by Naomi Beckwith, Donatien Grau, Jennifer Higgie and Lynette Yiadom-Boakye

http://28blacks.com/charles-dawson-14.html
11 Best Black Art Books of 2014
by VICTORIA L. VALENTINE on Dec 21, 2014
1. "Ellen Gallagher: Don't Axe Me," Edited by Gary Carrion-Murayari with a foreword by Lisa Phillips (New Museum, 122 pages). | Published Jan. 31, 2014
2. "Isaac Julien: Riot," By Giuliana Bruno, Paul Gilroy, Stuart Hall, and bell hook with Isaac Julien, et al. (Museum of Modern Art, New York, 248 pages). | Published Feb. 28, 2014
3. "Kerry James Marshall: Painting and Other Stuff," Edited by Navi Haq, with Dieter Roelstraete and Okwui Enwezor (Ludion, 192 pages) | Published April 30, 2014
4. "Charles Gaines: Gridwork 1974-1989," Edited by Naima J. Keith, with a foreword by Thelma Golden and contributions by Courtney J. Martin, et al. (The Studio Museum in Harlem, 168 pages). | Published Aug. 31, 2014
5. "Nick Cave: Epitome," By Andrew Bolton, Elvira Dyangani Ose and Nato Thompson with Nick Cave (Prestel USA, 288 pages) | Published Sept. 22, 2014
6 "LaToya Ruby Frazier: The Notion of Family," by LaToya Ruby Frazier, with Dennis C. Dickerson, Laura Wexler and Dawoud Bey (Aperture Foundation, 156 pages) | Published Sept. 30, 2014
7. "Black Artists in British Art: A History Since the 1950s," By Eddie Chambers (I. B. Tauris, 288 pages) | Published Oct. 2, 2014
8. "Chris Ofili: Night and Day," Edited by Massimiliano Gioni with Gary Carrion-Murayari and Margot Norton (Skira Rizzoli, 240 pages) | Published Oct. 28, 2014
9. "The Image of the Black in Western Art, Volume V: The Twentieth Century, Part 2: The Rise of Black Artists," Edited by David Bindman and Henry Louis Gates Jr. et al. (Belknap Press, 368 pages) | Published Oct. 31, 2014
10. "Prospect.3: Notes for Now," By Franklin Sirmans, with contributions by Rita Gonzalez, Christine Y. Kim, Mary A. McCay, et al. (Prestel, 176 pages). | Published Nov. 11, 2014
11. "Lynette Yiadom-Boakye," with contributions by Naomi Beckwith, Donatien Grau, Jennifer Higgie and Lynette Yiadom-Boakye

IAJDSS: The **International Association and Journal of DESign Science** is dedicated to research, prototyping and publication of results in the field of form generation, metaphysics, design theory and creativity with applications of the concepts, information technology and manufacturing technology to real world solutions in a holistic methodology know as D_ESign Science. Original works are expected from artists, industrial and academic contributors. Solutions, inventions and concepts must be presented as solutions to practical problems relevant to needs of industries.

The areas of scope include, but not limited to, solid modeling, geometric modeling, geometric processing, computer graphics, computational geometry, computer-aided design, computer-aided manufacturing, computer-aided engineering, feature technology, concurrent engineering, collaborative engineering, computer integrated manufacturing, Internet-based CAD/CAM, rapid prototyping, assembly modeling, product data exchange, intelligent CAD, user interaction techniques, human machine and human computer interaction, RFID application, FEM/BEM, mesh generation, virtual reality, scientific visualization, CAPP, NC programming, and topological optimization, Artificial Intelligence and computer science and engineering etc.

ADDENDUM:
Thoth was **Seshat**, goddess of writing, the keeper of books, and patron goddess of libraries and librarians who was alternately his wife or daughter.
Christ's appearance in radiant glory to three of his disciples (Matthew 17:2, Mark 9:2–3, Luke 9:28–36).

The *Ogdoad, Hehu* or Infinites, *'celestial rulers'* of a cosmic age might have come long before the Egyptian religious system currently recognized, the Ogdoad were concerned with the *preservation and flourishing of the celestial world*, and the *'formation' of the human race*. If we look at this as the creation story the characters could be very transmutable. They could be GODS, Forces, principles or other phenomena represented in a Spiritual Sciences tradition. The characters in our cosmic accounts could very well be the same as the Celestial rulers of the Egyptian era and way back as well. There is evidence in sites like Gobekli Tepe that could be the reference to the 'coming long before' of the Infinites. Whether it is before, during or after which will correspond with our time the process is eternal, the characters and their languages change. Threads of evidence are pointing to the constant creation model presenting us with parallels that might confirm this hypo or epi-the/i/sis.

The transfiguration of GODS into forces or energies and the reverse, seems to be a spiritual technology in Egypt. Transfiguration; noun: the Transfiguration a complete change of form or appearance into a more beautiful or spiritual state. Ancient Egypt has a history with more of it hidden that's left to the curious to imagine, investigate and interpret to discover the meaning and reason for being and becoming. From the Pre-Dynastic Period (c.6000- 3150 BCE) to the Ptolemaic Period (323-30 BCE), the last dynastic era of Egyptian history was influenced by the civilization building began with the GOD Toth or Hermes Tresmegistus. His wisdom, knowledge and spiritual influence inspired the wisdom tradition, schools and temple cultures. Thoth's Egyptian name was Djehuty (also *dhwty*) meaning "He Who is Like the Ibis". His name would be taken by kings of Egypt.
Tuthmoses "Born of Thoth", scribes, and priests is one example. Toth is commonly depicted as a man with the head of an ibis or a seated baboon with or without a lunar disc above his head. Were these interpretations fact, spiritual knowledge or magic?

Worshiping Thoth began in Lower Egypt most likely from the Pre-Dynastic Period (c. 6000-3150 BCE) to the Ptolemaic Period (323-30 BCE). Thoth's veneration was among the longest of the Egyptian gods or any deity from any civilization. The Aquarian Gospel tells us of the Transfiguration of Jesus the hierophant into 'The Christ'. After intense studies, 'training' and testing by the priests in the temple he becomes a 'Master'. The knowledge Toth imparted to the priest sect became the 'canon' of the wisdom tradition.

Naqada III or King Menes; 3200 to 3000 BC, ruled in the last phase of the Naqada culture. This **Protodynastic Period** was characterized by an ongoing process of civilization and political unification. Very little is known of the mystery tradition and role of the Pharaonic and temple cultures played in the transfigurations of Egyptian pharaohs into GODS. Were they GODS. Menes founded Memphis as the capital of ancient Egypt in the north, near the apex of the Nile River delta which became the dominant metropolis in ancient Egypt.

Djer was the second or third pharaoh of the first dynasty of Egypt, which dates from approximately 3100 BC. *Uncertainty over the first pharaohs of this dynasty, Menes or Narmer, and Hor-Aha, and possible confusion with the final ruler of the Protodynastic Period, makes the numbering of the First Dynasty problematic.* Thoth is the Egyptian God of *writing, magic, wisdom, and the moon.* He was one of the most important Gods of ancient Egypt. He was *self-created* or born of the *seed of **Horus** from the forehead of **Set**.* What do we find at the forehead that could be the seed? Mystery traditions know this as the third eye. It is connected to the Hypothalamus where the Pineal and Pituitary glands are also considered the '*seat of the soul*'. Horus and Set are essences of consciousness with principles involved in the living dynamic and connections to source. As the son of these two deities, who represented order and chaos respectively, he was also the god of equilibrium and balance and associated closely with both the principle of ma'at (divine balance) and the goddess Ma'at who personified this principle (and who was sometimes seen as his wife). Another of his consorts was the goddess Nehemetawy ('She Who Embraces Those In Need") a protector goddess. In his form as A'an, Thoth presided over the judgment of the dead with **Osiris** in the Hall of the Truth and those souls who feared they might not pass through the judgment safely were encouraged to call upon Thoth for help.